THE HISTORIOGRAPHY OF THE CHEMICAL REVOLUTION: PATTERNS OF INTERPRETATION IN THE HISTORY OF SCIENCE

THE HISTORIOGRAPHY OF THE CHEMICAL REVOLUTION: PATTERNS OF INTERPRETATION IN THE HISTORY OF SCIENCE

BY

John G. McEvoy

Routledge
Taylor & Francis Group

LONDON AND NEW YORK

First published 2010 by Pickering & Chatto (Publishers) Limited

Published 2016 by Routledge
2 Park Square, Milton Park, Abingdon, Oxfordshire OX14 4RN
711 Third Avenue, New York, NY 10017, USA

First issued in paperback 2015

Routledge is an imprint of the Taylor & Francis Group, an informa business

BRITISH LIBRARY CATALOGUING IN PUBLICATION DATA

McEvoy, John G., 1942–
The historiography of the chemical revolution: patterns of interpretation in the history of science. 1. Chemistry – History – 18th century – Historiography. 2. Chemistry – History – 18th century. I. Title 540.7'22-dc22

ISBN-13: 978-1-138-66126-4 (pbk)
ISBN-13: 978-1-8489-3030-8 (hbk)

Typeset by Pickering & Chatto (Publishers) Limited

CONTENTS

For Marilyn, Tommy and Scott

who thought it would never end

ACKNOWLEDGEMENTS

This book has been an intermittent, but ever present, object of my attention for longer than I care to remember. It originated over two decades ago in response to scholarly criticisms of my earlier work on Joseph Priestley and the Chemical Revolution. These criticisms focused on historiographical rather than straight-forward historical issues and problems. They marked the beginnings of a period of historiographical ferment in the discipline of the history of science, during which time this book took on a life of its own. Part of the furniture of my mind for so long, it inevitably bears witness to broader intellectual and personal influences which I wish to recognize with gratitude at this juncture.

I am deeply indebted to Ted McGuire, Larry Laudan, John Nicholas and the late Gerd Buchdahl and Satish Kapoor who, though they may hesitate to admit it, shaped the basic contours of my identity as a historian and philosopher of science. I am also indebted to the works of Jan Golinski, Simon Schaffer and Steven Shapin, which guided me through the minefield of issues, problems and opportunities generated for historians of science by the burgeoning discipline of the sociology of scientific knowledge. The sociological dimensions of my own historiographical model were also shaped by the interventions of Bob Young, the *Radical Science Journal* and the Marxist tradition that sustained them. Among members of the broader scholarly community who have supported this project with insights, discussions, collegiality and friendship, I would like to thank Chris Lawrence, John Brooke, Arthur Donovan, Anna-K. Mayer, Brain Easlea, Dan Anderson, Bernadette Bensaude-Vincent, Marco Beretta and the late Bill Smeaton.

Work on this project benefited from invaluable intellectual and material inputs at my home institution, the University of Cincinnati, which has been a generous supporter of my research and intellectual endeavours for nearly forty years. Travel and research for this project was facilitated by grants from the Charles Phelps Taft Research Center, while the book itself was written during a period of extended academic leave facilitated by a Taft Center Fellowship. The Philosophy Department, held together by the firm but gentle hand of 'super secretary' Arlene Bridges, has been an integral part of my life these many years; those among its faculty and graduate students who have been particularly helpful to my research on this project include Bob Faaborg, Larry Jost, Bob Richardson, Ted Morris, Koffi Maglo, John Bickle, Kathy Russell, Kerry Walters, Joe Cronin, Arthur Morton and, especially, Julian Wuerth who took it upon himself to make sure I didn't balk at the final fence. I only wish that my dear friend and mentor, Don Gustafson, who hired me into the department, had lived long enough to see the fruits of a labour he did so much to encourage.

I owe a special word of thanks to Scott Stewart who spotted some of the more egregious errors in the manuscript: the ones remaining can only be called my own. Scott also sustained me over the years with a stalwart friendship and the warm hospitality of his – and Laura's – home nestled in the beautiful Blue Ridge Mountains of Virginia.

None of this would have been possible without the enduring love and support of my family, on both sides of the Atlantic. It was only the determination and sacrifices of my late mother, Gertrude McEvoy, and my elder brother, James ('Jimmy') McEvoy, that enabled me to be the first member of my working-class family to entertain seriously, let alone pursue, the path of higher education in England in the early 1960s. This path eventually brought me to Cincinnati, where my sons, Tommy and Scott, grace my world with a happiness I never thought I'd know and my wife Marilyn makes sure the centre holds and that thinking about the past doesn't prevent me from taking care of the present, which includes the pressing canine needs of Sherwood and Gracie. I dedicate this book to them as a small token of my love and affection.

This book draws in part on some of my previously published articles. I thank the relevant institutions and individuals for permission to use here work previously published. The core thesis of the book was first outlined in 'In Search of the Chemical Revolution: Interpretive Strategies in the History of Chemistry', *Foundations of Chemistry*, 2 (2000), pp. 47–73. Chapter 1 is a modified version of 'Positivism, Whiggism and the Chemical Revolution: A Study in the Historiography of Chemistry', *History of Science*, 35 (1997), pp. 1–33, by permission of the editor. Parts of Chapters 2, 3 and 4 draw on 'Postpositivist Interpretations of the Chemical Revolution', *Canadian Journal of History*, 36 (2001), pp. 453–69 and 'Modernism, Postmodernism and the Historiography of Science', *Historical Studies in the Physical and Biological Sciences*, 37 (2007), pp. 383–408, by permission of the editors. Chapter 7 incorporates extracts from 'The Enlightenment and the Chemical Revolution', in R. S. Woolhouse (ed.), *Metaphysics and the Philosophy of Science in the Seventeenth and Eighteenth Centuries: Essays in Honor of Gerd Buchdahl* (Dordrecht: Kluwer, 1988), pp. 307–25, by permission of Kluwer Academic Publishers; 'Continuity and Discontinuity in the Chemical Revolution', in A. Donovan (ed.), *The Chemical Revolution. Essays in Reinterpretation* (Osiris, 2nd Series, Volume 4) (Philadelphia, PA: History of

Science Socirty, 1988), pp. 195–213, by permission of the History of Science Society; and 'The Chemical Revolution in Context', *The Eighteenth Century: Theory and Interpretation*, 33 (1992), pp. 198–217, by permission of Texas Tech University Press.

PREFACE

In 1989 Francis Fukuyama famously proclaimed the 'end of history'. Expressing the unalloyed triumphalism of American liberalism at the end of the Cold War, Fukuyama's controversial declaration also gave voice to postmodernist aspirations to escape from the fixed identities, traditions and institutions of the past and live in a pluralistic present of self-fashioning and unfettered desire. This aspiration seemed to place in jeopardy humanity's historical sense, threatening an unexpected nihilistic outcome to twentieth-century debates about the nature and significance of that sense and its place in the formation and maintenance of social ties and cultural bonds. In an important paper of 1976, the British historian Lawrence Stone examined these debates from the perspective of the professional historian. His tone was both celebratory and cautionary. While looking to the future with diminished expectations and some apprehension, Stone identified the 'new history' of the previous forty years as a period of unsurpassed creativity 'in the whole history of the profession', brought on by 'borrowings from the social sciences'.[1] The interaction between the traditional narrative form of history, professionalized between 1870 and 1930, and the neighbouring disciplines of sociology, anthropology, economics, psychology and demography issued in the formation of new fields of historical inquiry, including demographic, cultural and social histories, as well as histories of mass culture, science and the family, which, at the time of Stone's writing, were still in 'their heroic phase of primary exploration and rapid development'. This is still the case with the history of science which, contrary to Stone's gloomy expectations, has continued to ferment under the influence of extrinsic disciplinary pressures. The present volume is an attempt to understand the disciplinary ferment in the history of science since World War Two in a way that throws light on the formation and mutation of our historical sensibilities, of our sense of time, place and heritage.

This book explores the 'coming of age' of the discipline of the history of science – or more particularly, the history of chemistry – and the numerous identity problems associated with that tumultuous process. It avoids the perils of analytical irrelevance, and anchors abstract historiographical reflection to the concrete practices of historians, by approaching a set of interpretive issues and problems shared by the contemporary history of science community as a whole from a perspective generated by an analysis of existing interpretations of a single significant event in the history of science: the Chemical Revolution of the eighteenth century. Traditionally associated with the French scientist Antoine Lavoisier and the oxygen theory of combustion, the Chemical Revolution occurred towards the end of the Enlightenment. It embroiled some of the finest scientific minds of Europe – including Lavoisier's arch-rival, the British chemist Joseph Priestley, who upheld the phlogiston theory of combustion – in a debate of lasting significance for subsequent generations of scientists and historians of science. Indeed, the Chemical Revolution is generally regarded as the birth of modern chemistry and as the very paradigm of a scientific revolution. Some scholars even regard it as part of a 'Second Scientific Revolution' which took place between 1778 and 1839 and which rivalled, in its scope and implications, *the* Scientific Revolution of the sixteenth and seventeenth centuries. In a broader context, the conjunction of the American and French Revolutions with the Chemical Revolution marked the beginning of the 'Age of Revolution', a period of cultural and political turmoil and upheaval that swept Lavoisier to his tragic death at the hands of the Terror and Priestley to his lonely political exile in rural Pennsylvania. So viewed, the generative phase of modern chemistry is one of the most compelling and rewarding subjects in the history of science. The recent resurgence of interest in the Chemical Revolution, fuelled by the growing ferment in the history of science, makes this an apt topic for a reflexive study of the kind proposed here.

Mainly the amateur and gentlemanly preserve of scientist-historians throughout the eighteenth, nineteenth and early-twentieth centuries, the history of science emerged as a recognizable profession with a clearly defined academic identity only after World War Two, eventually becoming a focus of intense interest and activity for journalists and popularizers in the closing decades of the twentieth century. Variation in the practitioners of the discipline brought with it changes in its identity, as interests and perspectives rooted in more established disciplines, such as the natural sciences, theology, philosophy, sociology, literary theory and political history, sought to stipulate and regulate the interests and inquiries of this newly emergent discipline. Always a contested terrain, a site of intense philosophical and ideological intervention, in recent years the history of science has become an intriguing maze of contending and conflicting interpretive strategies, procedures and perspectives. While the proliferation of alternative research strategies is a welcome development in any discipline, there comes a time when further gains can best be made by reflecting critically on what has already been achieved. Not surprisingly, the history of science community has taken a reflexive turn in recent years, with many of its practitioners reflecting on the methods as well as the objects of their inquiries. With this situation in mind, this study offers a critical account of extant patterns or models of interpretation in the history of science, designed to elucidate their philosophical origins and presuppositions in a way that throws light on their proper scope and validity, as well as their relative strengths and weaknesses. It also uses this analysis to outline a new interpretive strategy and identity for the discipline. Besides making this intriguing and exciting discipline accessible to students, interested readers and specialists in other disciplines, it is designed to stimulate further reflection and discussion among its current and future practitioners.

Although interest in the Chemical Revolution has enjoyed something of a renaissance among recent historians of science, compared with its more venerable predecessor – the Scientific Revolution of the sixteenth and seventeenth centuries – and its more notorious successor – the Darwinian Revolution of the nineteenth century – it is still a relatively neglected affair, sorely in need of further historical analysis and historiographical reflection. This reflexive study responds to this unsettled and unsatisfactory historiographical situation in a number of different ways. Addressing the needs and concerns of the scholarly practitioner, it is designed to provide the neophyte with a useful guide through the intricate labyrinth of fascinating philosophical and historiographical issues associated with past and present interpretations of the Chemical Revolution, thereby keeping new and current historians in touch with the important results of earlier scholars, which are in danger of being overlooked or poorly understood in a historiographical environment unsympathetic or even hostile to the methods and procedures used to procure them. By providing a critical exploration and evaluation of existing historiographies in this area of inquiry, it is also designed to assist established practitioners and scholars in the endless task of honing their skills and refining their practices. The survey of the secondary literature (and extensive bibliography) inevitably involved in a study of this kind should also benefit the general student and reader interested in the fields of the history and philosophy of science and science studies. Finally, by relating specialist interests and issues in the history and philosophy of science and science studies to broader philosophical, sociological and historiographical movements and themes, this study also hopes to attract the interest and attention of a wide range of scholars working in the humanities and social sciences.

But this study has broad philosophical and historiographical, as well as practical, implications and benefits for our understanding of the Chemical Revolution. It provides information and perspectives for a broader 'historical' comparison of competing interpretations of the Chemical Revolution in particular and definitions of science in general which, as seems to be agreed by both philosophers and sociologists of science, should be used to supplement, or even replace, traditional 'logico-epistemological criticism'.[2] In the presence of conflicting and contending historiographical strategies, historians of the Chemical Revolution also need to preserve and incorporate the important insights of past and present interpretive strategies into a more balanced and comprehensive historiographical framework. Faced with the historiographical 'other', historians of science can reflect on their own identity and history, placing the emergence and development of their discipline in a broader historical framework. Some historians and commentators have already remarked on this framework, drawing parallels between, for example, the progressivist 'Whig interpretation of history' and the fortunes of British imperialism, between the static structures of post-war structuralism and postpositivism and the 'immobilism' of the Cold War, and between the deconstructionist tendencies of postmodernism and 'the post-1968 rejection by many left-wing intellectuals of any perspective of global social transformation'.[3] The interpretive model outlined in the final chapter of this study entails but, for want of time and space, does not explore this reflexive entanglement. Clearly, an adequate appreciation of 'the history of science as history', expressed in R. G. Collingwood's notion of 'historical criticism', requires closer and more extensive contacts between historians of science and general historians intent on both the examination of the past and their examination of it. Without guaranteeing a true account of an immutable past, history and historiography, taken together, can move forward in a self-corrective manner.

Notes

1. L. Stone, 'History and the Social Sciences in the Twentieth Century', *The Past and the Present* (Boston: Routledge and Kegan Paul, 1981 (1976)), pp. 3–44, on p. 30. See also J. C. D. Clark, *Our Shadowed Present. Modernism, Postmodernism and History* (London: Atlantic Books, 2003), pp. 1–12.

2. See I. Lakatos, 'History of Science and Its Rational Reconstructions', in C. Howson (ed.), *Method and Appraisal in the Physical Sciences. The Critical Background to Modern Science* (Cambridge: Cambridge University Press, 1976), pp. 1–39; S. Shapin, 'History of Science and Its Sociological Reconstructions', *History of Science*, 20 (1982), pp. 157–211.

3. G. S. Jones, 'The Pathology of English History', *New Left Review*, 46 (1967), pp. 29–43, on pp. 30–31; E. P. Thompson, 'The Poverty of Theory', in *The Poverty of Theory and Other Essays* (New York: Monthly Review Press), pp. 1–210, on pp. 73–75; A. Callinicos, *Against Postmodernism. A Marxist Critique* (London: Pluto Press), p. 85.

INTRODUCTION: THE PHILOSOPHICAL AND HISTORIOGRAPHICAL TERRAIN

More than twenty years ago, the late Carleton Perrin likened the current state of our scholarly understanding of the Chemical Revolution to the parable of the blind men and the elephant. While historians of this complex event have a shared sense of being in the presence of a great beast, they mistake the part each of them has touched for the whole thing and hence cannot agree on its nature or identity.[1] As a historical event, the Chemical Revolution is readily identified. It occurred towards the end of the eighteenth century and involved some of the finest scientific minds of Europe in an upheaval of considerable scope and consequence. What is not so easy to determine is the meaning or significance of this event, both for its participants and for subsequent commentators. Nineteenth and early twentieth-century historians of chemistry identified the Chemical Revolution with the conflict between the English natural philosopher Joseph Priestley and the French chemist Antoine Lavoisier over the nature of combustion, with Priestley defending the traditional view that burning substances emit 'phlogiston' (the principle of inflammability) against Lavoisier's innovative suggestion that they absorb oxygen. But the issues joined in this debate went well beyond the question of the empirical adequacy of competing scientific explanations, encompassing methodological, epistemological, ontological, linguistic and institutional issues that related to the very identity of chemistry as a scientific discipline. While generations of historians of chemistry have been united in the belief that the birth of modern chemistry involved a fundamental break with previous chemical theory and practice, they have failed to arrive at any consensus on the identity of the offspring and the nature of its gestation. Did this act of parturition, which brought forth modern chemistry, hinge upon an experimental discovery, a theoretical insight, a methodological reform, an epistemological reorientation, or an ontological purification? Or did it involve the coming of reason to an arcane corner of experimental knowledge, or merely the machination of local sociological forces? The aim of this study is to explore these different interpretations of the Chemical Revolution, elucidate their underlying historiographical principles and philosophical presuppositions, and propose a new interpretation that recognizes the complexity and temporality of this perplexing and elusive historical event. As an exercise in the historiography of science, it is concerned not so much with the Chemical Revolution as an object of historical inquiry as with the different ways in which it has been described and interpreted by its participants and subsequent historians.

Until recently, the Chemical Revolution was the Cinderella of 'scientific revolutions' in the discipline of the history of science.[2] Demurely wedged between her noisier and more noticeable sisters, the Scientific Revolution of the seventeenth century – which involved the birth of modern science – and the Darwinian Revolution of the nineteenth century – which evoked passionate debate about the origin of life and human destiny – the more prosaic issues associated with the Chemical Revolution attracted the interest of only a handful of historians and historically-minded chemists. This situation changed in the 1960s and 70s however, and the intervening years have witnessed almost as many studies of the Chemical Revolution as occurred in the preceding century. This resurgence in scholarly interest coincided with a historiographical ferment in the discipline of the history of science, in which the long-held positivist and Whig view of science as a teleologically structured body of experimental knowledge gave way, first in the 1960s to the postpositivist view of science as theory akin to speculative metaphysics, and then in the 1970s and 80s to the postmodernist view of science as a sociological entity shaped by the contingent constraints of specific agents practising in local contexts. Current interpretations of the Chemical Revolution, which run the gamut from the materialist view of an unchanging past to the idealist sense of its continual 'reconstruction' or 'renewal' by our attempts to understand it, are shaped by the complex intermingling and sedimentary layering of these interpretive styles.

This period of historiographical ferment and renewal is exciting and invigorating for specialist historians of chemistry, leading them into a deeper and more extensive understanding of the arcane details of their object of inquiry. But without the assistance of texts designed to offset the difficulties in communication that accompany the fragmentation involved in increased specialization, the resulting situation can be daunting for the student and intelligent lay-reader and problematic for the general scholar. This study tackles the problem of intelligibility and communication by revealing and exploring, beneath the welter and multiplicity of diverse specialist interpretations of the Chemical Revolution, the presence and influence of a small number of interpretive patterns, forms or styles, grounded in clearly defined sets of historiographical strategies and philosophical themes intelligible equally to the student, lay-reader, scholar and specialist. But this reflexive activity is not only an exercise in clarification and communication, intended to make accessible to a wider audience the results of more narrowly conceived scholarly interpretations of the Chemical Revolution; it is also designed to further these scholarly inquiries by exploring their philosophical foundations and, hence, their proper scope and implications. Hopefully, it will also whet the growing appetite among historians, philosophers and sociologists of science for historical and critical accounts of their own disciplines as an end in itself. With these considerations in mind, this study offers, in tandem with a history of the history of the Chemical Revolution, an in-depth account of the broader philosophical and cultural movements that encapsulated these histories.

This kind of inquiry will be received differently by philosophers or sociologists of science, who are concerned with general historiographical issues and problems, and historians of science, who seldom explore the conceptual and methodological foundations of their work. Largely an empirical discipline, the history of science is marked by an 'individualism and particularism' that discourages explicit theorizing and which defines the task of the historian of science primarily in terms of specific bodies of empirical material and their accompanying narratives. This defensive empiricism characterized the reaction of an older generation of historians of science to the hegemonic inroads of philosophers of science like Imre Lakatos, who assigned to the historian of science the derivative task of providing 'rational reconstructions' of past science in terms of preconceived philosophical models of scientific knowledge and rationality. More recent historians have moved away from this defensive posture, manifesting a greater understanding of the inescapable presence of philosophical and historiographical presuppositions in their empirical inquiries and revealing a greater familiarity with the views of philosophers and sociologists of science such as T. S. Kuhn, Bruno Latour, and Michel Foucault, who in turn showed a greater willingness to learn from the history of science.[3] Nevertheless, the majority of historians of science continue to treat philosophy and historiography instrumentally, reaching into the toolbox of interpretive devices and strategies to illuminate or solve the empirical problems under scrutiny, but paying little or no attention to the philosophical intent or overall historiographical implications of the chosen analytical devices. Whether practising historians should, in the course of their everyday inquiries, become more historiographically self-conscious or continue in their instrumentalist ways is a question that is not addressed in this study. Whatever form the workaday relation between historical inquiry and historiographical reflection takes – whether they are actively mingled in the work of each investigator or divided among a community of specialists – this study aims to open up more cogent and compelling lines of communication between these distinct, but equally necessary, aspects of historical inquiry. Moving at times far afield from the paradigmatic domain of the history of chemistry, the basic aim of this study is to provide a clear and systematic account of the philosophical and historiographical dimensions of past and present interpretations of the Chemical Revolution which will not only sharpen our appreciation of the real significance, relative worth and scope of available and competing interpretations of this crucial event in the history of chemistry, but will also accelerate its future comprehension. The burden of this introduction is to map out the more prominent features and contours of the philosophical and historiographical terrain occupied by these interpretive patterns, forms and strategies.

Discipline of the History of Science

Interpretations of the Chemical Revolution have been shaped by interpretive strategies associated with the emergence and development of the discipline of the history of science. Since its inception in the Enlightenment, the discipline of the history of science has occupied a contested intellectual terrain, shaped by philosophical and ideological forces generated by the development and cultural entanglements of science itself.[4] In the eighteenth century, Jean d'Alembert and Joseph Priestley linked the discipline of the history of science to Enlightenment programmes of social and intellectual reform, while Adam Smith viewed it as a species of 'philosophical history' designed to elucidate the universal principles of the human mind.[5] During the last two centuries, a broad spectrum of religiously minded historians of science, including Joseph Priestley, Pierre Duhem and Stanley Jaki, also treated the history of science as 'a focal point of debate in the conflict between secular and religious cultures'.[6] While such religion-based historiographies played a significant role in shaping past and present interpretations of the Scientific Revolution, which involved a break with medieval religiosity, and the Darwinian Revolution, which rejected Christian accounts of the creation, they played no discernible role in the formation of past and present interpretations of the Chemical Revolution. This is decidedly not the case with positivist historiographies predicated on the repudiation of religion (and metaphysics); these historiographies played a crucial role in the work of nineteenth- and early twentieth-century historians who linked the Chemical Revolution to the emergence of chemistry as the first 'positive' science.

Reinforced by Whig views of political and cultural progress, nineteenth-century positivism developed philosophies of history designed to uphold the Enlightenment view of the progressive disentanglement of science from non-science. Positivism's view of the 'cognitive monopoly of science' shaped the influential historiography of science developed by George Sarton in the opening decades of the twentieth century. Calling for the development of a 'new humanism', which tied human progress to scientific progress and sought to humanize science in a way that increased its present and future progress, Sarton appropriated the discipline of the history of science to the justificatory and celebratory needs of science itself. The resulting historiographical sentiments played an important role in the efforts Sarton, Charles Singer and William Osler made in the first part of the twentieth century to establish the history of science as a recognizable and useful academic discipline and profession.[7] Somewhat eclipsed by the long and gloomy shadow of World War Two, these celebratory sentiments resurfaced among some historians of science in the 1980s and 1990s, as can be seen, for example, in Bill Brock's *Fontana History of Chemistry*, according to which 'the history of chemistry not only informs us about our great chemical heritage but justifies the future of chemistry itself'.[8] More recently still, a group of historians of biology, eager to resurrect Sarton's original perspective, have advocated the study of the history of science as a means of 'illuminating science and making it better', claiming that the adoption of this perspective 'on a larger scale' would 'transform not only science but the discipline of the history of science as well'.[9] The philosophical underpinnings and historiographical implications of positivism's justificatory and celebratory approach to the history of science and its impact on the historiography of the Chemical Revolution will be explored in Chapter 1.

The tumultuous years of the 1960s and early 70s saw the deconstruction of the positivist hegemony in the history and philosophy of science, together with the emergence of 'postpositivist' sensibilities in the works of W. V. O. Quine, Karl Popper, T. S. Kuhn, Imre Lakatos, Paul Feyerabend and Larry Laudan, among others. As will be seen in Chapter 4, postpositivism shared positivism's modernist heritage, generating justificatory and celebratory analyses of scientific knowledge and its unique rationality. But, in keeping with the speculative tenor of the time, postpositivism also adopted a less certain and more independent epistemic stance towards science. Besides replacing the positivist model of empirical certainty in science with fallibilistic accounts of its theoretical structure and status, postpositivist philosophers of science sought to project onto the history of science philosophically derived models of progress and rationality. Instead of appropriating the history of science to the justificatory and celebratory needs and interests of science itself, postpositivist historians and philosophers of science appropriated the history of science to the aims and interests of the philosophy of science. They sought the 'rational reconstruction' of the history of science in terms of philosophical models of rationality. The philosophical roots of these models and their influence on our understanding of the Chemical Revolution will be discussed in Chapters 2 and 3. This account will offset the tendency to downplay the historical significance and specificity of postpositivism that results from treating it as either a minor modification of positivism or a brief prelude to the emergence of postmodernism.[10] Despite its brief ascendancy in the pantheon of interpretive styles, postpositivism had a profound and productive influence on the historiography of the Chemical Revolution, and this study should enable us better to understand and appreciate its complex, variegated and important results.

But the positivist thesis of the epistemological and utilitarian pre-eminence of science did not go unchallenged by philosophical and cultural forces and movements designed to combat the hegemonic impulse of modern science, as can be seen in the emergence and development of Romanticism, German Idealism and American Transcendentalism in the nineteenth century, and Continental Phenomenology, Hermeneutics and Critical Theory in the twentieth century.[11] A similarly critical and sceptical stance towards modern science and its positivist boosters shaped sociological accounts of the development of science which came to the fore in the wake of postmodernism in the 1980s and 90s. As science developed in the nineteenth and twentieth centuries into a force of daunting proportions in the economic, political and cultural affairs of modern society, and as scientific beliefs and practices became more and more insinuated into the complex structures and institutions of those societies, so science itself became an unavoidable object of inquiry for anyone trying to understand those structures and institutions, as well as an obvious focus of opposition and concern for anyone adversely affected by them. Although this levelling stance was absent from the well-known Marxist historiographies of science that burst upon the Western scene in the 1930s, the perception by philosophical and ideological friends of science that 'by displaying its banausic and practical origins', these materialist historiographies devalued science strengthened the hand of idealist and postpositivist historiographies of science that flourished in the United States during the anti-communist fervour of the 1950s and 60s.[12] As will be seen in Chapter 2, idealist and postpositivist philosophers of science enjoined historians of science to focus on the 'internal', conceptual content of science and to shun, or downplay, any reference to its 'external', social or psychological, circumstances, thereby countering the threat to the autonomy and rationality of science they perceived in Marxist, Freudian and other naturalistic accounts of science.[13] But, as will be seen in Chapters 4 and 5, sociologists of scientific knowledge working in the 1970s and 80s, in turn, rejected these celebratory and idealist historiographies of science, adopting instead a more neutral naturalistic approach towards the epistemological and utilitarian aspirations of science. These scholars treated science not as a privileged body of knowledge, but as a social activity on a par with other social activities. Challenging the normative distinction between the internal (cognitive) content of science and its external (material) circumstances, they deployed the concepts and methods of the social sciences, including sociology, psychology and anthropology, to describe and explain, rather than reconstruct and justify, scientific thought and practice. They replaced 'rational reconstructions' with 'sociological reconstructions' of the history of science. These developments in the sociology of scientific knowledge had a profound and lasting impact on the discipline of the history of science, and as will be seen in Chapter 6, this impact left its mark on the historiography of the Chemical Revolution.

The historian of science I. Bernard Cohen summarized the state of the profession and the multitude of historiographical options available to the discipline of the history of science in a 1977 paper entitled 'The Many Faces of the History of Science – A Font of Examples for Philosophers, a Scientific Type of History, an Archaeology of Discovery, a Branch of Sociology, a Variant of Intellectual or Social History – *Or What?*'[14] Partly rhetorical and certainly designed to celebrate the rich potential and interdisciplinary nature and significance of the discipline of the history of science, Cohen's question has lost none of its initial relevance and significance in the intervening years. On the contrary, subsequent developments in science studies have, if anything, served to render it more pressing and significant than when Cohen first formulated it. This study responds to Cohen's question with an exploration of the different disciplinary sources and pressures that have shaped past and present interpretations of the Chemical Revolution and, in Chapter 7, it answers his rhetorical 'Or What?' with 'History!' Instead of subsuming the history of science under the disciplinary hegemony of science (positivism), philosophy (postpositivism), or sociology (sociology of scientific knowledge), the interpretive model developed in Chapter 7 – 'robust contextualism' – emphasizes the historicity and, hence, temporality, of these appropriating disciplines.[15] Robust contextualism integrates the history of science in general and the Chemical Revolution in particular into a dynamic, open-ended historical ontology based on the specificity and irreducible complexity of *historical* events. In offering a substantive alternative to the interpretive models of the Chemical Revolution discussed in the rest of the book, Chapter 7 pays more attention to the Chemical Revolution as an object of inquiry, but like the previous chapters its main concern is with the method, sources and presuppositions of that inquiry.

History of Science as History

As a metadiscipline, the object of which is the discipline of history, historiography is concerned with the interpretive presuppositions and principles that make possible and give coherence to historical inquiry. Like the objects and activity of that inquiry, historiography is itself an object of history, and shares in its multiplicity and complexity. Indeed the turbulent history and methodological variability of the discipline of the history of science lends credence to R. G. Collingwood's claim that 'no historical problem should be studied without studying ... the history of historical thought about it'.[16] The inextricability of history and historiography, expressed in Collingwood's notion of 'historical criticism', or 'the history of history', informs the current interest that many historians of science have in 'the history of science as history'.[17] These reflexive sensibilities have surfaced recently among analytical philosophers and philosophers of science who hope to revitalize the hybrid discipline of the history *and* philosophy of science by stimulating a renewed cooperation between historians of science and philosophers (and sociologists) of science estranged in the wake of Kuhn's relativistic account of the history of science.[18] Adopting the spirit of this renewed endeavour, this study calls upon historians of the Chemical Revolution to pay close and critical attention to the current state, future development and philosophical status of their interpretive enterprise. It canvasses scholarly, critical and hermeneutic consideration to support this reflexive turn.

Reflecting on the state of current scholarship, a number of historians of eighteenth-century science and the Chemical Revolution have noted how in the absence of more synthetic interpretive considerations, the proliferation in the last forty years of studies dealing with particular problems, issues and individuals in the Chemical Revolution has produced a state of uncertainty verging on 'crisis', characterized by a breakdown of communication and the proliferation of dissensus in the historical community.[19] While dissensus itself is not a bad thing, clear guidance and critical reflection on the central themes and problems of the discipline would not only facilitate better communication and coordination among its practitioners, but also make the job of the student and lay-reader a lot easier and more productive. Some steps have been take in this direction; but they are incomplete and partial, consisting in a few interpretive articles and anthologies summarizing the state of scholarship on selected problems and issues. However, there is still no systematic and coherent account of past and present interpretations of the Chemical Revolution, detailing and assessing their problems, methods and presuppositions. This study is designed to fill that gap.

As an exercise in the history of ideas, or intellectual history, the main body of this study shows how, since World War Two, the discipline of the history of science has passed through three historiographically distinct, but temporally overlapping, stages, yielding significantly different interpretations of the Chemical Revolution. The first stage, which stretched from the early nineteenth century to the middle of the twentieth century, projected the positivist-Whig view of science as a unitary and progressive body of experimental knowledge grounded in an algorithmic method of inquiry. This perspective gave way in the 1960s to the postpositivist identification of science with theory and the history of science with the exfoliation of global theoretical doctrines and research traditions. No sooner had postpositivism established itself in the community of historians and philosophers of science than it was challenged, in the 1970s, by the sociology of scientific knowledge, which treated science as a social activity rooted in the self-interested activities, or practices, of specific agents in local contexts. The result is a reflexive narrative of the Chemical Revolution that draws attention to its multiple sources and variegated perspectives.

The critical dimension of this study links this survey of past and present interpretations of the Chemical Revolution to the formation of an interpretive model that weighs and balances the strengths and weaknesses of available perspectives and offers an alternative approach that is both novel and fruitful. It uses the logic of 'conjecture and refutation', associated with Collingwood's notion of 'historical criticism', or 'the history of history', to link the description of past and present historiographical practices to the prescription of future historiographical strategies. Thus it describes different theories of history in the temporal order in which they occurred, while relating successive theories to attempts to remove the empirical and conceptual difficulties encountered by earlier ones. More particularly, in the last chapter, it explores and evaluates accounts of the Chemical Revolution based on the disciplinary directives of the disciplines of science, philosophy and sociology from the perspective of a *historical* model of the Chemical Revolution. But it does not posit a logic of history, based on the elimination of error and the discovery of a greater truth. Rather, it envisages a terrain of historiographical and philosophical diversity and multiplicity, reflexively mapped to reveal its objective contours and thereby to facilitate an easier passage for those who care to make this interesting and fruitful intellectual journey.

The success of this reflexive study of the history of the history of science requires further consideration of the familiar topic of the Whig interpretation of history and its powerful, pervasive effect on historians of science. Over forty years ago, in his epoch-making work *The Structure of Scientific Revolutions*, Kuhn highlighted the aspirations of a new generation of historians of science when he drew attention to an ongoing 'historiographical revolution', in which the traditional Whig concern with 'the permanent contribution of an older science to our present vantage' began to yield to contextualist attempts 'to display the integrity of the science in its own time'.[20] The notion of a historiographical struggle between the competing forces of whiggism and contextualism governed subsequent reflections on the state of the discipline of the history of science, with most commentators supporting Kuhn's view that the Whig interpretation of history was fast approaching oblivion. But, as will be seen in the course of this study, this conclusion was premature in the extreme. Not only did Whig (and positivist) sensibilities remain implicit, but undiminished, in a number of ostensibly anti-Whig interpretations of the Chemical Revolution generated by postpositivist and sociological historians of science, some scholars in the 1980s and 90s still insisted that, given the uniquely progressive nature of science, 'the whiggish idea of progress had inevitably to be built into the history of science'.[21] As the historian Herbert Butterfield noted over sixty years ago, whiggism is a highly viscous, intractable and alluring ideology. It is, according to Butterfield, 'the historian's pathetic fallacy', which involves 'an unexamined habit of mind historians may fall into' when they momentarily step back from their detailed research to communicate their results to a wider public.[22] Ironically, later in his own life, Butterfield succumbed to this fallacy when he abandoned the contextualist historiography of his 'misguided' youth in order to 'celebrate this Whig inheritance of ours with a robust but regulated pride'.[23]

This is not to deny that there has been real movement among recent historians of science along the axis from whiggism to contextualism. But, as this study will show, this movement has been more ambiguous, incomplete and confused than is usually supposed. It is furthermore accompanied by an impoverished (reflexive) understanding of whiggism, which overflows into a narrow and distorted assessment of the nature of the contextualist challenge and alternative historiographies. The tendency to identify the Whig interpretation of history with a retrospective, celebratory view of the past ignores the more abiding and significant ontological, epistemological and axiological underpinnings of this historiography. As will be seen in Chapter 1, an appreciation of the 'deeper' philosophical principles and presuppositions, concerned with the nature of agency, the role of structure and the place of values in history, is essential to a proper understanding of the different forms of whiggism (and positivism), its persistent hold on the imagination of historians and the variety of contextualist alternatives formulated more recently.

Similarly, the transition from positivist to postpositivist historiographies of science can be understood and evaluated only in relation to the more fundamental transformation in ontological, epistemological and methodological conceptions of science brought about by the emergence of theoreticism. Theoreticists replaced the positivist ideal of certain knowledge, grounded by algorithmic methodological procedures in the indubitable world of sense experience, with the notion of an inherently fallible knowledge, underdetermined by methodological principles and procedures, and dependent for its meaning and significance on prior theoretical presuppositions and commitments. As will be seen in Chapter 2, fallibilism encompassed a range of historiographical strategies generated by lively debates between realists and idealists about the ontological significance of scientific theories, between rationalists and relativists about the epistemological status and standing of scientific knowledge, and among methodologists who disagreed about whether to prioritize the conceptual (foundational) or the empirical dimensions of science.

Chapter 4 will show how the ensuing clash between postpositivist and sociological interpretations of the Chemical Revolution involved a 'deeper' polarity between theoreticism and sociologism, between a realist or globalist philosophy of science, which emphasized the unity of science and history, and the nominalism of postmodernism, which stressed the irreducible multiplicity and diversity of scientific practices and historical contexts, including those of the sociology of scientific knowledge itself. An eclectic mixture of Continental philosophical influences, emanating from phenomenology, ethnomethodology, hermeneutics and poststructuralism, coalesced with Anglo-American philosophical sensibilities, associated with pragmatists, Wittgensteinians, Kuhnians, New Historicists, Cultural Marxists and the 'new empiricists', to shape the sociological notion of 'science as practice'. This notion generated a bewildering array of interpretations of the Chemical Revolution, depending on whether 'practice' was identified as material, instrumental, experimental, discursive, investigative, didactic, rhetorical or cultural.

Chapter 4 will also consider the role played by deeper or broader philosophical and cultural changes, associated with the transition from modernism to postmodernism, underlying the replacement of philosophical – positivist and postpositivist – with sociological interpretations of the Chemical Revolution. The struggles and divisions between philosophers of science and sociologists of scientific knowledge had a profound effect on the historiography of the Chemical Revolution, bearing witness to the more epoch-forming conflict between the modernist problematic of autonomy and legitimation, embedded in the notions of formal reason, representational thought and progressivist history, and postmodernist naturalism and nominalism, with its 'spacialized' sense of a fragmented and static world of fleeting simulations and interventions immune to the coherent movement of modernist 'temporality'.

The critical dimension of this study also points to the importance of 'deeper' philosophical principles and presuppositions in the formation and development of historiographical strategies and interpretive models. It argues that existing accounts of the Chemical Revolution are not *historical* to the extent that they subsume history under the disciplinary interests and strategies of science, philosophy or sociology. Instead of grasping the Chemical Revolution as a product of history, a specific mode of temporality, they view it as a scientific discovery, a moment of rationality, or a matrix of social interests that happen to have occurred in the past. In order to evaluate these disciplinary intrusions, and to answer Cohen's question 'or What?' with an unequivocal 'History', this study highlights the need to develop a clear sense of the priority and irreducibility of history and the methods used to study it. Instead of approaching history scientifically, philosophically or sociologically, it treats science, philosophy and sociology historically. Instead of viewing chemistry as a well-defined science, 'which has a history one can choose to study or ignore', it 'envisages this science as the *product* of history ... [as] a history in progress'.[24] Recalling the famous claim of Marx and Engels that there is only one science, 'the science of history', the final chapter of this study deploys consideration derived from historical materialism to articulate its sense of the priority and irreducibility of history.

The claim that existing accounts of the Chemical Revolution are not *historical* is intended not as an evaluation of their intellectual worth or scholarly competence, but as an assessment of the historiographical styles or patterns of interpretation that inform them. Following the path delineated by Louis Althusser and Michel Foucault in the 1960s, Chapter 7 argues for the priority and irreducibility of history by linking it to the concept of 'complexity'. The complexity involved here is not one of detail, or 'incidental circumstances', which can be captured in more adequate and complete philosophies or sociologies of science.[25] Historical complexity is fundamental; it is rooted in temporality and is irreducible to the unfolding of scientific experience, the instantiation of formal structures, the realization of material interests, or any simple conjunction thereof. To insist on the complexity of historical events in this sense is to reject the idea that the Chemical Revolution had a defining essence, or crucial moment, which ordered its different aspects into a unitary historical process. But this is not to defend a pluralistic conception of historical events as the simple conjunction of absolutely autonomous factors. To insist on the complexity of historical events is to treat them as 'decentered totalities', the parts of which display a 'relative autonomy'.[26] This study approaches the positive characterization of the notion of historical complexity and its applicability to the Chemical Revolution laid out in the last chapter by first examining existing histories of the Chemical Revolution and the way in which this concept eludes them.

The interest of historians of science in the history of the history of science is reinforced by hermeneutical considerations that make the consequences and subsequent interpretations of a historical event integral to the identity and meaning of the event itself. The notion that history is a narrative, or a story, which has a plot with a beginning and an end, supports the idea that an event becomes a historical event only in relation to later events, and that historical descriptions of an event are richer than empirical observations made at the time of the event. Unlike natural events, which have no history and are completely available for scientific analysis, historical events, as integral parts of patterns of historical change, are forever incomplete, dependent for their identity and meaning on subsequent events and developments.[27] Thus, when we claim that World War Two started in 1939, we consider the events of the year 1939 from the retrospective view of a war that ended in 1945. This view is shaped by events subsequent to 1939, and hence is beyond the interpretive focus of someone living in 1939. The identity and meaning of a historical event, in this case World War Two, is open to modification by subsequent events and consequences and the interpretations that accompany them. On this view, the goal of the historian is not the traditional unrealizable goal of a final objective knowledge of an independently existing past, so much as the construction and reconstruction of its ever evolving identity and meaning. Merging history with the history of history, some recent historians of the Chemical Revolution have tried to display not the mythical original identity and meaning of this event so much as its unfolding identity and meaning through time.[28]

Historiography and the Historian

As a metadiscipline, which takes as its object of study the discipline of history, historiography identifies or enunciates the philosophical commitments, ideological contexts and technical problem-field of a given historical analysis. A particular historical analysis can be philosophically classified as materialist, idealist, empiricist, historicist, structuralist or some hybrid of these and other philosophical styles and strategies, depending on its overall view of history and the nature and place of evidence, concepts and objects in historical inquiry.[29] Given the disciplinary and historical proximity of the history of science to other disciplines, especially philosophy and sociology, and given the ferment in this disciplinary matrix, it is not surprising to find considerable philosophical variety among historiographies of science. But this variety and multiplicity is not without some underlying unity and uniformity, provided, in part at least, by a pervasive polarity between, on the one hand, idealist and relativist philosophies of science and, on the other hand, materialist and realist philosophical sensibilities. As this study will show, idealist and relativist sensibilities, which prioritize scientific concepts over practices, assimilate objects to their interpretations, and question the capacity of science to represent an independently existing reality, have had a long and lasting impact on the historiography of science in general and the Chemical Revolution in particular. But the final chapter of this study will opt for the other side of the philosophical ledger, offering a historiography of the Chemical Revolution predicated on the representational function of scientific ideas and their embodiment in the material practices of history. More generally, this study registers the fluidity and variability of such philosophical terms as 'realism', 'relativism', 'idealism' and 'materialism', which include local variations among their core connotations.

In its articulation of the ideological and political dimensions of the discipline of the history of science, historiography stresses mutual implications rather than invariable correlations between ideologies and philosophies. Thus, the variety of philosophical and historiographical expressions and manifestations of Marxist politics and ideology, and the different political uses, both conservative and radical, to which Kuhn's philosophy of science can be put, support the more general claim that the relation between the ideological and philosophical dimensions of a historical analysis is historically variable, not conceptually necessary. While these ideological considerations point historiography in the direction of broader issues and considerations than those involved in the philosophical dimension, the technical dimension involves a narrowing of focus. In its narrower technical dimension, the historiography of science articulates and prioritizes the issues and problems relevant to 'particular interpretations of a tightly-defined object of historical investigation', indicating, for example, the relative weight to be given to the role of matter theory and phlogistic chemistry in the Chemical Revolution, the part played by Neoplatonism in the genesis of the Copernican worldview or the validity of the Merton thesis concerning the relation between science and society in the seventeenth century.[30] Although the philosophical, ideological and technical dimensions of a historiography are closely intertwined, the main body of this study will focus on the philosophical and technical dimensions of past and present interpretations of the Chemical Revolution. But this restriction is not intended to downplay the significance of the ideological dimension of these historiographies. On the contrary, the historiographical model delineated in the last chapter of this study emphasizes the integral role of ideological and political parameters in the emergence and development of science and, by implication, its historiography. This narrowing of focus is necessary, however, to keep a vast and highly complex inquiry within manageable proportions. Thus, by and large, this study will consider political and ideological parameters and issues only to the extent that they exerted a direct and specific influence on the formation of philosophical and technical strategies articulated by the historiographies under scrutiny. A more thorough exploration of the complex web of mediated connections between the ideological, philosophical and technical dimensions of the historiographies of the Chemical Revolution considered in this study must await a future inquiry focused more on the political and ideological dimensions of the historiography of science. This strategy is entirely consistent with the historiographical model delineated in Chapter 7, which emphasizes the 'relative autonomy' of these dimensions and, hence, the need to recognize and characterize their 'specificity' before considering their relational identity, or connectedness.

The following characterization and classification of the historiographical principles and practices of historians of science is replete with 'isms'. It is more concerned with systems or patterns of beliefs, doctrines, practices and strategies than with their specific instantiations in particular texts or authors. Going against the grain of the prevailing sociological hegemony in the history of science community, which emphasizes the specificity and contingency of local practices and agents, this study revives an earlier interest in the determining role of temporally extended traditions and 'movements' in history. But this concern with general historiographical patterns rather than specific interpretive practices is not without its problems, many of which will be rehearsed in the course of this study. Among these problems is the vague and somewhat ambiguous nature of the language of 'isms'. Thus, there is no general consensus about the precise meaning of such terms as 'positivism', 'postpositivism', 'realism' and 'relativism', which can vary from discourse to discourse and even within the same discourse. This is notoriously the case with the terms 'modernism' and 'postmodernism', as it is, for example, with the question of Foucault's identity as a 'structuralist' and Kuhn's relation to 'relativism'. Given these semantic vagaries and ambiguities, this study foregoes a systematic definition of its key philosophical terms, allowing them to be characterized more or less precisely as they emerge in the ensuing narrative.

While mainly concerned with historiographical styles and traditions, this study also has something general to say about their instantiation in particular interpretive texts and practices. Thus, it posits a relation of instantiation between a sequence of interpretive styles and a sequence of texts that is complex and uneven. Concerned with the specificity and multiplicity of historical events, working historians adopt an instrumentalist, or opportunistic, approach to the toolbox of available interpretive devices, choosing and using them more in accord with their narrative needs than with the requirements of doctrinal purity and philosophical rigour. When applied to the Chemical Revolution, this narrative focus produces texts which are mixtures or blends of different interpretive styles and modes of exemplification, so that each of the historiographical stages distinguished in this study are characterized by a dominant rather than an exclusive interpretive style. The temporal fit between historiographical paradigms and practitioners is also skewed by the uneven nature of historical development. While some scholars lag behind the curve of history, appearing outdated and old-fashioned in their historiographical affiliations, others stay ahead, playing the role of pioneers and precursors in the development of new historiographical styles and strategies. The relation between texts and styles, scholars and strategies, is further complicated by the interdisciplinary nature of the discipline of the history of science, which enables and encourages its practitioners to draw interpretive inspiration and guidance from such diverse disciplines as philosophy, sociology, literary theory, political history, religion, anthropology and science itself. Given these complexities, this study will focus on the description, evaluation and prescription of historiographical styles and strategies, using individual texts and authors primarily for purposes of exemplification, and leaving to another, longer inquiry a more detailed analysis of the relations between historiographical paradigms and their practitioners. This strategy is supported by the fact that no matter how mediated the relation is between style and text, strategy and scholar, paradigm and practitioner, historiographical styles, strategies and paradigms are real and effective forces in the formation of the interpretive texts and disciplinary identities of individual historians.

Reflexivity and the Priority of History

Historiography is essentially reflexive. It involves historians reflecting on the methods as well as attending to the objects of their inquiry. The intense interest of contemporary historians of science in the nature of their discipline, in questions and recommendations about how and why it should be practised, derives not only from concerns about disciplinary identity and boundaries produced by the encroachment of other disciplines, but also from the reflexive orientation intrinsic to these disciplines. Predicated as they are on general theories of belief formation, the disciplines of the philosophy of science and the sociology of scientific knowledge imported into the historiographical arena the tenet of reflexivity, an unavoidable concern with the application of theories of belief formation to themselves. Latent and implicit in the philosophy of science, this commitment came to the fore among sociologists of scientific knowledge, though it dominated the interests of only a minority of these scholars. While not a major focus of this study, a brief consideration of the variable meaning and significance of 'reflexivity' in the scholarly community serves to highlight significant issues and differences between the philosophical, sociological and historical approaches to the history of science considered in the following pages.

As Larry Laudan noted, any 'perfectly *general* theory' of belief formation, scientific or 'avowedly *non-scientific*', must, on pain of inconsistency, 'necessarily be self-reflexive'.[31] In this vein, Lakatos replaced 'logico-epistemological' criticisms of competing models of scientific rationality with a historiographical one, claiming that his theory of scientific rationality, which identified scientific progress (rationality) with progressive scientific research programs, was progressive (rational) to the extent that it constituted a 'progressive "historiographical" research programme'.[32] But Lakatos's nod in the direction of reflexivity was something of an afterthought, tacked on to the end of his 'History of Science and Its Rational Reconstruction', but frequently omitted from anthologized versions of this seminal paper. David Bloor, on the other hand, made the tenet of reflexivity central to the sociology of scientific knowledge, identifying it as one of four necessary conditions for the scientificity of the 'Strong Programme' in the sociology of knowledge.[33] Whereas the proponents of rational theories of belief formation took the tenet of reflexivity in their stride, viewing it as a source of added strength and reinforcement for their position, sociologists of scientific knowledge encountered it as a problem, which some ignored while others pursued. As both proponents and opponents of the sociology of scientific knowledge agreed, 'if it were true that *all* beliefs were not the result of rational deliberation and enlightened evaluation', but were 'simply determined by the social situation of the believers', then the beliefs of the sociologists of scientific knowledge would themselves 'have no relevant rational credentials', and the whole enterprise would be 'self-indicting'.[34]

Not surprisingly, philosophical critics of the sociology of scientific knowledge were happy with this conclusion, noting that in order 'to avoid being hoisted by his own petard', the sociologist of scientific knowledge must recognize that 'some beliefs are rationally well-founded, rather than socially determined'.[35] Equally unsurprisingly, sociologists of scientific knowledge vigorously resisted this conclusion and its destructive consequences for their discipline, adopting a variety of postures and attitudes – including denial, indifference and pursuit – towards the problem of reflexivity. Those sociologists who pursued the matter transformed the problem from a destructive dilemma into a heuristic challenge, treating it as an opportunity 'to pursue some fruitful lines of exploration which are opened up by a fresh attitude towards reflexivity'.[36] So motivated, these scholars deployed novel textual and rhetorical devices – such as mixed typographies, biographical statements and subversive paradoxes – all designed to provide an account of the world while drawing critical attention to the constructedness of the account and its relation to alternative accounts. As will be seen in Chapters 4 and 5, these sociological responses to the problem of reflexivity endorsed the more general postmodernist sense of history and historiography as fundamentally 'ironic', concerned more with narratives of 'self-concern' and 'self-exploration' than with representations of an independently existing reality, past or present. These chapters will also relate the difference between philosophers of science, who construed reflexivity as an irresolvable dilemma for the sociology of scientific knowledge, and sociological theorists, who grasped it as a research opportunity, to the deeper divide and wider gulf between the problematic of modernism, which valorized criticism and reform, and postmodernist naturalism, which sought to describe and explain rather than criticize and excoriate.

The historical model outlined in Chapter 7 offers a 'third way' of dealing with the issue of reflexivity, which eschews the formalism of philosophers and the relativism of sociologists in favour of a realist and materialist sense of its historicity. It locates the source and solution to the problem of reflexivity in neither the formal adequacy of ahistorical theories of rationality nor the heuristic practices of sociological inquiry, but in the specific historical conjuncture which, while transcending the scope of formal theories and individual narratives, governs and legitimates the production and dissemination of these theories and narratives. Thus, hermeneutical reflection makes conscious the tacit connection between the interpretive concepts and categories of historians and a prior understanding that is rooted in their sociohistorical context. In keeping with the hermeneutical idea that the possibility and objectivity of historical inquiry and understanding hinges on the objective context of dynamic historical traditions, which are embodied in the objects of historical inquiry and critically engaged by the interpretations of those objects, the historical model, in its fullest development, neither reproduces nor reconstructs the past, so much as reflexively mediates it with the present.[37] The vehicles of this mediation are neither formal theories nor individual perspectives, but the temporally extended, materially grounded, critically engaged and mutually interacting traditions and movements of history. Though not pursued in any detail in this study, this construal of reflexivity serves to prioritize history in the manner of the philosophy of historical materialism, according to which humans, nature and society are formed and modified by concrete processes of development through time, rather than by the isolated instantiations of abstract and formal principles and theories or the accumulation and aggregation of the specific practices of isolated individuals.

Scientific Change and the Chemical Revolution

The Chemical Revolution has generally been regarded as the very paradigm of a scientific revolution. It was recognized as a revolution in its own time, and has been so viewed by subsequent generations of historians and scientists. Spanning the last two decades of the eighteenth century, the Chemical Revolution occurred during a period of intense social and political upheaval in Europe, the dawn of a new 'Age of Revolution'.[38] The French Revolution was just beginning, the American Revolution was still unfolding, and the English Civil War was a distant but disturbing memory among the European elite when, in 1789, Lavoisier published the first systematic account of his new system of chemistry. These political events exerted a powerful influence on the thought of the day, leading to the development of new concepts of social, political and cultural revolutions, which were applied to the development of a new historiography of science. According to Cohen, throughout most of the eighteenth century there was 'some confusion and ambiguity about the sense of the word "revolution" in relation not only to science but to political events'. An 'older sense of "revolution" as a cyclical phenomenon ... or a repetition' intermingled with a new use of the term to describe 'a breach of continuity, or a secular change of real magnitude'.[39] This confused mode of thought was appropriate to the political compromise achieved by English society in the Glorious Revolution of 1688, providing contemporary historians and philosophers with a means of understanding Newton's towering contribution to the Scientific Revolution. After 1789, however, the new meaning of 'revolution' came to the fore, shaping the social, political and cultural sensibilities of the modern age. It is in the modern sense of the word that Lavoisier was the first scientist to refer to his work as 'revolutionary'. He associated his scientific activities with the political events of the day, denying the possibility of a return to the 'old order' in either politics or chemistry.[40] Similarly, Lavoisier's great adversary Priestley applied the concept of 'revolution' to the political and scientific realms. Although he championed the American and French Revolutions, he bitterly opposed Lavoisier's Chemical Revolution, regarding it as a 'check' on the chemistry of gases, which was previously enjoying 'the most rapid and promising state of growth'.[41] Nevertheless, Priestley shared Lavoisier's exhilarating sense of living in an 'age of revolutions, philosophical as well as civil'.[42]

Not only did the Chemical Revolution occur in a self-consciously revolutionary period in modern history, it was, prior to postmodernism's historiographical devaluation of temporality, deployed in the service of numerous philosophical theories of scientific change and cognitive development. The suddenness, brevity and pace of the Chemical Revolution, together with the burst of new discoveries and foundational conflicts that accompanied it, marked it in the minds of many commentators as arguably the best example of a classic revolution in the history of science.[43] In this vein, Priestley spoke for his contemporaries, as well as for subsequent commentators, when he said,

> 'There have been few, if any, revolutions in science so great, so sudden and so general, as the prevalence of what is now usually termed *the new system of chemistry*, or that of the *Antiphlogistians*, over the doctrine of Stahl, which was at one time thought to have been the greatest discovery that had ever been made in the science'.[44]

This view of the Chemical Revolution was given its definitive modern form at the end of the nineteenth century, when the chemist and historian of chemistry Marcelin Berthelot used the occasion of the Academy des Sciences's centenary celebrations of the French Revolution to eulogize Lavoisier as the father of modern chemistry.[45] The thesis that the generative phase of modern chemistry involved a fundamental break with previous chemical theory and practice served to unite scholars of diverse historiographical persuasions and otherwise incompatible philosophical sensibilities in the twentieth century.

This study highlights two fundamental 'mentalities' encountered in existing models of scientific change. Emphasizing conceptual permanence, cognitive continuity and cumulative progress in the development of science, the positivist model rendered change in science problematic, if not impossible, and treated an episode like the Chemical Revolution as a transition from a prescientific or nonscientific form of consciousness to a scientific mode of thought. In contrast, postpositivist philosophers of science treated conceptual change, rather than cognitive continuity and permanence, as the normal state of affairs in science, regarding the Chemical Revolution merely as a more prominent feature in a landscape of cognitive upheaval. On both the positivist and postpositivist models, however, the Chemical Revolution appeared as a moment of radical discontinuity, in which Lavoisier ushered in the age of modernity in chemistry by making a fundamental break with prevailing traditions and practices, whether scientific or philosophical. A corresponding incompatibility between the scientific sensibilities of Priestley and Lavoisier was an integral part of these scenarios. Although postmodernist scholars eventually broke with these Manichean sensibilities in their 'polymorphous and multipolar networks model[s]' of the Chemical Revolution, they did so only by abandoning the problematic of temporality and the reality and significance of scientific change associated with it.[46] In contrast, this study stays within the problematic of temporality, drawing attention to interpretations of the Chemical Revolution that offer more balanced accounts of the interrelatedness of the moments of continuity and discontinuity in scientific change and developing, in Chapter 7, a general interpretive model to sustain them. This model challenges prevailing monomial accounts of the Chemical Revolution, which mistake a single level or dimension of science for its complex, polynomial identity. Thus, whereas positivists identified *empirical* continuity with *scientific* continuity and postpositivists identified *theoretical* discontinuity with *scientific* discontinuity, the model outlined in Chapter 7 stresses the multidimensionality of science and the complexity of scientific change, insisting that continuity and cumulativitiy at one level is perfectly compatible with discontinuity at another level. This analysis undercuts the fashionable thesis of incommensurability, according to which competing and historically successive theories are insufficiently incongruous to rule out the possibility of comparison on a shared set of criteria. On the contrary, scientific change, like science itself, is a complex phenomenon, constituted by the intermingling and interconnectedness of contrary moments, such as the continuous and discontinuous, the gradual and revolutionary and the progressive and retrogressive.

Summary and Conclusion

This study offers an exegetical and critical survey of past and present interpretations of the Chemical Revolution, designed to lend clarity and direction to the current ferment of views and perspectives in the historiography of science. Aimed at a mixed audience of scholars, students and interested lay-readers, it adumbrates the philosophical presuppositions of these interpretations, and formulates an alternative interpretation in their stead. Concerned with patterns rather than particulars, with styles rather than the individuals who embody them, it divides the history of the history of science since World War Two into three distinct, though overlapping, stages, each characterized by a dominant, if not exclusive, interpretive style. It relates this sequence of interpretive styles – positivism, postpositivism and the sociology of scientific knowledge – to the emergence and development of philosophical and sociological models of science, and it shows how each of these styles marked the hegemony of science, philosophy or sociology in the historiography of science. It explores within this framework a range of different interpretations of the Chemical Revolution, noting conflicts and tensions between rationalist and relativist, realist and antirealist, materialist and idealist and essentialist and nominalist philosophical sensibilities. Stressing the hegemonic status of science in the modern world, it draws attention to the constitutive and critical role played by a broad spectrum of Anglo-American and Continental philosophical traditions in shaping our cultural image of science and its history; it also references the contours of a cultural upheaval, associated with the transition from the modernist problematic of temporality and normativity to the nominalist spatiality of postmodernist naturalism, which underpinned significant developments and changes in this image. Finally, it outlines an alternative, *historical* interpretation of the Chemical Revolution, based on the idea of the ineluctable complexity of historical events and the priority of history, vis-à-vis science, philosophy and sociology, in the constitution and comprehension of the history of science. The *historical* model of the Chemical Revolution integrates the monomial, linear temporality of modernism and the dispersed spatiality of postmodernism into a more balanced account of the interrelatedness of the moments of continuity and discontinuity, identity and difference, permanence and mutability, in the phenomenon of scientific change.

1 POSITIVISM, WHIGGISM AND THE CHEMICAL REVOLUTION

Logical Positivism was a dominant paradigm in twentieth-century philosophy: it shaped 'virtually every significant result obtained in the philosophy of science between the 1920s and 1950'.[1] Logical Positivism was itself a narrow technical expression of more general philosophical sensibilities associated with the Positivist Movement which first emerged in France at the beginning of the nineteenth century. This movement mingled with the English Whig tradition in political and general history, leading to the idea that the history of science consists in a progressive, teleological struggle between the inexorable agents of cognitive progress and their reactionary opponents. Whig historiographical sensibilities endorsed the positivist view of science as a unitary domain of value-free knowledge, hermetically sealed from metaphysics by the operation of an algorithmic method of inquiry. The mingling of these philosophical and historiographical sensibilities resulted in the hybrid, positivist-Whig historiography of science, which had a long and powerful influence on our understanding of the Chemical Revolution. This chapter relates this influence to variations, within a shared metaphysical framework, on the idea of a unique and defining method of scientific inquiry. It delineates the components of this hybrid historiography – positivism and whiggism – and relates them to deeper and broader philosophical influences associated with the essentialist doctrine of knowledge as inscribed in the nature of things and the historicist notion of an inherent logic of history. Reference to these deeper and broader considerations helps to explain the intellectual hold that the positivist-Whig interpretive model had on earlier generations of historians of science, as well as the difficulties more recent scholars have had with loosening its grip. These considerations also set up a more adequate interpretive framework for understanding the emergence and development of postpositivist and postmodernist interpretive strategies in the history of science in general and the Chemical Revolution in particular.

Whiggism

The term 'Whig history' was first used to refer to history interpreted from the perspective of the English Whig political party, which was formed in 1679. A clearly definable Whig historiographical tradition can be traced from Edward Coke and Paul de Rapin, in the seventeenth and early eighteenth centuries, to Henry Hallam, Thomas McCauley, William Stubbs, Edward Freeman and John Green, in the nineteenth century. The central themes of Whig histories concerned the development of civil and religious liberties, which they associated with Protestantism and the rights of Parliament over the King. Simple in outline, the Whig interpretation of English history encompassed a range of political and ideological variations on the basic theme of the triumphant struggle of the great men of history to secure for everyone the blessings and benefits of constitutional liberties and representational institutions. Towards the end of the nineteenth century, Whig historiographical sensibilities slipped their political moorings and 'the Whig interpretation became the national interpretation'; it survived into the early decades of the twentieth century until it no longer served the ideological needs of an imperial and oligarchic power in decline.[2]

English Whigs and French positivists traded in the Enlightenment currency of the progress of man, which was flexible and varied in its coinage. Celebrating the Glorious Revolution of 1688, English Whigs linked progress to the expanding constitutional liberties of the freeborn Englishman. In contrast, French positivists responded to the social dislocations in post-Revolutionary France by making progress and order 'two aspects, constant and inseparable, of the same principle'. While French positivists wrote philosophical histories designed to elucidate the universal principles of human nature and the general laws of society, English Whigs focused more on the individual, concerning themselves with practical problems and political issues endemic to their own society. They adopted a narrative style in which the facts of history were used not to formulate a general theory of society, but to instruct and edify citizens living in a specific constitutional setting. While positivism sought to establish history as a 'social science', English Whigs assimilated it to the 'moral sciences' of psychology and ethology.[3]

Nineteenth-century Whigs were united in a confident possession of the past, which they celebrated and revered as a source and sanction of what they valued in the present. They balanced conservative preferences for a fixed tradition and radical demands to adapt political institutions to changing socioeconomic circumstances in a philosophy of gradual, ordered progress, in which the living past was organically linked to the vital present. By identifying a 'teleological order in the past', Whig historians tried to 'create a tradition that demanded and inspired emulation in the present'. The study of history met the quasi-religious needs of the nineteenth-century Whig historian by providing 'a source of consolation, direction, and inspiration for his flock'.[4] But Whig (and positivist) faith in the progressive amelioration of the human condition foundered on the harsh realities of late-nineteenth and early-twentieth-century capitalism, collapsing almost completely in the wake of the universal catastrophe of 'the Great War'.

Not surprisingly, theories of historical progress were challenged in the early part of the twentieth century by Spenglerian notions of the 'decline of the west'. As fear and foreboding increased in the decades following World War One, Arnold Toynbee set to work on his multi-volume *A Study of History* (published between 1934 and 1954), which heralded the decline of western civilization. In a similar critical vein, Herbert Butterfield penned *The Whig Interpretation of History*, which was first published in 1931. Butterfield provided a definitive statement and trenchant critique of the 'historian's pathetic fallacy', or ingrained tendency 'to emphasize certain principles of progress in the past and to provide a story which is the ratification if not the glorification of the present'.[5] This document played a central role in shaping the postpositivist challenge to positivism and Whiggism, giving the term 'Whig history' its generally accepted meaning among twentieth-century historians of science.

Contrary to the way it is usually read, Butterfield's text is not primarily a 'negative essay', a monolithic critique of whiggism designed to tell historians 'what history should not be, not what it should be'.[6] Butterfield developed a positive as well as a negative, a prescriptive as well as a proscriptive, dimension to his analysis: he linked his criticism of Whiggism to an affirmation of contextualism. Although Butterfield's contextualist notion of understanding the past in its own terms would today be rejected as philosophically naïve and illusory, he put it to good rhetorical use in highlighting the ontological, epistemological and axiological assumptions that informed the Whig interpretation of history. He argued that whereas contextualism upheld the complexity of human actions and treated historical change as a complex process of 'interactions' in which 'nothing less than the whole of the [complex] past ... produced the whole of the complex present', Whig historians projected onto the past 'an enthusiasm for something in the present', and they sought the teleological thread connecting 'one thing' in the past with 'one thing' in the present. Interested in the agents rather than the processes of history, Whig historians replaced the contextualist view of history's relational unity and complexity with a Manichean duality in which 'the modern world emerge[d] as the victory of the children of light over the children of darkness'. Whig historians focused on a select number of individuals who participated in the transcendental subject, or *telos*, of history; they elevated these individuals above their frail and finite stations in life and placed them in the immortal ranks of the great men of history. Whereas contextualism questioned the efficacy of individual agents and the significance of historical origins, and viewed historical events as the unintended and unpredictable consequences of a complex 'clash of wills', Whig historians referred 'changes and achievements to this party or that personage'. These different and opposing approaches to history involved different and opposing attitudes towards values. Whereas contextualism rejected judgements of origins and values because they overlooked the complexity of historical causation and the relativization of worth to circumstances and consequences, Whig historians used history to formulate 'simple and absolute judgments' about the origins of historical events and the moral efficacy of historical agents, understood in relation to the unfolding *telos* of history.[7] According to this scenario, Whig historians developed an understanding of the past in terms of the present, which delineated the teleological lineage of history and described and evaluated the place of individual historical agents in relation to that lineage.

Positivism

Whig historiography encouraged historians of science to search for the origins of contemporary science in the actions and Manichean struggles of individual historical agents conceived in relation to the present state and historical *telos* of science. But it was only in conjunction with the positivist theory of scientific knowledge that whiggism yielded the more specific and substantive models of scientific change and development that dominated scholarly interpretations of the Chemical Revolution for so long. The emergence of positivism was one of the most significant episodes in nineteenth-century philosophy. It marked the transition from epistemology to the philosophy of science, from the analysis of science based on philosophical views of the nature and limits of knowledge to the identification of knowledge with the activities and achievements of science itself. Providing a 'utopian' celebration of the 'dazzling success of science', positivism's scientistic sense of the pre-eminence of science was what Continental philosophers had in mind when they characterized all Anglo-American philosophy as 'positivistic'.[8]

A full appreciation of the importance of positivism in the discipline of the history of science must take into account the complexity and variability of this protean philosophy. The term 'positivism' was first used by the Utopian Socialist Saint-Simon to designate the scientific method and its extension to the solution of problems in philosophy and society. Adopted by Auguste Comte in the 1840s, it came to designate a general philosophical and cultural movement which exerted a powerful influence on scholars and intellectuals in Europe and America for over a century.[9] The Positivist Movement found its philosophical inspiration in a blend of empiricism, rationalism and utilitarian philosophies developed in the previous century. Combining 'positive philosophy' with 'positive polity', positivism deployed these philosophical resources in an optimistic response to the social and cultural upheavals generated by the Industrial Revolution that linked progress and improvement to the explosive growth of science, technology and industry. The ideological debt that positivism owed to the Enlightenment was evident throughout the Positivist Movement, which defended the liberal and rational gains made by the Enlightenment against the atavistic thrust of the Counter Enlightenment, whether in the form of obscurantist metaphysics, reactionary theologies or conservative political movements. Although this Movement passed through a number of distinct intellectual and institutional stages, it is sufficient for current purposes to highlight and clearly distinguish its initial Comtean stage of development from its final manifestations in the form of Logical Positivism in Vienna in the 1920s and Logical Empiricism in Anglo-American philosophy of science in the 1940s and 50s. By assimilating positivism to Logical Positivism (and Logical Empiricism), twentieth-century positivists and their interlocutors overlooked important differences between the formalism of the Logical Positivists (and Logical Empiricists) and the historicized epistemology of the Comtean School. A just appreciation of these differences must be factored into an accurate assessment of the role of positivism in the development of the discipline of the history of science.

Comte's Historicized Epistemology

Nineteenth-century historicism – which interpreted man, nature and reason as developmental processes in time – and scientism – which identified knowledge with the methods and content of science – shaped Comte's historicized philosophy of science. Rejecting traditional models of science based on philosophical visions of the nature and limits of knowledge, Comte and his followers sought to explicate the norms and criteria of knowledge implicit in actual scientific practice. They replaced epistemology and metaphysics with the philosophy of science and articulated the meaning and significance of science in terms of the progressive nature of its historical origins, development and impact on 'the entire context of life'.[10] Assuming that the past progressed towards the present, positivist and Whig historians of science deployed the principles and criteria of modern science to describe and evaluate past science and to distinguish it from metaphysics, religion and the pseudo-sciences of magic, alchemy and astrology.

The point of departure for Comte's historicized epistemology was his famous law of the 'three stages of human progress', according to which the general history of humanity, the specific history of the individual and the particular history of each cognitive discipline, all pass through a theological and metaphysical stage before entering the era of the 'positive spirit'. These stages are distinguished by a characteristic mode of thought. The stage of the 'positive spirit', or the stage of science, was reached when earlier attempts to discover the hidden underlying causes of natural phenomena, whether gods or forces, gave way to the careful observation of phenomena and the formulation and predictive application of lawlike relations between them.[11] Comte used the law of the three stages of human development to outline a 'historical and systematic' classification of the basic disciplines of science, insisting that the hierarchy of astronomy, physics, chemistry, biology and sociology constituted a 'necessary and invariable subordination' which determined not only the order of their epistemic dependence and historical development, but also the order in which they were to be taught. Comte excluded mathematics and psychology from this order because he regarded the former as the basis of all the sciences and the latter, insofar as it used the method of introspection, as not a science at all.

Comte's developmental view of history was both deterministic and teleological, resulting in an ambiguous historiography of science. Comte's deterministic view of the organic development of knowledge and society encouraged a relativistic respect for the developmental stages of the different disciplines of science; but this respect was mitigated by his teleological view that all the sciences are moving towards the same basic mode of knowledge. The fundamental orientation of the positivist interpretation of science was teleological: it consisted in a presentistic historiography of progress and a strong sense of the unity of the disciplines of science and their essential demarcation from metaphysics.[12] Logical Positivists like Rudolph Carnap and Otto Neurath embraced these views with particular enthusiasm, linking the unity of science to the deployment of a single international method and language of science grounded in the theory-neutral protocols of individual experiences. The 'goals and rhetoric' of the Positivist Movement, from Comte to the Logical Positivists, 'dovetailed with the larger movements of architectural, literary, and philosophical modernism' which were concerned with the 'simple basis', 'international unity [and] progressive nature of their disciplines'.[13] The link between positivism and modernism will be explored more fully in Chapter 4, where its demise will be related to the celebration of complexity, multiplicity and specificity associated with the rise of postmodernism in the 1980s and 90s.

Positivism linked the 'positive spirit' to the methodological rules and procedures that guarantee 'scientific objectivity'. The positivist notion of objectivity combined empiricism, rationalism and Enlightenment epistemological principles and appropriated them to the goal of a methodological definition of science.[14] At the core of this definition was the distinction between the unreality of speculative metaphysics and the objectivity of factual science. Accepting the empiricist principle of the epistemological primacy of sensation and the Enlightenment notion of the methodological nature and unity of rational inquiry, positivists related the 'certainty' of scientific knowledge to the 'sense certainty of systematic observation' and the 'methodological certainty' of an algorithmic and 'obligatory unitary procedure'. In accord with rationalism's emphasis on theory, positivists called not only for the accumulation of facts but also for their theoretical systematization.[15] The rationalist orientation of positivism shaped the influential Covering Law Model of Explanation, developed after the World War Two by the Logical Empiricists, with its emphasis on the formal structure of scientific theories and arguments.[16] At the same time, however, positivists rejected the rationalist or metaphysical notion of an 'absolute' knowledge of the ultimate origins or essences of things. Like Locke, they replaced the metaphysical notion of 'real essences' with the notion of 'nominal essences', understood as objective lawlike relations or regularities among observable phenomena. They also insisted that, in contrast to metaphysics, the scientific investigation of the laws of nature was always unfinished and incomplete. In this non-epistemological sense, scientific knowledge is, for the positivists, 'relative' to the historical situation in which it is generated. Once generated, however, scientific knowledge is 'objective' and cumulative in its development.

Positivism found a middle ground between empiricism and rationalism in the method of hypothesis.[17] Comte rejected the naïve inductivism associated with Bacon and Newton, which insisted that theories are acceptable in science only if they are generated from the phenomena by an infallible logic of discovery. Comte argued that the origin of a theory is irrelevant to its credibility, which depends entirely on its deductive conformity with the phenomena. Embracing a Kantian notion of the theory-ladeness of observation, Comte claimed that science aims not at the observation and accumulation of facts, but at the discovery of laws and theories. Contrary to the spirit of rationalism, however, Comte refused to ascribe ontological significance to hypotheses about unobservable entities; he related the progress of science to the replacement of hypothetical links between facts with an exact representation of the facts themselves. While Comtean positivists embraced a range of views on the methodology of hypothesis formation and evaluation in science, they identified the theoretical with the general and upheld an epistemological distinction between the observational and theoretical languages of science.

This distinction dovetailed nicely with positivism's devaluation of the cognitive significance of metaphysics. Positivists claimed that while metaphysics revels in its own hypothetical constructs, even the most elaborate theoretical constructions of science deal with, or refer to, the real world, though somewhat indirectly. Thus, whereas the observational terms of science, such as 'carbon', 'blue' and 'length', refer to observable entities in the world and observational statements describe their observable properties, theoretical terms, such as 'electron', 'gene' and 'resonance hybrid' have no direct referential significance. According to positivism, the theoretical terms of science either acquire cognitive significance indirectly from the 'correspondence rules' that connect them to observational terms; or they function as heuristic instruments, bereft of any representational significance, for the systematization of observations. The epistemological and ontological devaluation of theoretical terms is integral to the positivist doctrine of sense certainty and is clearly expressed in the doctrine of a theory-neutral observation language. Postpositivism based its opposition to positivism on the thesis of the theory-ladenness of observations and observation statements.

Logical Positivism

Within a shared epistemological and ontological framework, Logical Positivism rejected Comte's view of an integral relation between the history of science and the philosophy of science. As a combination of logicism (the programme to reduce arithmetic to logic) and positivism, Logical Positivism sought to do justice both to the central role of mathematics, logic and theoretical physics in science and to the general positivist programme for the elimination of metaphysics through the articulation and application of an empiricist criterion of factual (scientific) meaning. In keeping with the scientistic sensibilities of their Comtean predecessors, they did not try to shore up the uncertainties of science with a foundational philosophy (empiricism), but to establish the certainty of science and 'secure the scientific status of philosophy'.[18] To this end, Logical Positivists drew a clear and distinct line between, on the one hand, the domain of the logical, the formal, the analytic and the *a priori,* and, on the other hand, the domain of the empirical, the factual, the synthetic and the *a posterior.* Since the statements of metaphysics, theology, and ethics are neither analytic nor experimentally testable, Logical Positivists dismissed them as 'meaningless or linguistically impossible'.

This Manichean research strategy shaped the view that Logical Positivists held about the nature and function of the philosophy of science and its relation to the discipline of the history of science. Since the philosophy of science is clearly not an empirical discipline, Logical Positivists were forced by the threat of meaninglessness to construe it as a logical or formal activity. Identifying the philosophy of science with the logic of science, they developed the techniques of formal logic to characterize and analyse scientific theories as ahistorical, abstract formal structures constructed according to universal criteria of adequacy and a unitary logic of formation. This formalist orientation encouraged the belief that there is only one logic of science, which further supported the more general positivist notion of the 'unity of science'. Within this dualistic framework, Logical Positivists upheld a categorical distinction between the philosophical study of the logic of science and other disciplines, such as the history, psychology and sociology of science, concerned with the formation of actual scientific beliefs and practices. They mapped this categorical division of labour onto the distinction between the context of discovery, in which historians of science describe the historical origins and genesis of particular scientific beliefs and practices, and the context of justification, in which philosophers of science are concerned with formal considerations of the adequacy and acceptability of scientific theories. While Logical Positivists were more than willing to use the history of science to illustrate and explicate formal arguments, as a matter of principle they drove a disciplinary wedge between the philosophy of science and the history of science. They rejected the view, held by their Comtean predecessors and postpositivist successors, that the history of science is an appropriate source or testing ground of philosophical theories of the nature of science and its cognitive development. By the early 1960s, 'historians thought it inappropriate to address epistemological questions in their research' and 'philosophers saw no need to "stoop" to consult the historical record'. Scholars who 'muddled historical origins with logical justification' were accused of committing the 'genetic fallacy', of lapsing into prohibited states of 'psychologism' and 'sociologism'.[19]

George Sarton and the New Humanism

Postpositivist historians and philosophers of science in the 1960s launched a concerted assault on the positivist distinction between the context of discovery and the context of justification. The 'historical school' of postpositivists returned to the Comtean view that an adequate understanding of science involved an appreciation of 'what science had been (hence the historical component) and what it ought to be (the philosophical element)'.[20] Linking the emergence of the philosophy of science to the end of epistemology and the rise of historiography, Comtean positivists used the notions of progress, unity and the epistemological pre-eminence of science to link the history of science to the general history of mankind. These historical sensibilities united scholars across a range of cultural contexts, national boundaries and political interests; they coalesced into a coherent disciplinary and academic identity under the leadership of George Sarton, who articulated the historiographical implications of the positivist philosophy. Upholding Comte's faith in the unity and continuity of humanity, Sarton presented the history of science as 'the most important historical discipline and the very heart of the history of civilization'.[21] Positivists like Sarton looked to the history of science not only as a source of insight into the progressive unity of science, but also as an essential ingredient in the 'new humanism' which, by bridging the gap between the sciences and the humanities, sought the further improvement of science, society and the species.

Sarton used the Comtean doctrine of 'the three stages of human progress' to emphasize the unity of science. Held together like branches on a tree, the disciplines of science are demarcated from nonscientific disciplines by their epistemological, methodological and developmental unity. Linking the unity of science to the unity of nature and the unity of mankind, the positivist historiography of science emphasized links between the epistemological foundations and methodology of science and its international character and chronological development. Embracing the Enlightenment faith in scientific and secular progress, Sarton pursued the 'internal history' of science along the path of 'accumulated knowledge and the discoveries of individual men of genius'. In giving priority to the 'internal', intellectual factors over the 'external', social factors of science, Comtean positivists registered their view of the pre-eminence of science as a spiritual, intellectual and material force in the history of humanity. On this view, the internal development of science is the motive force, not the derived effect, of its external, social and material, circumstances. While internalism remained a cornerstone of postpositivism in the 1960s and 70s, sociologists of scientific knowledge in the 1980s banished it from the history of science altogether.

Within an 'internalist' historiographical framework, positivist historians of science painted portraits of isolated and unyielding heroes 'fighting for Truth against the forces of darkness'. The positivist historiography of genius grew out of the Enlightenment deification of the 'incomparable Mr. Newton' and reinforced the link that eighteenth-century historians drew between the development of science and the emergence of individual freedom and creativity.[22] In accord with Comte's view of the inexorable movement of progress, the positivist notion of 'genius' served to shift attention away from the historical specificities of the work and achievements of individual scientists and towards their lasting contributions to an unfolding body of permanent knowledge. Like the Whig historiography of the great men of history, the positivist historiography of genius involved a revelatory and foundational epistemology, according to which knowledge consists in a pre-established harmony between thought and reality, subject and object, made transparent in the receptive minds of a few gifted individuals. Upholding the eureka-moment notion of a scientific discovery, understood as a 'single event of individual labor', positivism identified the context of discovery with the singular revelatory experiences of individual luminaries and the context of justification with the disciplines and traditions generated by the collective memories and discursive articulations of these singular moments of enlightenment.[23] Understood as the exfoliation of an underlying *telos*, history consisted in the discoveries and achievements of the great men of science.

For Sarton and like-minded positivists, the ultimate value of the history of science, with its story of heroic acts of individual discovery, lay in its service to the 'new humanism': 'The history of science is the history of mankind's unity, of its sublime purpose, of its gradual redemption'.[24] Like the Renaissance humanists, with whom they compared themselves, the new humanists sought to increase knowledge, expand understanding and elevate intellectual and moral standards in order to 'deepen our understanding of human beings and their nature'.[25] Unlike the Renaissance humanists, however, who looked to the past for their inspiration, the new humanists set their sights on the future. They were interested in the past only to the extent that it could be used to shape and influence the future. It followed that the history of science was to be approached as a source of ideas and methods that could be put to some current or future scientific or metascientific use. The positivist historiography of science was presentistic in a twofold sense: both the movement of past science and the historian's comprehension of that movement were oriented towards the present, which was in turn viewed as dynamically linked to the future. Given the positivist notion of the historical unity of the sciences, the history of any scientific discipline involves the same sequence of epochal stages, which 'belong essentially to that science'. On this view, 'the past history of science, its genealogy, is the proof of its validity'.[26] Ultimately, for Sarton and his fellow positivists, the history of science served the interests of modern science.

The instrumentalist view of the history of science as a means for the improvement and development of present and future science reinforced the orientation towards method already present in the positivist doctrine of the unity of science and its demarcation from other intellectual disciplines. The non-critical, instrumental view of rationality favoured by positivism further strengthened the emphasis on method: since the aims and ends of science are teleologically predetermined, the problem of rationality is to figure out the appropriate means or method for the realization of these ends, which are themselves in no need of rational appraisal or justification. Not surprisingly, throughout the nineteenth- and early twentieth century, positivistically inclined historians and philosophers of science tried to 'distil the method by which previous discoveries were made, on the assumption that its continued use would insure future success'. In this manner, Sarton viewed the history of science as 'the history not so much of discoveries as of the method which made them possible'. Similarly, Charles Singer, a twentieth-century pioneer of the discipline of the history and philosophy of science, insisted that the history of science should teach 'the method by which knowledge has been gained'.[27] Like Sarton, Singer placed the history of science in the vanguard of the new humanism. The legacy of methodism was also evident in the *Harvard Case Histories in Experimental Science*, published in the 1950s under the editorship of James B. Conant and designed to provide the reader with 'a unique opportunity of learning at first hand about the methods of science'.[28]

In the spirit of the new humanism, the historiographical focus on method functioned not only to appropriate past science to the internal interests of present science, but also as a strategic bridge between science and its cultural milieu. Both Sarton and Conant presented the history of science as a way of bridging the growing cultural gap between modern science and the educated public. They envisioned the history of science as a means of giving humanists a deeper insight into the scientific method and scientists a keener sense of the cultural significance and implications of their work. They feared that without this shared historical perspective, science would degenerate into a shallow technical enterprise devoid of educational or cultural value, while the humanities would lapse into a rigid and reactionary opposition to progress and all things scientific. Only the history of science, with its vision of the developmental unity of science, society and nature, could prevent this cultural dissolution and its dire consequences for civilization. Comtean positivists regarded the development and institutionalization of the discipline of the history of science as an essential component of a modern civilized society, and the humanist vision of positivism played a significant role in the formation and institutionalization of the discipline of the history of science in the 1920s and 30s.[29] Unfortunately, as will be shown in Chapters 2 and 4, the positivist sense of the progressive unity of science, society and nature did not survive relativist and postmodernist accounts of the perspectival status of scientific knowledge, its ontological separation from nature and its hegemonic function in society.

Essentialism, Historicism and Idealism

Positivism and whiggism acquired their paradigmatic force, thematic unity and cultural hegemony from a set of philosophical principles and assumptions associated with essentialism, historicism and idealism. These assumptions grounded knowledge in the nature of things, posited an inherent logic of history and upheld the identity of thought and things. This philosophical legacy passed through Hegel and Marx and had its origin in Descartes's *cogito,* which grounded knowledge in the certainty of the unmediated presence of the self to itself. Identity philosophy shaped the 'subject-centred' historiographies developed by Continental phenomenologists and existentialists, which varied according to whether they viewed the subject as the source of meaning or the centre of being. The assumption of an identity between the knowing subject and the known object also informed empiricism, which anchored knowledge in the object of experience, rationalism, which located it in the subject, and pragmatism which referred it to a determinate interaction between subject and object. The broad sweep of identity philosophies from Descartes to Hegel validated the power of reason by identifying the laws of thought with the laws of things.[30] An underlying identity between thought and things, subject and object, enabled the scientific mind to discover the laws and regularities that inform the apparent chaos and confusion of immediate experience. This identity was encapsulated in Sarton's claim that the existence of science required 'at one and the same time the unity of knowledge and the unity of nature'.[31] The assumption of the identity of thought and things also informed the positivist notion of 'the methodological certainly of obligatory unitary procedure', as well as the associated distinction between states of cognitive transparency, such as truth, knowledge, illumination, objectivity and rationality, and states of cognitive opacity, such as error, ignorance, prejudice, superstition and darkness, in which reality is occluded or distorted by the unreceptive mind.[32] The dialectic between identity and difference, transparency and opacity, played a central role in the positivist-Whig historiography of the Chemical Revolution, with its view of a Manichean struggle between Lavoisier, the agent of illumination and progress, and Priestley who languished in a reactionary and obfuscating state of ignorance, prejudice and superstition.

The essentialist cast of the positivist theory of science is immediately evident in Sarton's representation of the 'trinity' of 'the unity of nature, the unity of knowledge and the unity of mankind' as the 'dispersion of a fundamental unity' which is itself beyond our cognitive grasp.[33] The principle of expressive causality, according to which nature, society and the mind are the visible manifestations or expressions of an invisible cause, essence or centre of things shaped Comte's notion of the epochal unity of the three stages of human progress and conditioned the more general positivist and Whig belief in an unfolding logic, or *telos*, of history.[34] The idealist identification of reality with thought further reinforced the idea of history as logic unfolding in time. Thus, Whigs and positivists alike reduced the specificity of a historical event to that of a mere moment in a dynamic totality unfolding according to the logical dictates of a predetermined end. The unfolding essence or subject of history imparted continuity to historical development and homogeneity to the developing parts.

In accord with the notion of an unfolding originary essence, historicism involved the postulates of genesis, continuity and totalization.[35] The basic problematic of historicism was that of 'periodization', or the division of history into different periods or epochs representing different stages in the realization of the emergent historical essence. The problematic of 'periodization' shaped Comte's law of the three stages of human progress, as well as the Whig interpretation of the Renaissance, the Reformation and the Enlightenment as progressive stages in the evolution of political freedom from its glimmerings in the medieval Dark Ages to the high noon of Whig constitutionalism. The same historicist problematic informed Butterfield's claim that the Scientific Revolution, rather than the Renaissance or the Reformation, was 'the real origin both of the modern world and of the modern mentality'.[36] More generally, the idealist and essentialist ontological underpinnings of historicism predisposed positivist historians of science to favour a form of historiographical idealism which upheld the image of science as essentially an intellectual search for the truth independent of practical needs and sociocultural circumstances.

The postulates of genesis, continuity and totalization, which are integral to the historicist view of history, implied the homogeneity and continuity of historical time. As expressions of an underlying essence or totality, the stages of development of one section of a society are perfectly coordinated with those of every other section.[37] On this view of historical time, there is an absolute synchronization of the tempo, rhythm and duration of the different aspects of historical development, so that no one aspect outstrips or lags behind any of the other aspects. When positivist and Whig historians did recognize differences in temporalities – such as those between the different stages and rates at which different scientific disciplines pass through the 'three stages of human progress', or those between the different stages and rates at which different cultures and countries adopt Whig constitutional practices and institutions – they viewed them as different levels on the same ladder or scale of history.

The term 'historicism' is not without its ambiguities; it is used in the current literature to denote two contending and mutually inconsistent historiographies. The term is used here to denote the idea that history has an inherent logic or objective rationality, which can be used to formulate historical predictions and evaluations. This is consistent with Popper's influential use of the term.[38] Other authors use the term to emphasize the idea that ways of thinking are embedded in specific historical contexts and can be understood and rationally evaluated only relative to these contexts. Comte's developmental view of history involved the consistent use of both senses of the term by incorporating a relativistic respect for the developmental stages of history into a teleological account of their sequential lawfulness. Since the positivist historiography is fundamentally teleological, the term historicism will be used in this study to denote the idea of an inherent logic and rationality of history; the terms 'contextualism' and 'relativism' will be used to characterize those views that reject the search for objective historical laws and reasons in favour of analysing and evaluating each historical situation in its own terms. This usage will be blurred somewhat in Chapter 4, however, which examines how the New Historicists used the term as a synonym for contextualism.

Positivist-Whig Historiography of Science

Compelled by the essentialist and historicist assumption of the teleological orientation of the past towards the present, positivist-Whig historians of science developed a retrospective, or presentistic, view of the progressive unfolding of past science towards present science. Identifying science with 'the constellation of facts, theories and methods' collected in 'up-to-date science textbooks', they chronicled the 'successive increments' in the 'growing stockpile' of 'scientific technique and knowledge' and 'the obstacles that have inhibited their accumulation'.

In accord with the doctrines of sense certainty, methodological certainty and the demarcation between science and nonscience, positivist-Whig historians created a Manichean division of 'thinkers into two categories variously described as right and wrong, scientific and superstitious, open-minded and dogmatic, observer of fact and speculant'. Emphasizing the cognitive efficacy of individual agents and the eureka-moment notion of discovery, they pursued two main objectives: to determine when, how and by whom 'each contemporary scientific, fact, law and theory was discovered or invented' and to delineate the 'congeries of error, myth and superstition that have inhibited the more rapid accumulation of the constituents of the modern science text'.[39] Problems of chronology, priority and authorship were an integral part of the historicist problematic of periodization, with its concern to delineate carefully and precisely the different stages in the unfolding logic of history. In relating the logic of history to the lives and achievements of the great men of science, positivist-Whig historians identified crucial moments of unity and progress when human beings, through a shared knowledge of nature, moved closer to their common humanity.

In accord with the thesis of the demarcation and unity of science, positivist-Whig historians argued that 'in science there was only one revolution', the Scientific Revolution of the sixteenth and seventeenth centuries. This originary event involved a 'revolution against prejudice and superstition which started the smooth development' of modern science. Articulating the problematic of periodization, with its associated notions of the homogeneity and linearity of historical time and the developmental unity of the disciplines of science, positivist-Whig historians claimed that subsequent revolutions in eighteenth-century chemistry and nineteenth-century biology were either watered-down versions or 'postponed' extensions of the scientific sensibilities that first emerged in astronomy and physics in the sixteenth and seventeenth centuries.[40] They also made use of the theses of the methodological certainty and unity of science to claim that once the revolutionary break was made with prescientific modes of thought, scientists merely had to deploy the appropriate scientific method to ensure a state of gradual and cumulative cognitive progress. Positivist-Whig historians saw in the history of science a clear reflection of the unique rationality of science and its progressive disentanglement from the irrational.

Positivist-Whig historians linked the gradual ascent of past science towards present science to the purposive and highly moral activity of the agents of progress – the 'children of light' – who, in the service of the high ideals of science and humanity, used the one true scientific method to overcome the intrusion into science of the inherently nonscientific modes of thought and practice emanating from the 'children of darkness'. Accordingly, they treated science as an inherently 'consensual activity' in which persistent 'dissensus' marks the intrusion of alien, ideological forces into science and not, as the postpositivists were

later to claim, an underdetermination in the method of science and its application to the problem of theory choice. Positivist-Whig historians were impressed
by the 'high degree of agreement in science' compared with the constant debates
and disputes that characterize such 'ideology-laden fields' as philosophy, sociology and political theory. Stressing the fact that 'consensus' formed in science
around a rapidly changing body of theoretical opinions, they posited 'agreement among scientists at a deeper level', viewing consensus as 'the by-product
of a prior methodological and axiological compact'. In accord with the theses
of sense certainty and methodological certainty, positivist-Whig historians
restricted rational scientific debates to algorithmically resolvable disagreements
over the relative abilities of competing theories to explain theory-independent
facts. This view of the 'linear and cumulative' development of science guaranteed 'that no sensible person could resist the appeal of the new theory'.[41] When
such resistance did arise, positivist-Whig historians traced it to the intrusion of
'external', ideology-laden modes of dissent into the domain of scientific rationality. Although positivist-Whig historians recognized and documented the role of
'external' circumstances in scientific development, they concentrated their energies on the more fundamental task of describing the 'internal' exfoliation of the
facts, methods and theories of science.

Positivist-Whig historians of science distinguished themselves from one
another by adopting significantly different construals of the consensus-producing
method of science. Three versions of this method were particularly important:
inductivism, falsificationism and conventionalism. On the inductivist view,
scientific discoveries consist in 'facts' and their inductive generalizations, and
scientific revolutions involve the transition from states of mere belief or pseudo-
science to a state of proven scientific knowledge. The inductivist 'thesis of instant,
certain truth' in science dominated methodological thought for fifty years after
the publication of Newton's *Principia* in 1687. This thesis was undermined by
eighteenth-century skepticism and the emergence of hypothetico-deductive sensibilities, however, which fostered the less ambitious goal of reaching the truth in
the long run. Unable to show how the method of hypothesis selects truer theories
even in the long run, philosophers embraced the methodology of falsificationism, which focused on the role of facts in disproving theories or hypotheses.[42]
But, as Pierre Durham noted, falsificationists ignored the holistic character of
scientific tests, which renders refuting instances inconclusive for the specific
theory being tested. Treating theories as conceptual schemes for the organization of facts into coherent wholes, conventionalists regarded genuine epistemic
progress in science as cumulative and factual, but treated theoretical progress in
terms of convenience or simplicity and not in terms of truth.[43] Whereas modern conventionalists (i.e. postpositivists and constructivists) relativize both the
empirical base and the criteria of simplicity and convenience to the prevailing

scientific theory or practice, positivist-Whig conventionalists posited an infallible empirical base and shared criteria of simplicity and convenience for the evaluation of competing theories.

In accord with the doctrines of sense certainty and methodological certainty, positivist-Whig historians represented the history of science in terms of direct and objectively resolvable conflicts between the empirical claims of competing theories. While inductivists interpreted 'crucial experiments' as proving the new theory and disproving the old one, falsificationists interpreted them as refuting the old theory and leaving the new one as the unrefuted alternative, and conventionalists interpreted them as establishing one theory as unequivocally more simple than its alternatives. These different methodological perspectives upheld a shared notion of 'instant rationality', according to which the rational credentials of a given scientific proposition are established once and for all in a given crucial experiment and cannot be altered in any subsequent court of appeal. The interrelated notions of crucial experiments and instant rationality harmonized with the Manichean and historicist sensibilities of the positivist-Whig historiography of science, which viewed scientific progress in terms of the inevitable victory of the forces of truth, rationality, objectivity and genuine science over the forces of ignorance, prejudice, superstition and pseudo-science. These notions were also reinforced by the presentist and teleological tendencies of the positivist-Whig historiography, which picked out crucial experiments as 'the parting of the ways from which only two roads lead, a true one leading up of course to us, and a false one leading back into the realm of error and superstition'.[44] Incorporated into the positivist-Whig ontology of agency, this view of the purposive actions of specific individuals and their crucial experiments enabled positivist-Whig historians to judge the theories, facts and practices of past scientists as either good or bad, rational or irrational, scientific or nonscientific, according to the extent of their lasting and unequivocal contributions to the present state of scientific knowledge. In this manner positivist-Whig historians of science succumbed to the psychological imperative behind 'all the fallacies of the Whig interpretation of history': the 'passionate desire to come to judgments of value, to make history answer questions and to decide issues and to give the historian the last word in a controversy'.[45]

Positivist-Whig Accounts of the Chemical Revolution

Manichean Dualism and Cognitive Inversion

The teleological orientation of the positivist-Whig historiography is immediately evident in the idea of the Chemical Revolution as the origins of modern chemistry, as the originary moment in the exfoliation of the facts, theories and methods to be found in modern textbooks of chemistry. Thus James R. Partington claimed in 1937 that Lavoisier's *Traité* read 'like a rather old edition of a modern text-book'. Manichean sensibilities informed the tendency to interpret the Chemical Revolution primarily as a struggle between 'two giants', Antoine Lavoisier and Joseph Priestley, viewed not as fully contextualized historical figures, but as archetypal representatives of the unfolding *telos* of the development of science.[46] Integral to this Manichean teleology was a cognitive inversion, or transition from the 'looking glass' chemistry of phlogiston to the real world of oxygen. In viewing the phlogiston theory as an inverted perspective, which turned 'everything upside down' and was 'as difficult to overcome as any left-handed habit', Butterfield and Charles C. Gillispie expressed the positivist sense of a profound cognitive discontinuity between the phlogiston theory and the oxygen theory. The metaphor of cognitive inversion shaped the claim of nineteenth-century chemists and historians of chemistry that Lavoisier 'did not revolutionize chemistry, he founded it'. The Founder Myth traced Priestley's refusal to recognize Lavoisier as 'the father of modern chemistry' to the illegitimate intrusion into his scientific discourse of nonscientific, ideological factors. Positivist-Whig historians deployed a variety of rhetorical polarities, such as truth and prejudice, reason and authority, science and metaphysics, rationality and irrationality, to dramatize their sense of the sudden and extensive cognitive upheaval that accompanied the teleological emergence of modern chemistry.[47]

Viewing science as the systematization of facts, positivist-Whig historians of science placed at the core of the Chemical Revolution competing and contrasting explanations of the phenomena of combustion and the calcination (oxidation) of metals. Deploying metaphors of inversion, they noted that whereas Stahl and his followers viewed combustion and calcination in terms of the *release* of phlogiston (the principle of inflammability) from the combustible into the atmosphere, Lavoisier and his cohorts viewed the same phenomena in terms of the *absorption* by the burning substance of oxygen in the atmosphere. Stressing the consensual, algorithmic resolution of factual debates in science they tied the choice between these competing hypotheses to Lavoisier's quantitative experiments on the gain in weight of metals and nonmetals during combustion and calcination. They placed at the heart of the Chemical Revolution Lavoisier's experiments in the fall of 1772 and spring of 1773, in which he used his observation of the increase in weight of sulphur and phosphorous during combustion and of lead and mercury during calcination to conclude that combustion involves the gain of oxygen and not the loss of phlogiston. Positivist-Whig historians eagerly contrasted Lavoisier's rational deployment of Newton's principle of the conservation of mass with the failure of 'Phlogistians' to appreciate the rational force of Lavoisier's crucial quantitative experiments. Robert Multhauf spoke for many of his fellow historians when he claimed that Lavoisier ushered in the age of modern chemistry with his 'unrelenting use of the balance' and his clear recognition of 'the significance of the concepts of the conservation of mass and the constancy of chemical composition.'[48]

In accord with the myth of the given and the eureka-moment notion of discovery, positivist-Whig historians linked Lavoisier's experiments on the gain in weight in metals and nonmetals during combustion to 'a vision of something much bigger than a theory of combustion, a revolution in chemistry and physics'.[49] This revolution extended beyond the theoretical and methodological dimensions of inquiry to the epistemological and ontological domains. Emphasizing Lavoisier's link with the corpuscular and experimental philosophies of the seventeenth century, and especially with Boyle's critique of the 'rarified essences' and 'hypostatized qualities' of Aristotelian and Paracelsian chemistry, many historians, including Hermann Kopp, T. L. Davis and Pierre Duhem, drew attention to the way in which the French chemist established for chemistry an ontology of 'simple substances', such as carbon, oxygen and hydrogen, which are isolable in the laboratory and detectable by the balance.[50] On this perspective, the experimental, theoretical and methodological revolution that Lavoisier brought to chemistry entailed an equally discontinuous transition from the classificatory logic and ontology of 'principles' to an ontological schema, developed by David Hume and Etienne Bonnot de Condillac, based on the causal contiguity and concomitance of 'simple substances'.

Other scholars, including George F. Rodwell, John H. White and Douglas McKie emphasized the epistemological dimensions of the Chemical Revolution, claiming that Lavoisier replaced a metaphysical and speculative mode of discourse with a scientific factual mode of inquiry.[51] The doctrine of sense certainty shaped the contrast that positivist-Whig historians drew between Lavoisier's open-minded and objective attention to the facts of the matter and the tendency of his phlogistic contemporaries to allow 'prejudice and superstition' to blind them to the empirical shortcomings of their own hypotheses. In accord with the historiography of the great men of history and the Manichean division between the 'children of light' and the 'children of darkness', positivist-Whig historians drew a sharp contrast between the genuine scientific sensibilities of Lavoisier and the fundamentally religious cast of Priestley's thought. The positivist-Whig metaphor of inversion implied a relationship of cognitive discontinuity between the phlogiston theory and the oxygen theory, the chemistry of Priestley and the chemistry of Lavoisier, which was as radical as that evoked by the later postpositivist thesis of the incommensurability of contending paradigms. Thus, as will be seen more clearly in Chapter 3, postpositivist theories of scientific change involved a modification of their positivist heritage rather than, as is usually supposed, a complete break with their philosophical past.

Founder Myth

The 'Founder Myth' stressed the demarcation between science and nonscience and the determining role in the progress of science of individual men of genius and their eureka-moments of discovery. As the 'founder of modern chemistry', the mythical Lavoisier broke with his immediate chemical past and, in a moment of fiery illumination evoked by his experiments on combustion, established an 'immutable and eternal' order for the future development of chemistry. In keeping with the historicist sense of the homogeneity and unity of historical time, proponents of the Founder Myth presented nineteenth-century chemistry as a continuous 'extension of the Lavoisian enterprise', dismissing the problems with Lavoisier's original synthesis as 'minor excesses' and viewing the nineteenth-century addition of un-Lavoisian concepts, such as 'atoms' and 'affinities', as mere modification or completions of Lavoisier's original insights. In accord with their sense of the role and significance of agency in the development of history, positivist-Whig historians used the refusal of Phlogistians to enter the Lavoisian promised land not to 'raise questions about the nature of science', but to make 'judgments of individuals'.[52]

Despite Lavoisier's early recognition of the continuity of his work with the chemistry of his day, his *Traité* was an important source of the Founder Myth. Most of his colleagues and supporters enthusiastically accepted and promulgated this myth, and the cult of Lavoisier reached its apotheosis in the historical writings of J. B. Dumas in the 1830s. Contrary to the internationalism of its positivist underpinnings, the cult of Lavoisier took on nationalistic overtones during the Franco–Prussian War and the First World War, with Hermann Kolbe, Jacob Volhard and Wilhelm Ostwald objecting vigorously to the claim made by Adolph Wurtz and Duhem that chemistry was a 'French science', with a French father.[53] In the midst of this turmoil, Marcelin Berthelot provided a more balanced and definitive statement of the Founder Myth in *La Revolution Chimique, Lavoisier,* which was published in 1889. By no means the prerogative of French historians of chemistry, the Founder Myth was eagerly embraced by their anglophone counterparts in the nineteenth and twentieth centuries, many of whom proclaimed Lavoisier to be 'one of the immortals of modern science' and the 'Father of Modern Chemistry'.[54] Even Henry Guerlac, who did a lot to shift the interpretive focus of historians of the Chemical Revolution away from Lavoisier and towards the chemical traditions that shaped his work, concluded that even if Lavoisier 'did not create a new science *ex nihilo,* as some earlier writers believed', he changed chemistry so radically that 'the science as we know it today seems to have been born with him'.[55] The wider circulation of the Founder Myth among European intellectuals in the late nineteenth and early twentieth century is evident in the writings of A. N. Whitehead and Frederick Engels, who claimed that Lavoisier 'was the first to place chemistry, which in its phlogistic form had stood on its head, squarely on its feet'.[56] The Founder Myth exerted a tenacious hold on the imagination of scholars and intellectuals for so long because it functioned as a focal point for a nexus of doctrines associated with the positivist-Whig historiography of science, including the myth of the given, the historiography of great men, the Manichean demarcation between science and nonscience and the historicist sense of the homogeneity and linearity of historical time.

The Founder Myth did not portray Lavoisier as acting entirely alone. While it viewed him as breaking completely with his immediate predecessors and contemporaries in the eighteenth-century chemical community, it also linked his revolutionary strategies to the upheavals associated with the Scientific Revolution of the sixteenth and seventeenth centuries. This association secured Lavoisier's rightful place among the 'giants' and 'immortals' of a mythic scientific past. In accord with the problematic of periodization, and the associated notions of the homogeneity and linearity of historical time and the unity and demarcation of scientific thought, positivist-Whig historians explored the historical connections and intellectual affinities between the Chemical Revolution and the Scientific Revolution. Although they tried to explain why 'the emergence of modern chemistry should come at so late a stage in the story of scientific progress', in general, they agreed that in both the Scientific Revolution and the Chemical Revolution, the same principles of enlightenment triumphed over the same forces of obscurantism: In both of these originary events, the scientific principles of analysis, experimentation, quantification and the mechanization of nature replaced metaphysical predilections for speculation, animism and occult principles.[57] In this process of expanding illumination, Lavoisier took up where Robert Boyle and the other scientific luminaries of the seventeenth century left off. Stressing the consensual nature of scientific progress, positivist-Whig historians related the intervening 'Dark Ages' in chemistry to the intrusion into the scientific domain of the inherently nonscientific modes of discourse associated with the phlogiston theory. According to White, for example, the intrusion of the phlogiston theory into chemistry made the transition from Boyle to Lavoisier 'more difficult not more easy'. On the assumption that 'certain workers' in the seventeenth century were 'already more than half-way towards a solution' to the problems of combustion, calcination and the nature of heat and light, the phlogiston theory appeared as a 'step in the wrong direction', as a turn of events in which 'intelligent chemists were sidetracked into a cul-de-sac instead of following the broad way of truth to its logical conclusion'. Butterfield adopted this perspective on the Chemical Revolution when in a notorious reversal of his earlier anti-Whig position, he included in *The Origins of Modern Science, 1300–1800* a chapter entitled 'The Postponed Scientific Revolution in Chemistry'. As White noted, with the advent of the phlogiston theory, 'a number of important results and conclusions dropped out of sight and had virtually to be rediscovered a hundred years or so later'.[58]

Proponents of the Founder Myth offered a number of different perspectives on the generation of modern chemistry depending on their view of the specific historical connections that linked the Chemical Revolution to the Scientific Revolution. Scholars like Arthur Donovan and Simon Schaffer claimed that as part of a 'Second Scientific Revolution' which occurred between 1775 and 1830, the Chemical Revolution brought to practical fruition the disciplinary and institutional ideals of the Scientific Revolution.[59] While these scholars claimed that chemistry, understood as an integrated, coherent and independent science, was a generalized product of the Scientific Revolution, other proponents of the Founder Myth identified one or other of the 'founding fathers' of modern science as Lavoisier's true spiritual guardian and intellectual mentor. But even the promulgators of the new chemistry could not agree to which 'founding father' they owed their patrimonial allegiance. According to Guyton de Morveau, for example, it was by bringing the spirit of Cartesian scepticism to bear on the whole of chemistry that Lavoisier was able to rebuild chemistry 'on a firm and secure foundation' provided by a clear and distinct mode of thought and practice. In contrast, Lavoisier himself invoked the name of Condillac, and by association Locke and Newton, to justify the claim, inherent in the Founder Myth, that genuine chemical knowledge is not derived from received opinions and judgments but from the direct experience of nature. Besides tracing the ultimate source of the Chemical Revolution to the philosophical and methodological theories of Locke and Newton, Lavoisier's version of the Founder Myth also reflected the more general Enlightenment belief in the liberation of the human mind from error and confusion through the clear and precise apprehension of nature.[60]

Chemical Revolution and the Scientific Revolution

The positivist-Whig view that the Chemical Revolution was rooted in the achievements of seventeenth-century philosophers and scientists was part of a broader historiographical picture that emphasized the overarching influence of Newton on eighteenth-century thought and that traced the immediate ancestry of the Enlightenment mind to the scientific philosophies of the seventeenth century. Within this overall historiographical framework, some positivist-Whig historians emphasized a more specific and direct influence of Newton on Lavoisier, representing Lavoisier's principle of the conservation of mass as the methodological corollary of 'Newton's idea of hard permanent atoms'. On this account, 'it was the unrelenting use of the balance and the assumption of conservation' that made Black's quantitative experiments on 'fixed air' (carbon dioxide) so important for Lavoisier and that placed both thinkers in the Newtonian tradition of 'physico-mathematical' chemistry. This version of the Founder Myth incorporated the lineage of Black, Cavendish, and Lavoisier into 'a distinctly "Anglo-Saxon" view of the history of chemistry as the work of great men led by Boyle, Newton, and Dalton'.[61]

Despite the dominating and pervasive influence that the historiography of Newtonianism exerted on earlier historians of eighteenth-century science, positivist-Whig historians interested in the continuity of the Chemical Revolution with the Scientific Revolution turned their attention more frequently to the figure of Robert Boyle. They noted how, in the closing decades of the seventeenth century, Boyle used the mechanical and corpuscular philosophy to establish chemistry as a science independent of medicine and alchemy, to eliminate from chemistry its 'ancient and mystical residues' and to formulate the first clear statement of the problems and concepts of modern theoretical inorganic chemistry.[62] While one school of thought focused on the way in which he deployed quantitative experiments on the calcination of metals to disprove the qualitative ontologies of the Aristotelians and the Iatrochemists and to support the quantitative ontology of 'matter in motion', another line of inquiry drew attention to the way in which Boyle replaced the Aristotelian and Paracelsian notion of a few universal principles of matter with the modern definition of an element as the end-point of laboratory analysis. Either way, it was regarded as unfortunate for the progress of chemistry that Boyle 'did not know how to use in his experimentation his own ideas'. Consequently, he failed to have a significant impact on the development of chemistry.[63]

On these accounts, it was left to Lavoisier, in the early 1770s, to reformulate independently 'many of Boyle's fundamental chemical concepts' and to obtain 'their almost immediate and entirely general acceptance'. Besides giving theoretical and empirical substance to Boyle's abstract definition of an element, Lavoisier repeated Boyle's quantitative experiments on the calcination of lead and tin. According to Multhauf, it was 'by weighing the sealed calcination vessels before and after the process' that 'he proved quantitatively that the gain in weight [of the metals] could not be attributed to the accession of fire particles', as Boyle had supposed. On this account, 'Boyle could have made a hundred years progress in one step', if only he had refrained from opening the vessel – and thereby allowing outside air to rush in and replace air in the container absorbed by the metal – before reweighing it after calcination. In accord with the positivist-Whig historiography of periodization, and the associated interest in origins, priorities and authorship, positivist-Whig historians portrayed Boyle as an isolated 'precursor' of Lavoisier, as someone who failed to become the 'father of modern chemistry' he might so easily have been because the 'time was not ripe'.[64]

Rise and Fall of the Phlogiston Theory

The positivist-Whig presupposition of the essential continuity of the Chemical Revolution with the Scientific Revolution informed the view that Boyle's work evoked little response and theoretical chemistry stagnated until the time of Lavoisier because chemistry in the early eighteenth century became engrossed in the phlogiston theory of combustion, first formulated by J. J. Becher in 1669, and elaborated and further developed by George E. Stahl in the first three decades of the eighteenth century. Positivist-Whig presuppositions of the methodological unity and demarcation of science also shaped the claim that the phlogiston theory functioned as an obstacle to the otherwise inevitable formation of a progressive consensus among eighteenth-century chemists because it was either a methodologically flawed piece of science or an essentially nonscientific mode of thought.[65] This view of the obstructionist role of the phlogiston theory did not prevent a minority of positivist-Whig historians from recognizing, in accord with the deterministic sense of the developmental stages of the history of science, its positive function in eighteenth-century chemistry. Thus, some scholars noted that the phlogiston theory exerted some 'influence for good' by introducing a 'certain amount of order ... among a vast chaotic mass of chemical facts'. Indeed, as White and Gillispie recognized, it was by providing 'the first really unifying principle' that was of any assistance to the experimental chemist that the phlogiston theory 'bridged the gap that existed between empirical chemistry and modern chemistry'. These concessions notwithstanding however, Rodwell insisted that although the phlogiston theory provided a means of 'escape from mystic science', a bridge from arcane to modern chemistry, it was a 'bad bridge, and built upon shifting sands'.[66] A chorus of scholarly voices in the nineteenth and twentieth centuries insisted that the phlogiston theory was experimentally, methodologically and ontologically flawed; it was based on hasty generalizations; it disregarded the new knowledge of gasses; it 'neglected the quantitative aspects of chemical changes ... and paid little attention to the atomic theory'. On this interpretation, the problems and deficiencies of the phlogiston theory were built into its initial formulation at the beginning of the eighteenth century, and they only became more pronounced and insurmountable as the century unfolded. According to Partington and McKie, it was only a matter of time before such an 'imaginary structure built upon loose ties' would 'fall or rather vanish like a phantom'. As Butterfield concluded, 'the interposition of the phlogiston theory' made the transition from 'the chemistry of Boyle and Hooke' to that of Lavoisier 'more difficult rather than more easy'. Progress returned to chemistry not through modifications to the flawed phlogiston theory but by the return to pristine origins involved in Lavoisier's deployment of Boylean concepts and methods.[67]

Positivist-Whig historians located three areas of research and deployment in eighteenth-century chemistry that led to the eventual downfall of the phlogiston theory. Gillispie and A. Rupert Hall noted that despite its early heuristic value, the phlogiston theory 'became a hindrance ... after 1765, when the accommodation of the findings' of gas chemistry began to complicate rather than sophisticate the theory. A number of other historians emphasized how the well-known phenomenon of the gain in weight of metals during calcination revealed that the phlogiston theory was not only a heuristic liability, but an empirically flawed and internally inconsistent hypothesis as well. According to McKie, among others, it was the experiment on the calcinations of mercury in 1775 that was particularly crucial in showing that the phlogistic generalization was 'too hasty and was insufficiently supported by facts'. The fact that mercury became a calx when heated and that the calx reverted to mercury when heated further was inconsistent with a fundamental tenet of the phlogiston theory, according to which the calcination of a metal involved the expulsion of phlogiston and the revivification of the resulting calx required the addition of a phlogiston-rich substance, such as charcoal or inflammable air. A number of positivist-Whig historians used this experiment to stress the general methodological difference between the qualitative basis of the phlogiston theory and the quantitative orientation of the oxygen theory.[68]

The distinction between the qualitative chemistry of the phlogiston theory and the quantitative science of the oxygen theory constituted the high watermark of the positivist-Whig interpretation of the Chemical Revolution. Thus Gillispie noted that although with phlogiston, chemistry became a science, it became 'a qualitative rather than a material one'. In keeping with the historiography of periodization, Gillispie regarded the phlogiston theory as typical of the early stages of any science, 'during which it explores effects as real qualities out there in nature'. According to Gillispie and Crosland, the qualitative ontology of property-bearing principles led the Stahlians to devalue gravimetric analysis and to ignore Boyle's mechanistic programme for the reduction of qualities to 'matter in motion'. Noting that the weight-gain in metals during calcination led some Stahlians to ascribe 'negative weight' to phlogiston, Crosland emphasized how the gradual acceptance of the 'Newtonian tradition of mass' made it 'easier for unprejudiced minds to interpret combustion as a gain of oxygen rather than a loss of phlogiston'. If the Manichean sensibilities of the positivist-Whig historiography conditioned the idea that Lavoisier ushered in the age of modern chemistry by 'cutting phlogiston ... right out of consciousness', then the problematic of periodization, with its view of the developmental inevitability of the events of history, suggested to Multhauf that when Lavoisier 'attended regularly to the weight balances exhibited by chemical reactions, he hardly rejected phlogiston: it simply disappeared'. For positivist-Whig historians of science, the transition from the qualitative phlogiston theory to the quantitative oxygen theory was the very essence of the Chemical Revolution: 'Far more intimately in chemistry than in any other science, the foundations of objectivity were embedded in a quantitative conscience'.[69]

Crucial Experiments

Stressing the myth of the given, the eureka-moment notion of discovery and the cognitive efficacy of individual scientists, positivist-Whig historians analysed the Chemical Revolution in terms of the doctrine of instant rationality, according to which science progresses through a series of 'crucial experiments', or 'crucial moments', in which the great men of science replace older and less adequate observations and theories with new and more adequate ones. The resulting accounts of the Chemical Revolution varied according to the moments in Lavoisier's career that were identified as crucial and the kind of decision that made them so. Thus, whereas Harold Hartley thought that the 'the critical moment for Lavoisier' came in 1772 when he observed the absorption of air and the increase in weight of sulfur and phosphorous during combustion, McKie located it in his 1775 experiment on the calcination of mercury and the 'discovery of oxygen'.[70] More typically, positivist-Whig historians viewed the Chemical Revolution as a *series* of crucial experiments. They argued that Lavoisier started his voyage of discovery in the fiery illumination of the combustion of sulphur and phosphorous, consolidated his initial insight in the experiments on the calcination of lead and tin, and reached the pinnacle of his analytical activities in his quantitative analysis of the calcination of mercury. However, it was only in 1783–4, with the analysis and synthesis of water, that he came to a successful and comprehensive theoretical conclusion.[71]

Positivist-Whig accounts of these defining moments in Lavoisier's career varied according to whether they were shaped by inductivist, falsificationist or conventionalist construals of their algorithmic structure. Treating Lavoisier's experiment on the calcination of mercury as 'a classic example of a crucial experiment, which clearly refuted the phlogiston theory', falsificationists like Hartley and Rodwell ascribed the refusal of the Phlogistians to recognize 'the merits of the new system' and 'the utter falsity of their own' to a 'blind adherence' rooted in 'the conservatism inherent in the human mind'.[72] Conventionalist historians rejected this interpretation however, noting that Phlogistians availed themselves of a number of subsidiary hypotheses for dealing with the weight-gain problem, each one of which suggested that the emission of phlogiston by the burning metal was accompanied by the absorption of some other material from the metal's surroundings. Nevertheless, conventionalists like Conant still insisted that the experiment on the calcination of mercury was the crucial moment in the Chemical Revolution. Claiming that Lavoisier's explanation of the calcination of metals only involved two ingredients – the metal and the oxygen that combined with it – and the phlogiston theory involved three ingredients – the metal, the phlogiston it emitted, and the substance it gained – Joshua C. Gregory argued that it was 'the greater simplicity of Lavoisier's system, and its greater coherence', that 'prevailed over the complexities of phlogistic chemistry'. For conventionalism, the Chemical Revolution illustrated a recurring pattern in the history of science: scientific progress consists in the replacement of empirically adequate but increasingly complicated and ad hoc conceptual schemes with conceptual schemes of greater explanatory power and simplicity.[73]

Intermingled with falsificationist and conventionalist versions of the positivist-Whig interpretations of the Chemical Revolution was the inductivist view that the emergence of modern chemistry involved the replacement of 'premature and erroneous speculation by proven truth'. According to the inductivists, Lavoisier transformed practical 'lore' into theoretical 'science' when he replaced a speculative, erroneous and essentially nonscientific hypothesis with a genuine understanding of the experimental discoveries of his time, arrived at by the judicious deployment of the scientific method and a proper appreciation of quantitative procedures. According to Rodwell, the 'impossibility of experimenting with' the unisolable and imponderable phlogiston meant that the phlogiston theory 'was not founded upon direct experimental evidence'. Rather, it was 'essentially and completely a syncretistic theory', which was 'built upon *idola theatri* collected from various sources' and 'cemented together by the *idola specus* of Becker and Stahl'. When compared with the 'path of true discovery' opened up by Boyle, Hooke and Mayow, and followed by Lavoisier, the phlogiston theory appeared to the inductivists as a 'jungle' of error and superstition, a testimony to 'the perversity of the human mind' and 'an obstacle ... which had to be swept away in the triumphant march towards truth'.[74] The inductivist version of the Founder Myth of the Chemical Revolution, with its Manichean division between the scientific oxygen theory and the nonscientific phlogiston theory, reached its apotheosis in Gillispie's view of Lavoisier's role in Chemical Revolution:

> With a clarity of mind which cannot be too much admired, Lavoisier saw what had to be done to create a modern science of chemistry out of the chaotic legacy in which were jumbled the Greek doctrine of elements, old Stoic vestiges of fire and flux, Paracelsian principles of salt, sulfur and mercury, alchemistical distillations and purifications, mineralogical lore, and the proliferation of laboratory discoveries of his own time. And to clear his mind – perhaps, too, to fix it – he wrote down in his laboratory register what he meant to do for the rest of his life.[75]

Priestley and Lavoisier

Similar Manichean sensibilities informed the explanation positivist-Whig historians gave for the persistent loyalty of some eighteen-century chemists to the phlogiston theory long after Lavoisier's supposed demonstration of the superiority of the oxygen theory. Thus they traced the failure of scientists such as Joseph Priestley to respond appropriately to the moments of instant rationality and scientific consensus involved in Lavoisier's crucial discoveries to the disruptive influence of essentially nonscientific states of mind, such as prejudice, superstition and obstinacy, on their scientific reasoning. Linking the positivist distinction between the scientific oxygen theory and the metaphysical phlogiston theory to a moralistic and Manichean vision of the significance and polarity of human agents in history, positivist-Whig historians transformed the cognitive conflict between the progressive oxygen theory and the regressive phlogiston theory into a titanic moral struggle between the agents of progress and their reactionary opponents. Treating Priestley and Lavoisier as 'giants' of this confrontation, positivist-Whig historians took the cognitive differences that existed between their competing research traditions and strategies and exaggerated, polarized and hypostatized them in a nexus of polar oppositions between the 'intellectual personalities' of the 'protagonists of the Chemical Revolution'. On this account, chemistry progressed because Lavoisier, the theoretician, understood what Priestley, the experimentalist, could only discover.[76] In this vein, positivist-Whig historians accepted at face value the claims made by Priestley and his son that his discoveries were 'accidental', the result of 'no previous theory, unlooked for and unexpected'.[77] Although some early nineteenth-century commentators, such as Humphry Davy and Thomas Thomson, saw Priestley's theoretical naiveté as a scientific asset, enabling him to respond imaginatively and sympathetically to the unpredicted and the unexpected in nature, most positivist-Whig historians saw it as a shortcoming or weakness, robbing his work of real meaning and significance.[78] As Gillispie noted, despite Priestley's untutored prowess in the discovery of new gases, he failed to appreciate the 'Cartesian imperative towards order and unity of doctrine' which determined Lavoisier's more educated and sophisticated focus on the theoretical reform of chemistry.[79]

Positivist-Whig historians deployed a variety of strategies to uphold this moralistic view of Priestley's role in the Chemical Revolution. Praising Lavoisier for using the scientific method to unearth the enduring facts and theories of chemistry, they criticized or patronized Priestley for allowing the inherently non-scientific influences of prejudice, superstition and dishonesty to intrude into his science and to prevent him from appreciating the obvious validity of Lavoisier's arguments and conclusions. The historiography of progress and the associated interest in origins, authorship and priority informed the judgment of many positivist-Whig historians that, despite his many contributions to the development of experimental techniques and the chemistry of gases, Priestley's claim to be 'one of the fathers of modern chemistry' was obviated by his failure to maintain a strict distinction between his scientific interests and his nonscientific concerns. This historiography praised Lavoisier for adhering to the positivist's sense of the cognitive autonomy of scientific theory and practice, while chastising Priestley for the way in which he linked chemical theory to religious and political issues. The positivist-Whig interpretation of the Chemical Revolution reached its apotheosis when it presented Priestley's response to the oxygen theory in terms of a failure to appreciate the progressive significance of Lavoisier's achievement, rooted in a pathological deviation from reason inherent in the unwarranted intrusion of metaphysical, theological and political forces into the domain of scientific discourse and debate.[80] Positivist-Whig historians did not present Priestley and Lavoisier as fully contextualized historical figures, enmeshed in the complex circumstances of their individual lives, but as the transcendental, almost mythical, great man and his shadow, archetypal representatives of the conflicting forces engendered by the unfolding *telos* of the history of science.

Summary and Conclusion

The positivist-Whig interpretation of the Chemical Revolution was well established when the discipline of the history of science became an independent field of inquiry in the decades immediately following World War Two. It presented the practitioners in this new academic discipline with an impressive and coherent account of the origins of modern chemistry. On this account, modern chemistry arose, phoenix-like, from the ashes of the fallacious phlogiston theory. At the heart of this conflagration was the 'genius' of Antoine Lavoisier, who turned his back on the traditional theories and practices of chemistry and attended to the teachings of nature. Extending the quantitative and mechanical principles of the Scientific Revolution to chemistry, Lavoisier took up where Boyle left off. In a series of crucial experiments on combustion and the calcination of metals, Lavoisier deployed the quantitative method of inquiry to refute the phlogiston theory and to establish the central role of oxygen in the workings of nature. Like astronomy and mechanics before it, chemistry broke with its nonscientific past and moved inexorably and methodically towards the present state of scientific knowledge. This 'break' involved a cognitive inversion, in which the imaginary world of phlogiston gave way to the real world of oxygen. Phlogistians like Priestley failed to appreciate the new chemistry because their minds were clouded by the intrusion of alien, metaphysical and religious, modes of thought into their scientific inquiries. Lavoisier grounded chemistry in a uniform experience and a unitary method of inquiry, which formed the basis of subsequent progress in chemistry. This progress was linear, cumulative and autonomous. It marked the victory of the 'children of light' over the 'children of darkness'. The scientific struggle between the agents of light and the agents of darkness had significant moral and social implications. It was the responsibility of the positivist-Whig historian to determine the crucial moments, pivotal discoveries and individual contributions to this struggle, and to apportion praise and blame accordingly. Postpositivist historians and philosophers of science challenged this mandate in the 1960s and 70s with empirical evidence against the positivist-Whig interpretation of the Chemical Revolution and with a new interpretive framework within which to interpret this crucial event in the history of chemistry. As will be seen in the next two chapters, the resulting clash between tradition and innovation was a productive one; it energized scholars who developed a variety of new interpretations of the Chemical Revolution.

2 POSTPOSITIVISM AND THE
HISTORIOGRAPHY OF SCIENCE

Historians and philosophers of science first voiced their dissatisfaction with the positivist-Whig interpretation of the Chemical Revolution in the 1950s and 60s. In his seminal study *Lavoisier – The Crucial Year,* published in 1961, Henry Guerlac challenged the prevailing opinion that Antoine Lavoisier was 'the father of modern chemistry' because it overlooked 'the most significant ingredient of the Chemical Revolution', which concerned Lavoisier's scientific heritage and not his creative genius. Guerlac argued that:

> in the person of Lavoisier two largely separate and distinct chemical traditions seem for the first time to have merged. At his hands, the pharmaceutical, mineral, and analytical chemistry of the Continent was fruitfully combined with the results of the British 'pneumatic' chemists who discovered and characterized the more familiar permanent gases.[1]

Following Guerlac's lead, subsequent scholars developed 'thematic analyses of the Chemical Revolution from the perspective of larger developments in eighteenth century science'.[2] New interpretations of the Chemical Revolution appeared when the positivist-Whig interpretative framework gave way to interpretive schemata associated with the rise of postpositivism, which shifted the epistemological centre of gravity of science from individual experimentalists to theoretical traditions.

Postpositivism emerged in the 1960s and 70s as a critical philosophy designed to modify or replace positivism. The dialectic between innovation and tradition was a productive one, which yielded in two or three decades almost as many studies of the Chemical Revolution as appeared in the previous century. The resulting situation was complex and contradictory, with some scholars clinging to tradition and others modifying or rejecting it. Struggling to replace the positivist model of empirical certainty with a view of the fallibility of theoretical knowledge, postpositivist scholars generated a range of historiographical strategies which reflected the stresses and strains of shifting philosophical foundations. Tensions emerged in the resulting interpretive framework between the positivist view of science as rationally progressing towards certain knowledge of the external world and the relativist and antirealist tendencies nurtured by theoreticism. Further tensions within post-positivism developed between the followers of Karl Popper, who wrote normative historical narratives based on philosophical evaluations of past science, and scholars inspired by Ludwig Wittgenstein's view of philosophy, which suggested that the history of science is a purely descriptive discipline that takes the past as it finds it. Postpositivist narratives fluctuated between the scientistic strategies of positivism, which eschewed philosophical reasoning in order to explicate the norms and procedures of scientific practice, and more traditional epistemological practices which brought science to the bar of philosophical reason. These contrasting philosophical moments provided 'ideal types' of interpretive strategies, which were exemplified in concrete historical narratives in partial, incomplete, distorted, but nevertheless significant ways. Within this tensile interpretive framework, postpositivist historians of science developed new and innovative accounts of the history of science in general and the Chemical Revolution in particular.

Postpositivist historians of science seldom entered the philosophical fray associated with the global issues of realism, relativism and normativity, and they frequently reacted with disdain to the flagrant disregard for the historical record they encountered in the 'normative' narratives of some of their philosophical colleagues, who replaced history with its 'rational reconstructions'. But this disdain for philosophical meddling did not prevent these historians from using interpretive devices and strategies derived from postpositivist philosophical sources to fashion innovative views of the nature of science and its historical development. This chapter will explore the philosophical sources of these innovative historiographical strategies, while the next chapter will show how historians of the Chemical Revolution used them to re-examine new and familiar historical texts, issues and materials. The first part of this chapter will provide an account of the core epistemological principles and broad historiographical strategies associated with postpositivism, while the second part will focus on interpretive models and maneuvers that were particularly relevant to the postpositivist historiography of the Chemical Revolution. The identity of postpositivism as an international philosophical movement, encompassing Continental and Anglo-American philosophy, will be clarified in this chapter, but the historiographical implications of its French variant will be explored in subsequent chapters concerned with the sociology of knowledge. Postpositivism will first be approached through a consideration of its criticisms of the positivist-Whig historiography of the Chemical Revolution.

The Postpositivist Challenge

Postpositivist sensibilities shaped the criticisms of the positivist-Whig historiography of the Chemical Revolution that emerged in the 1950s, 60s and 70s. In 1952, Kuhn showed that the historicist assumption of a direct and progressive link between the Scientific Revolution and the Chemical Revolution was incompatible with the cognitive dissonance that existed between the reductionist thrust of Boyle's corpuscular chemistry and Lavoisier's antireductionist research programme. Eighteenth-century chemists rejected Boyle's attempt to derive the chemical properties of substances from the mechanical arrangement of their constituent particles because it precluded their interest in 'the isolation of elements and the determination of composition'. On Kuhn's analysis, Boyle failed to exert an immediate influence on the course of chemical theorizing not because he was ahead of his time, but because of 'the scientific shortcomings of his chemical doctrines, shortcomings which markedly differentiated his chemistry from the later doctrines of Lavoisier'.[3]

Kuhn also emphasized the contribution of Boyle's phlogistic opponents to the progressive development of chemistry in the early eighteenth century. According to Kuhn, 'the true progenitors of the Chemical Revolution were necessarily among Boyle's opponents'.[4] As can be seen in the works of Henry Cavendish, Richard Kirwan and Joseph Priestley, the phlogiston theory played a crucial and productive role in the development of pneumatic chemistry and was no less compatible than Lavoisier's chemistry with emerging quantitative sensibilities.[5] New evidence also showed that contrary to the positivist-Whig idea of a revolutionary break with the past, Lavoisier remained attached to the traditional view of chemical elements as the ultimate, universal constituents of matter, while some of his phlogistic rivals worked with the newer notion of simple substances.[6] Pointing to the existence of considerable cognitive parity between the oxygen theory and the phlogiston theory, some postpositivist scholars opposed positivist-Whig tendencies to construct a Manichean division between the scientific oxygen theory and the metaphysical phlogiston theory. Resisting attempts to reduce the Chemical Revolution to the emergence and application of a unique and determining scientific method, postpositivist scholars refused to assimilate Lavoisier's chemistry to modern chemistry.

Stephen Toulmin rejected the positivist-Whig notion of 'instant rationality' according to which, in 1775, Lavoisier used the crucial experiment on the calcination and revivification of mercury by heat alone to discredit the phlogiston theory and establish the oxygen theory.[7] As Toulmin noted, Priestley responded to this experiment in 1783 with a crucial experiment of his own. Priestley appealed to the reaction between minium (lead oxide) and inflammable air (hydrogen) over water as a 'counter demonstration', in which the phlogiston needed to convert minium into lead came from the observed diminution in the volume of inflammable air. Questioning the logical efficacy of crucial experiments in scientific debates, Toulmin supported the postpositivist thesis that an experiment is crucial only relative to a particular theoretical framework.

Highlighting Humphry Davy's refutation at the end of the eighteenth century of Lavoisier's claim that all acids contain oxygen, Alan Musgrave rejected the inductivist claim that the Chemical Revolution involved a transition from the realm of metaphysical error and superstition to the domain of scientific truth and rationality. He also noted that contrary to the claims of falsificationism, Lavoisier's quantitative experiments on the calcination of metals did nothing to falsify the phlogiston theory, which posited not only the release of phlogiston, but also the absorption by the metal of a substance in its surroundings. He further argued against the conventionalist view that Lavoisier's explanation of the calcination of metals was simpler than those developed by his phlogistic rivals. Just as the Phlogistians posited the release of phlogiston and the absorption of an 'aerial' substance by the metal, so Lavoisier thought that when a metal decomposed oxygen gas, it seized its 'oxygen principle' and liberated caloric in the form of heat.[8] Musgrave's line of reasoning is supported by the fact that in some of its ontological implications, the phlogiston theory was even simpler than the oxygen theory. While most Phlogistians reduced all substances to a few universal principles, such as the Aristotelian four elements or the Paracelsian three principles, Lavoisier and the oxygen theorists analysed each laboratory substance into a small number of an indefinite number of 'simple substances'. On the other hand, Phlogistians were forced to appeal to differences in the 'mode of combination' of a few generic principles in order to account for the multiplicity of chemical appearances – a complication that the oxygen theorists did not need.[9] Whatever form they took, positivist-Whig notions of 'instant rationality' and 'crucial experiments' could not capture the complexities and nuances of scientific thought during the Chemical Revolution.

Postpositivist scholars argued that the positivist-Whig view of a Manichean division between the progressive 'children of light' and the reactionary 'children of darkness' ignored or downplayed the similarities and continuities that existed between the 'intellectual personalities' of Priestley and Lavoisier. These scholars showed how Priestley's methodology was more systematic and sophisticated, his scientific training more adequate and complete and his critical powers more developed and acute than positivist-Whig historians had supposed. Priestley was well attuned to the primacy of quantitative considerations, and he put the phlogiston theory to excellent heuristic and explanatory use; he functioned in the Chemical Revolution not as a dogmatic and entrenched defender of the phlogiston theory, but as a valuable critic of Lavoisier and his supporters.[10] Although postpositivist historians joined their positivist-Whig predecessors in stressing the role of theological, metaphysical, epistemological and political parameters in the formation of Priestley's science, they did not do so in order to question his worth as a scientist. Postpositivist scholars replaced the positivist demarcation between science and nonscience with an awareness of the place of the Chemical Revolution in the broader field of eighteenth-century thought. They assimilated the history of science to the history of ideas.

Postpositivism

These empirical criticisms functioned as a force for historiographical innovation only when they were incorporated into the interpretive framework provided by postpositivism. Distinct from the radical antipositivist sensibilities of poststructuralism and postmodernism, which came to the fore in the 1970s and 1980s, the term 'postpositivism' denotes an earlier, narrower, more ambiguous philosophical culture, associated with a loose coalition of Anglo-American historians and philosophers of science.[11] This coalition was held together in the 1960s and 70s by the views and writings of W.V.O. Quine, Karl R. Popper, Norwood Russell Hanson, Paul K. Feyerabend, Thomas S. Kuhn, Imre Lakatos, Larry Laudan and Stephen Toulmin. These philosophers rejected the positivist doctrine of empirical certainty and maintained instead that knowledge is fallible and theoretical. Louis Althusser and the young Michel Foucault developed similar views in France in the 1960s. Althusser and Foucault drew on the French tradition of structuralism and the 'philosophy of the concept' developed in the 1930s and 40s by Gaston Bachelard and George Canguilhem to maintain the autonomy of theoretical knowledge from both the knowing subject and the known object. The roots of postpositivism can also be traced to the idealist historiographies of science developed in France in the 1920s and 30s by Emile Meyerson, Alexandre Koyré and Hélène Metzger. Under the tutelage of Koyré, leading Anglo-American scholars in the 1950s and 60s – including Henry Guerlac, A. Rupert Hall and Richard S. Westfall – assimilated science to theory and the history of science to the history of ideas. Popper also wrote the *Logik der Forschung* in the 1930s, but it languished in relative obscurity until the 1960s, when it was translated into English and attained an oracular status among postpositivists. This chapter will focus on the Anglo-American postpositivists because they had a more immediate impact than their French counterparts on the historiography of the Chemical Revolution. It will explore the French tradition in order to provide a more complete picture of postpositivism as an international philosophical movement, as well as to identify the way in which it evolved into a model for a later generation of sociologists of knowledge who eventually broke with it. Kuhn and Foucault were key figures in this transitional process, playing important roles in both the formation and deconstruction of postpositivism. While Chapter 4 will show how some sociologists of knowledge used the philosophies of Kuhn and Foucault to generate historiographies of science that were at odds with postpositivism, this chapter will consider only the postpositivist credentials of Kuhn and Foucault.

Fallibilism and Theoreticism

Postpositivism emerged as a critical force of opposition in two distinct and antagonistic philosophical traditions which dominated philosophy in the first half of the twentieth century. While Anglo-American positivists focused on the conceptual analysis of the natural sciences, phenomenologists and existentialists in Europe used philosophies of the human subject to elucidate the 'ontology' of the human and social sciences. This division ran deep; it involved antagonistic conceptions of the aims and methods of philosophy. Whereas positivists focused on epistemological and methodological issues raised by the logical structure and evidential basis of scientific theories, phenomenologists grounded scientific knowledge in metaphysical theories of the ultimate nature of man and his world. But this antagonism belied an underlying unity formed in response to a common enemy. While positivists like Moritz Schlick and Frederick Waismann avoided the cognitive excesses of speculative metaphysics by grounding knowledge in the passive reception of sensory experience, phenomenologists like Edmund Husserl and Martin Heidegger short-circuited metaphysical speculation by anchoring philosophy to an unmediated encounter with pure Consciousness or Being. Regarding scientific concepts as abstractions from the concrete 'given' of immediate experience, positivists valorized the objectivity of science, while phenomenologists stressed the incompleteness of scientific abstractions and their epistemological and methodological dependence on the 'life world'. While Heideggerian phenomenology grounded science in the secure and certain results of philosophy's encounter with Being, Logical Positivists like Moritz Schlick used the phenomenalist reduction of the concept of science to the immediate data of experience to ground philosophy in the achievements and continuities of the exact sciences. In their different ways, positivists and phenomenologists looked to foundationalist accounts of science to avoid the cognitive excesses and uncertainties of nineteenth-century metaphysics.[12]

As a form of 'fallibilism', postpositivism rejected as futile the search for certainty associated with foundationalism. The postpositivist rejection of certainty was part of a broader philosophical movement against the doctrine of immediate knowledge, or 'the myth of the given', which gathered strength in the second half of the twentieth century. According to the doctrine of immediate knowledge, the mind has no need of discursive intermediaries to refer to or know the external world: knowledge is directly and naturally derived from the privileged and incorrigible protocols of immediate experience. Positing a metaphysical gulf between man and nature, thought and experience, ideas and objects, fallibilism rejected the assimilation of thought to perception or experience. Denying the positivist claim that genuine knowledge 'corresponds to conceptually unmediated facts', as well as the phenomenologist's notion of an authentic realm of being that 'precedes conceptual articulation', fallibilism insisted upon the constitutive role of concepts and ideas in the formation of knowledge. Questioning the immediacy of 'facts', fallibilism took many more or less complete forms throughout the twentieth century, including the Hegelian Marxism of George Lukács; the Critical Theory of the Frankfurt School; Jurgen Habermas's 'quasi-transcendental' theory of interest-laden knowledge; structuralist and Wittgensteinian notions of rule-governed language, thought and practice; and constructivist accounts of the social *construction* of scientific knowledge.[13] Any vestiges of the 'myth of the given', of the idea of immediate, certain knowledge, were thoroughly expunged by poststructuralists, such as Giles Deluze, Jacques Derrida and Michel Foucault, who 'stressed the fragmentary, heterogeneous, and plural character' of reality, thought and the subject.[14] Accepting an activist view of the knowing mind, fallibilism generated a variety of perspectives on the nature of concepts and the way in which they mediate between thought and reality.

Postpositivist philosophers of science distinguished themselves from other fallibilists by identifying the conceptual with the theoretical. They defended the theoreticist claim that '[t]heory dominates the experimental work from its initial planning up to the finishing touches in the laboratory'. Underdetermined by formal reason or by empirical evidence, theory choice 'depends upon basic statements whose acceptance or rejection, in its turn, depends upon our decisions'.[15] In place of the positivist distinction between the certainty of theory-neutral experientially grounded observation statements and the uncertainty and endless mutability of theoretical statements, postpositivism insisted that all statements, observational and theoretical, are open to criticism and revision. Postpositivist scholars replaced the notion of the cognitive unity of the knower and the known, man and nature, mind and matter, which was inherent in the positivist doctrine of immediate knowledge, with a fallibilistic distinction between the known object and the real object, between thought and being, concept and nature. Like Theodore Adorno, Popper linked the (metaphysical) separation of thought and being inherent in the concept of conjectural theoretical knowledge to freedom and autonomy. Just as Adorno traced 'Heidegger's absorption into Hitler's "Führestaat"' to a 'philosophic attitude of mind in which the "Führer" and the dominant power of being were virtually identified', so Popper argued that 'the rationality of science lies not in the habit of appealing to empirical evidence in support of its dogmas', but in the 'critical use ... of empirical evidence' to correct our mistakes and in a never-ending process proceed 'to better theories'.[16] Whereas Adorno linked the critical method associated with fallibilism to the Marxist critique of capitalism, Popper tied it to the political liberty of the individual. Popper's abiding opposition to the forces of totalitarianism shaped his methodology of falsificationism and conditioned his intense opposition to other postpositivists, such as Kuhn, who emphasized the role of dogma in science, and Feyerabend, who nullified the critical force of science by undermining its epistemological status.

Propositions and Paradigms

Popper, Quine and Wittgenstein laid the philosophical groundwork for the critical assault on Logical Positivism that resulted in the formation of postpositivism. Quine evoked theoreticist sensibilities when he attacked the 'dogmas of empiricism' in 1951. Rejecting the positivist (semantic) distinction between observational terms and statements and theoretical terms and statements, Quine developed a holistic theory of science, according to which 'the unit of significance' is not an observation statement but 'the whole of science'. He claimed that any statement in science's 'network of meaning' can be 'held true come what may, if we make drastic enough adjustments in the system of science'.[17] Dubbed the 'Duhem–Quine thesis', the idea of the underdetermination of scientific decisions by data and method replaced the consensual model of science with a scenario of dissensus and endless conflict. Quine's holistic theory of science provided a backdrop for postpositivist philosophers of science who took their main inspiration from Popper and Wittgenstein. Shaped by numerous and diverse philosophical influences yet to be fully identified, the prominent contours of postpositivism can be triangulated from reference points provided by the philosophies of Popper and Wittgenstein.

Popper's Critical Rationalism

Popper provided the definitive theoreticist account of the rationality of scientific knowledge. In place of the positivist doctrine of sense certainty, which viewed knowledge as a direct relation between subjects and objects, persons and propositions, Popper identified scientific knowledge with the 'objective logical content' of 'theories, propositions, and statements', which are (logically) independent of the objects they denote and the minds that apprehend them.[18] Popper rejected the positivist model of knowledge because it conflated the justification of knowledge, or the 'logic of discovery', which involves logical relations between propositions, and the 'psychology of research', which provides causal accounts of the apprehension of these propositions by the knowing mind. Popper insisted that '[e]xperiences can motivate a decision, and hence an acceptance or a rejection of a statement, but a basic statement cannot be justified by them – no more than by thumping the table'.[19] Like the Logical

Positivists he otherwise opposed, Popper used the distinction between the context of discovery and the context of justification to identify the philosophy of science with the logic of science.[20]

Popper's propositional theory of knowledge tied objectivity to fallibilism:

> if we adhere to our demand that scientific statements must be objective, then those statements which belong to the empirical basis of science must also be objective, i.e. inter-subjectively testable. Yet inter-subjective testability always implies that, from the statements, which are to be tested, other testable statements can be deduced. Thus if the basic statements in their turn are to be inter-subjectively testable, *there can be no ultimate statements in science* ... [21]

Equating objectivity with intersubjective testability, Popper criticized (what he understood as) the research programme of the Vienna Circle – to derive scientific theories from basic statements grounded in nature and justified by immediate experience – because it introduced a private 'tacit' dimension ('feelings of finality') into science, accepted the 'fallacious' notion of an inductive logic, ignored the constitutive role of universal concepts in the descriptive vocabulary of science and failed to recognize the metaphysical distinction between nature and convention. Noting that 'statements' (propositions) can be (logically) derived and acquire their meaning only from other statements, and not from nature or experience, Popper related the validity and meaning of the 'basic [observation] statements of science' to a prior 'decision' to accept or reject a theory, or 'system of propositions'. Popper's propositional theory of knowledge and the associated distinction between nature and convention shaped the influential philosophies of science developed by Paul Feyerabend, Lakatos and other postpositivists. In place of the positivist distinction between observation statements, rooted in nature and based on experience, and theoretical statements, based on conjecture, Popper and his colleagues maintained that all scientific statements – basic postulates as well as 'basic statements' – are theoretical, conventional and fallible.[22]

Popper's theoreticism undermined the positivist-Whig idea that the history of science involves the accumulation of 'discoveries' generated by unmediated interactions between nature and individual men of genius. Popper's 'propositional' model of knowledge replaced subject-centred epistemologies with the concept of 'objective knowledge', or 'epistemology without a knowing subject'. The Popperian doctrine of 'objective knowledge' encouraged historians of science to shift their interpretive focus away from the experiences, actions and beliefs of individual scientists and towards the emergence and development of abstract theoretical doctrines and propositional systems. In this manner, postpositivist historians and philosophers of science appealed to the phenomenon of simultaneous discovery in science to undermine the positivist view of scientific discoveries as 'unproblematic mental events with obvious marks of identity'.[23] As will be seen in Chapter 4, sociologists of knowledge used the idea of 'knowledge without a knowing subject' to sustain the thesis that scientific discoveries are complex historical, social processes extended in time and across a range of individuals and groups.

Popper's reformulation of the positivist demarcation between science and metaphysics reinforced this historiographical shift. When Popper characterized metaphysics as unscientific because unfalsifiable, he did not conclude that it was therefore meaningless or worthless. On the contrary, he appealed to the histories of the atomic theory and the theory of evolution to justify the claim that conceptual refinement and alteration can transform metaphysical speculations into viable scientific concepts and theories.[24] Popper's (weaker) view of the demarcation between science and metaphysics encouraged historians of science to search for positive and productive links between modes of thought – the scientific and metaphysical – which positivist-Whig historians of science regarded as fundamentally alien and incommensurable. Popperian historiography replaced the positivist-Whig stress on the empirical foundations and methods of science with an interest in its theoretical structure, conceptual foundations and doctrinal development.

This does not mean that Popper and his postpositivist colleagues ignored the important role that facts and experiments play in the development of science. Advocates of the doctrine of the 'theory-ladeness of observation' did not deny the role of observation and experimentation in the development and evaluation of scientific theories. On the contrary, Popper made the experimental refutation of theories by facts the cornerstone of his theory of scientific rationality. What theoreticism did deny was the positivist-Whig view that experiments provide a direct unmediated access to nature, which yields incontrovertible facts for establishing, once and for all, the rational credential of competing theories. Upholding a fallibilist view of knowledge, postpositivist philosophers called attention to the ineluctable interpretive dimension of observations and experiments. They argued that new concepts and theories could challenge established facts and observations in the same way that new facts and observations test established theories. Unlike their sociological successors and adversaries, who focused on the practical skills and institutional networks involved in the performance and replication of experiments, postpositivist historians and philosophers of science tied the preparation, execution and interpretation of experiments to the dialectical development of conceptual structures and theoretical systems. They argued that without concepts and theories to interpret them, experiments are mute and facts have no cognitive content. Although science is about the external world, it begins and ends with our concepts and theories.

Popper developed a fallibilist account of the rationality and methodology of scientific knowledge. Accepting the validity of Hume's criticism of inductive logic, he denied that science progresses by deriving theories from facts or by confirming them with facts. Emphasizing the role of science in the Enlightenment struggle against prejudice and dogmatic modes of thought, Popper made 'falsifiability' the criterion of demarcation between science and nonscience. Although he recognized the real possibility of 'falsifying' a true theory with an erroneous, 'theory-laden' observation statement, he recommended 'methodological falsificationism', or 'critical rationalism', as the only alternative to scepticism, relativism and irrationality in science.[25] Like his positivist predecessors, Popper linked scientific rationality to scientific realism, claiming that science produces rational descriptions and explanations of a world that exists independently of the ways in which it is described and explained. He rejected the instrumentalist idea that theories are mere calculating devices with no ontological or epistemological significance, and he gave short shrift to the claims of idealists and constructivists who make the world dependent upon the way in which it is conceptualized or produced.[26] But Popper also rejected the 'naïve realism' of positivism which anchors the ontology and epistemology of science in its observational terms and statements. Popper's theoreticism and fallibilism constituted a form of 'critical realism', which rejects the goal of empirical certainty and locates the ontological and epistemological significance of science in its theoretical terms and statements. He was also a 'convergent realist', arguing that the aim of science is not to uncover the truth about the world so much as to move in that direction, even though it can never know for sure when it has reached its goal, or even got closer to it. Many of Popper's interlocutors had a hard time distinguishing between his noumenal notion of rationality, which renders reason unknowable, and relativism, which denies the relevance or existence of reason altogether. Indeed, while Laudan emphasized the historiographical problems with 'convergent realism', Feyerabend showed that 'sceptical fallibilism' is an ineluctable consequence of 'critical rationalism'. Other postpositivist philosophers, who were influenced by Wittgenstein and by 'global' views of scientific theories, embraced relativist and antirealist philosophies of science more directly.

Kuhn's 'Relativism'

Sceptical of the positivist-Whig programme for producing a theory-neutral observation language, Kuhn embraced Popper's theoreticism; but he rejected the propositional theory of knowledge associated with it. Resisting Popper's assimilation of the philosophy of science to the logic of science, Kuhn argued that an adequate theory of knowledge must combine 'syntactic' criteria, which concern logical relations between 'statements', with 'pragmatic' and 'semantic' considerations of how 'to relate sentences derived from a theory not to other sentences but to actual observations and experiments'.[27] While Popper excluded these considerations from the context of justification and the domain of the philosophy of science, Kuhn introduced the term 'paradigm' in *The Structure of Scientific Revolutions* to explain the role of 'concrete examples' rather than abstract rules in bridging the gap between the propositional content and experimental application of scientific theories.[28] Kuhn's focus on the pragmatic and semantic dimensions of scientific theories can be traced to the influence of Wittgenstein on Anglo-American philosophers of science, which is also evident in the earlier work of Norwood Russell Hanson and Stephen Toulmin.

In *Patterns of Discovery*, published in 1958, Hanson focused not so much on the theory-dependence of observation terms as on the 'theory-ladenness of observation itself'. Hanson accepted Wittgenstein's idea that languages and their concepts are instruments that acquire their 'sense' not from things in the mind, but from the functions they perform within the rule-governed activities of language games, or forms of life, involved in our everyday practical affairs.[29] While later sociologists of knowledge used Wittgenstein's concept of science as a form of life to support the view that science is a social activity grounded in social needs and interests, Hanson used it to assimilate perception and observation to theory and interpretation. According to Hanson (and Kuhn), observation and perception do not involve the 'passive' reception by the mind of unmediated perceptual data, but require an act of interpretation grounded in a given language game or linguistic activity. 'Interpretation is not something a physicist works into a ready-made deductive system', and theories are not 'simply man-made interpretations of given data'. A scientist explains the phenomena by 'perceiving the pattern' in the phenomena, by observing and theorizing at one and the same time. In contrast to Popper's realist distinction between language and reality, theories and the world, Hanson and Kuhn argued that theories, or 'pattern statements', are not only modes of interpreting the world; they are also ways of seeing or living in the world. Changes in theories, or 'pattern statements', do not depend on the accumulation of new and different data, but on the reorganization of existing data according to new conceptual schemes. According to Hanson and Kuhn, observation is 'theory-laden' and changes in theories are like 'gestalt switches', in which observers looking at the same object 'see different things' and gather different data, depending upon differences or alterations in their conceptual schemes, language games, or forms of life. According to Kuhn, after undergoing a gestalt switch, a scientist may be said to be living and working in a different world.[30]

Toulmin used Wittgenstein's instrumental theory of language, which links meaning to usage and locates usage in a 'form of life', to argue that scientific theories are not true or false descriptions or explanations of the external world, but rules for determining the functional meanings of theoretical concepts and governing the way in which inferences can be drawn between phenomena.

Drawing on R. G. Collingwood's theory of 'absolute presuppositions', Toulmin used the term 'paradigm' – later made famous by Kuhn – to refer to the global background presuppositions of scientific explanations and inquiries, which determine the meaning of the terms of a 'field of inquiry' as well as its domain of appropriate problems and range of acceptable solutions.[31] Understood as a form of life, as something – in the words of Michael Polanyi – 'to dwell in … as we do in our own body', a scientific discipline is 'essentially inarticulable' and cannot be evaluated from a neutral standpoint. Toulmin claimed that historians of science should not 'measure … scientific ideas against timeless logical ideals'; instead, they should describe the historical evolution of scientific disciplines and their specific canons of rationality.[32]

Following Wittgenstein, who identified learning a language with learning how to use it, Kuhn argued that scientists learn the terms of their discipline not 'exclusively by verbal means', but by acquiring 'words together with concrete examples of how they function in use'. Kuhn referred to these 'concrete examples' as 'paradigms', and he appealed to Wittgenstein's idea of 'family resemblances' to explain how they function. According to Wittgenstein, the meaning of a word is determined not by a set of characteristics that uniformly apply to all its concrete instances, but by the family resemblances, or concatenation of partial, overlapping and crisscrossing resemblances that exist between the different instances that fall under its conventionally established range of application. Similarly, Kuhn insisted that the 'problems and techniques' of a research tradition are not linked to a paradigm as instances of 'some explicit or even some fully discoverable set of rules' that define the tradition; rather, the 'coherence displayed' by a research tradition depends upon the ability of its practitioners to work directly from models and to know through their education and training 'what characteristics have given these models the status of community paradigms'. Kuhn insisted that the absence 'of a standard interpretation or an agreed reduction to rules will not prevent a paradigm from guiding research'. Arguing that paradigms are more complete, binding and fundamental than any 'rules for research' that may be derived from them, Kuhn argued that scientists could work without recourse to rules. According to Kuhn, 'nature and words are learned together' and science involves an element of 'tacit knowledge', which is learned 'by doing science rather than by acquiring rules for doing it'. While later sociologists of scientific knowledge took Kuhn's reference to 'tacit knowledge' to mean that science is not an autonomous rule-governed theoretical activity but a practical skill-based, socially controlled 'craft', Kuhn used the idea to endorse Hanson's claim that in order to interpret the world according to a certain theory, it is necessary to assimilate the 'time-tested and group-licensed way of seeing' the world associated with the theory.[33] Far from undermining theoreticism, Kuhn used the concept of a paradigm to endorse the theoreticist view of the priority and pervasiveness of theories in the determination of scientific knowledge *and* experience. Like Hanson, Kuhn treated scientific theories as 'gestalts', or ways of perceiving and experiencing the world, and he treated scientific changes as 'gestalt switches'.

The Philosophy of Science and the History of Science

Popper's Prescriptivism

The followers of Popper and Wittgenstein generated opposing views of the relation between the disciplines of the philosophy of science and the history of science. While Popperian philosophers of science used prescriptive models of scientific rationality to 'rationally reconstruct' the history of science, their interlocutors used Wittgenstein's view of the descriptive function of philosophy to argue that philosophers of science should describe how science is practised, not how it ought to be practised. Emphasizing the prescriptive function of philosophy, Feyerabend, in his early Popperian phase, assimilated the history of science to the justificatory concerns of the philosophy of science, using accounts of the '*actual* structure of science' as rhetorical aides – to be discarded in a more favourable philosophical climate – in his struggle to persuade unsympathetic philosophers to accept his 'abstract model for the acquisition of knowledge'. Despite his many appeals to the '*actual*' structure of science, Feyerabend insisted that 'this structure' must not be allowed to 'interfere with the models', which 'form a basis for the criticism as well as the improvement of what exists'.[34] Further articulating Popperian prescriptivism, Lakatos claimed that the 'history of science without the philosophy of science is blind'; he called upon historians of science to develop 'rational reconstructions' of the growth of scientific knowledge in terms of 'normative methodologies' derived from the philosophy of science.[35] Like Lakatos, Musgrave enjoined historians of science to formulate 'rational reconstructions' of the history of science in terms of models of science designed to control how science is practised by producing rules for how it ought to be practised.[36]

To this end, Lakatos formulated a significant and influential distinction between the 'internal history', or 'normative [rational] reconstruction', of the history of science and the 'external history' of its 'residual non-rational factors'. Lakatos recognized the need to supplement internal accounts of the unfolding logic of 'scientific research programmes' with causal accounts of those aspects of the history of science that deviate from this normative ideal, but he insisted on the priority of internal history. He argued that internal history determines the scope and problems of external history, which fills in the gaps left by internal history. The internal historian is also highly selective: 'he will omit everything that is irrational in the light of his rationality theory'. Lakatos allowed that in some cases the internal historian may offer a 'radically improved version' of the past, in which beliefs and concepts are ascribed to historical agents not according to their actual awareness of them, but on the basis of what 'fits naturally [logically] in the original outline' of their research programme. According to Lakatos, the way to write the history of science 'is to relate the internal history *in the text*, and indicate *in the footnotes* how actual history "misbehaved" in light of its rational reconstruction'.[37] Although postpositivist historians of science were outraged by Lakatos's philosophical high-handedness and unimpressed by his 'methodology of scientific research programmes', they still used the distinction between internal and external history to devalue sociological explanations of scientific knowledge, which they regarded as anathema to the intellectual worth and autonomy of science. Adherence to the internal–external distinction, or the 'arationality assumption', was a defining characteristic of postpositivist historians and philosophers of science in particular and 'internalist' historians of science in general. Sociologists of knowledge replaced this assumption with 'externalist' programmes for explaining all beliefs, rational and arational, in terms of their social causes.[38]

Kuhn's Descriptivism

While positivist-Whig historians of science judged past science in terms of present science, and Popperian historians of science reconstructed and evaluated the history of science in terms of modern philosophy of science, scholars influenced by Wittgenstein banished essentialism and prescriptivism from the historiography of science altogether. According to Wittgenstein, a philosopher determines the meaning of a linguistic expression not by comparing it with an abstract ideal or essence of meaning, but by describing the actual usage of this expression by speakers and interlocutors participating in a specific form of life. Philosophy does not criticize, improve or in any way 'interfere with the actual use of language; it can in the end only describe it ... it leaves everything as it is'.[39] Similarly, the task of the historian of science is not to evaluate the development of science in terms of normative models about how, as a rational enterprise, it ought to develop, but to describe the rational procedures that guide its actual development. While Lakatos wrote 'normative histories of science', in which he used Popper's view of 'the constitutional authority of an *immutable statute law* (laid down in his demarcation criterion) ... to distinguish between good and bad science', Toulmin appealed to the 'method of precedents' in 'common law' to argue that the description of the different ways in which rationality functions in different milieus is 'the sole but sufficient basis for determining how science derives its rationality'.[40]

In a similar vein, Kuhn contrasted his interest in the 'psychology of research' with Popper's focus on the 'logic of discovery [justification]'. Kuhn insisted that the historian of science 'must explain why science – our surest example of sound knowledge – progresses as it does, and ... must first find out how, in fact, it does progress'. Rejecting the influential distinction between the context of discovery and the context of justification, Kuhn questioned the assumption that history, construed as a 'purely descriptive discipline', has no relevance to issues in 'logic or epistemology'.[41] Although Kuhn was at times more equivocal than Toulmin, who consistently reduced the prescriptive to the descriptive, the essential to the nominal, he clearly opposed Popper's attempt to make history a tool of philosophy, insisting that 'when the historian of science emerges from the contemplation of sources and the construction of narrative', he or she has as much right as the philosopher of science 'to claim acquaintance with essentials'.[42] Eschewing Popper's normative interest in the 'logic' of justification, Kuhn enjoined historians of science to develop a 'psychological or sociological ... description of a value system [paradigm], an ideology, together with an analysis of the institutions through which that system is transmitted and enforced'. In order to avoid the circularity involved in the tendency to identify a paradigm as something the members of a scientific community share while simultaneously identifying the community as a group of individuals who share a paradigm, Kuhn argued that the historian of science should first identify and characterize 'the communal structure of science' and then map its various paradigms onto this structure.[43] Subsequent sociologists of knowledge have used passages like these to argue for the 'intrinsically sociological character of Kuhn's approach' to the history of science and to assimilate his analysis to the sociological programme for explaining scientific beliefs in terms of their social causes.[44]

But this interpretation does violence to Kuhn's real intention, which was to uphold a descriptivist, not a sociological, historiography. While Kuhn recognized the role of external, social causes in influencing the timing and tempo of scientific change and the significance and importance of some of its problems, he emphasized the 'conceptual insularity' of mature scientific communities.[45] Although Kuhn recognized the complementary nature of internal and external approaches to the history of science and called for a more cooperative relationship between their respective practitioners, he did not trace the paradigmatic conformity of a mature scientific community to its professional interests or wider social causes, but grounded its social cohesion in the attraction and achievements of its paradigm. Kuhn was not a proto-sociologist of knowledge, seeking to relate science to society, but a descriptivist philosopher of science, trying to articulate the internal mechanisms that do, in fact, generate scientific progress.

Despite their significant philosophical differences, postpositivist philosophers of science achieved considerable consensus in their decision to treat and evaluate scientific theories not as formal, static sets of abstract propositions about the world, but in terms of their cognitive use and development. Many of them rejected the distinction that Logical Positivists drew between the *historical* context of discovery and the *formal* context of justification, according to which the 'purely descriptive' study of 'what science has been' has no bearing on 'logic or epistemology', which is concerned with 'how it should be'.[46] While Logical Positivists criticized scholars who 'muddled historical origins with logical justification', postpositivist scholars allowed their understanding of how science has developed to play a decisive role in formulating their views about how, as a cognitive enterprise, it ought to develop. But they took this development to be an epistemological or conceptual one, focusing on the temporal unfolding of propositions and paradigms, rather than on the emergence of beliefs and their psychosocial causes. Sometimes referred to as 'the historical school', many postpositivist philosophers of science assimilated the philosophy of science to the history of science. Introducing a historical dimension into scientific appraisal, they combined historical evidence with philosophical reflection in order to show, in the manner of the earlier Comtean positivists, that what science has been is relevant to what it ought to be.[47] They required a successful theory of science to satisfy simultaneously the historical requirement of descriptive adequacy and the philosophical demand of prescriptive plausibility. Although postpositivist philosophers of science never fully realized this ambitious goal, the idea of a history of science with normative, epistemological and axiological, import shaped the hermeneutical strategies and fervour of a generation of historians of science.

But, despite the best efforts of the historical school, the postpositivist conception of history was not a genuinely *historical* one. While Laudan's research traditions added a dynamic dimension to the static synchronic structures associated with Kuhn's paradigms and Lakatos's 'research programmes', it still elevated abstract global structures above the concrete machinations, local practices and specific experiences of situated historical agents. In place of positivism's conception of a single, fixed scientific form or mode of rationality, postpositivism promulgated the conception of several relatively dynamic forms. In both cases, however, the forms of thought were regarded not as the products, but as the presuppositions, of historical activity. There was no conception in postpositivism of the empirical and theoretical characteristics of the global structures of science, whether dynamic or static, as emerging from the historical activities of their practitioners. On the contrary, experimental and theoretical practices were regarded as the mere result or application of underlying global structures. The specificity, complexity and agency of history eluded postpositivist accounts of the history of science.

Globalism

Joseph Agassi transformed Popper's 'methodological falsificationism' into a systematic historiography of science which enjoined historians of science to develop rational accounts of the growth of knowledge in terms of the formulation of 'bold' conjectures and their refutation or 'corroboration' in 'great negative crucial experiments'.[48] Arguing against the eureka-moment notion of factual discovery, which posits a direct, unmediated link between nature and the receptive mind of the scientist, Agassi insisted that behind every great factual discovery, there is a theory that the discovery refutes and from which it acquires its significance and importance. In this manner, Agassi refuted the positivist-Whig claim that Priestley's experimental discoveries were chance or 'accidental' events, involving neither the confirmation nor the refutation of a preceding theory. Besides loosening the grip that positivist-Whig sensibilities had on the historiography of science, Agassi's proposals had little or no substantive impact on the historical community. It was left to Feyerabend and Lakatos to incorporate Popper's philosophical ideas into more influential historiographical strategies, which shifted the interpretive focus from individual theories and conjectures to abstract, global theoretical structures. As Laudan noted:

> Much of the research done by historians and philosophers of science suggest that these more general units of analysis exhibit many of the epistemic features which, although most characteristic of science, elude the analyst who limits his range to theories in the narrower sense. Specifically, it has been suggested by Kuhn and Lakatos that *the more general theories*, rather than the more specific ones, *are the primary tool for understanding and appraising scientific progress.*[49]

Much to Popper's chagrin, this shift in epistemological focus diluted the Popperian notion of rationality sufficiently to allow Kuhn's image of the relativity of science and its historical development to come to the fore.

Popper's notion of 'epistemology without a knowing subject' and Wittgenstein's idea of science as a form of life shaped the 'globalist' movement among postpositivist historians and philosophers of science. Globalists, such as Kuhn, Lakatos, Laudan and Toulmin, distinguished between 'local', 'specific' or 'individual' theories, which provide specific predictions and detailed explanations of natural phenomena, and 'global theories', such as 'paradigms', 'research programmes', 'research traditions' or 'intellectual disciplines', which provide the ontological principles and methodological procedures necessary for the formulation of a variety of specific theories. While specific theories are evaluated for their empirical adequacy, their ultimate acceptance or rejection by the scientific community depends upon their place in a successful or failing global tradition. Less easily testable than the local theories they generate, global theories pass through a number of different formulations and have long histories extending through significant periods of time. For globalists, knowledge is not an instantaneous affair, associated with specific theories and individual scientists, but a 'complex temporal process' characterized by 'the patterns' that mark its historical development. Since global theories 'develop and change structure with time', their 'identification and evaluation ... are essentially historical in character'.[50] Globalist philosophers of science conflated the thesis of the historicity of scientific theories with the historical school's view of the essential role of the history of science in the philosophy of science. Like Hegel, globalist philosophers of science transformed epistemology into the philosophy of history.

Paradigms and Revolutions

Kuhn, Lakatos and Laudan developed globalist philosophies of science that influenced significantly the historiography of the Chemical Revolution. Regarded as a 'touchstone' by realists and antirealists, Kuhn's *Structure* was, in fact, a complex and unstable concoction of incompatible philosophical sensibilities, which deployed at one and the same time principles derived from rationalism and relativism, realism and antirealism, prescriptivism and descriptivism and internalism and externalism in the philosophy of science community. Kuhn based his globalist historiography on the concept of a paradigm, which, not surprisingly, was notoriously ambiguous. Although Margaret Masterman's claim that Kuhn used the term 'paradigm' in at least twenty-one different ways may be something of an exaggeration, Kuhn's usage of the term was sufficiently ambiguous to serve the antagonistic interests of postpositivist philosophers of science, bent on the intellectual reconstruction of the history of science, and their rivals in the sociology of knowledge, who sought the social causes of science.[51] Kuhn analysed a paradigm, or 'disciplinary matrix', into four components: ontological assumptions, methodological rules, explanatory strategies and procedures, and 'exemplars', or 'concrete problem-solutions', which show scientists 'how their job is done'.[52] As discussed earlier in this chapter, many of the ambiguities associated with Kuhn's philosophy of science reside in the conflicting interpretations that philosophers and sociologists of science placed on the idea of exemplars as forms of tacit knowledge.

Kuhn used the philosophical model of a paradigm as a tacit worldview to develop his influential account of scientific change. Viewing the history of science as a succession of competing and incompatible paradigms, or global worldviews, Kuhn claimed that when a scientific community accepts a paradigm, its members practice 'paradigm articulation', or 'normal science', in which they modify, falsify and even abandon individual theories, while placing the paradigm itself beyond the range of their critical activities. Challenging Popper's 'critical rationalism', Kuhn made dogma and entrenched commitment essential to the full realization of a paradigm's explanatory potential, equating normal science with 'puzzle solving', which tests the skills of the puzzle-solver, not the validity of the puzzle.[53] Kuhn argued that scientists question the dominant paradigm only after the accumulation of 'anomalies', or unsolved puzzles, plunges the scientific community into a state of 'crisis', when its members begin to take seriously alternative paradigms. A crisis-ridden community undergoes a 'scientific revolution' when it switches to a new paradigm for the further pursuit of normal science. But, Kuhn insisted, the spectre of relativism haunts the disputes over fundamentals that characterize these relatively brief and infrequent periods of 'revolutionary science'. According to Kuhn, rival paradigms are either imprisoned in their own worlds of 'incommensurable meanings', or, when they do share meanings and experiences, are too incongruous in their disciplinary problems, standards, aims and assumptions to be ranked on the same scale of criteria.[54] Since inter-paradigmatic debates are not subject to rational closure, Kuhn concluded that scientific revolutions are more like 'gestalt switches' in perception, or 'conversion experiences' in religion, than they are like the algorithms of rational decision-making.

Kuhn was both an 'epistemological relativist' and a 'convinced believer in scientific progress'. While he recognized that incommensurable paradigms could not be compared in terms of the traditional epistemological parameters of 'truth' and 'verisimilitude', he insisted that they could be ranked according to their 'puzzle-solving abilities'. Indeed, he maintained that later paradigms solve more puzzles more accurately than earlier ones. Kuhn suggested that 'the vexing problems associated with progress' would vanish if philosophers of science would replace traditional, epistemological evaluations of science with descriptive accounts of the scientific community as 'a supremely efficient instrument for maximizing the number and precision of the problems solved through paradigm change'. Most postpositivist historians of science ignored Kuhn's descriptivist programme, however, using his epistemological analysis to articulate relativist scenarios of scientific change and development. They were encouraged in this endeavour by Kuhn's occasional lapses into antirealism. In keeping with Wittgenstein's notion of theories as forms of life, Kuhn sometimes slid from the critical realist view that there is no theory-free access to reality to the antirealist, idealist, notion that there is no theory-free reality to be accessed. This slide is evident in Kuhn's discussion of the discovery of oxygen, where he claimed that as a result of this discovery, Lavoisier not only 'saw nature differently' but also 'worked in a different world'. Elsewhere, he argued that the practitioners of different paradigms, who 'have significantly different sensations on receipt of the same stimuli, do in *some sense* live in different worlds'.[55]

Kuhn's relativist and antirealist view of science rendered unintelligible the essentialist view of knowledge as inscribed in the nature of things and the historicist notion of a logic, or *telos,* of history. In place of the positivist-Whig vision of the unity, linearity and homogeneity of a single, absolute, historical time,

Kuhnian historians of science advanced the idea of a succession of distinct and different epochal times, each with its own unity, linearity and homogeneity. In this manner, Kuhnian historians of science rejected the problematic of periodization, origins, authorship and priorities, and replaced the positivist-Whig view of the progressive accumulation of knowledge with an image of the history of science as a succession of discrete, self-enclosed paradigms, or worldviews. The positivist-Whig idea that all revolutions in science are re-enactments of the break with metaphysics that first occurred in the Scientific Revolution of the seventeenth century contrasted with Kuhn's vision of the methodological diversity of scientific paradigms and their place in a discontinuous cognitive terrain. Kuhnian historians of science abandoned the positivist distinction between empirical knowledge and meaningless metaphysics. Equally theoretical, conjectural and enclosed, scientific paradigms are distinguished from metaphysical ones by their superior puzzle-solving capacities. Kuhnian historians of science replaced the positivist-Whig model of science as an inherently peaceful consensual, democratic activity, demarcated from metaphysics by the application of an algorithmic method of inquiry, with an image of science as a cognitive terrain characterized by a topology of dissensus, discontinuities and revolutionary fissures. Whereas positivist-Whig historians viewed the Chemical Revolution as a break with chemistry's nonscientific past, Kuhn treated it as an abrupt transition between incommensurable scientific paradigms. Thus, despite their considerable philosophical differences, positivist-Whig and Kuhnian historians of science viewed the Chemical Revolution as a moment of radical discontinuity in the history of science.

History as Rational Hindsight – or Not!

Imre Lakatos responded to Kuhn's assault on the citadels of objectivity and rationality in science by incorporating globalist themes into a 'sophisticated' version of Popper's 'methodological falsificationism', which treated scientific tests as 'three-cornered fights between rival theories and experiment' in which 'some of the most interesting experiments result ... in confirmation rather than falsification'.[56] According to Lakatos, a theory is falsified only in the presence of another theory which explains the success of the 'falsified' theory and has excess empirical content over its rival, some of which is corroborated. Lakatos used the methodological principles of 'empirical growth' and the 'proliferation of theories' in order to offset Kuhn's image of the paradigmatic conformity of normal science, to reduce the risk of falsifying a true theory, and to bring Popper's 'methodological falsificationism' more in line with the history of science. Adopting a historicized 'logic of discovery', Lakatos maintained that 'not an isolated *theory*, but only a series of theories can be said to be scientific or unscientific: to apply the term 'scientific' to *one single* theory is a category mistake'. He further argued that a 'series of theories' is held together by 'a remarkable [conceptual] *continuity*', which forms it into a 'research programme'. A research programme contains three elements: a 'hard core' (or 'negative heuristic') of fundamental assumptions, which cannot be abandoned without repudiating the research programme; a 'positive heuristic', which guides changes in specific theories; and a series of specific theories actually produced by the positive heuristic. Since a research programme is driven by its 'positive heuristic', and not by responding to empirical anomalies, the history of science enjoys a 'high degree' of theoretical autonomy, which, Lakatos insisted, cannot be explained by the 'naïve falsificationist's disconnected chains of conjectures and refutations'.[57]

Lakatos criticized the notions of 'instant rationality' and 'crucial experiment', and he rejected Agassi's attempt to link scientific discoveries to 'Popper's great negative crucial experiments'. Insisting that 'any theory can be defended progressively for a long time, even if it is false', Lakatos related the appraisal of a specific theory to its place in an unfolding research programme. According to Lakatos, a scientific decision is rational to the extent that a large-scale progressive, or predictive, programme 'overtakes a degenerating one', which 'gives only *post-hoc* explanations either of chance discoveries or of facts anticipated by or discovered in rival research programmes'. Lakatos viewed the probative relation between theory and experiment as ineluctably temporal and historical, arguing that 'no experiment is crucial at the time ... it is performed'. Further rejecting the doctrine of incommensurability, which rules out crucial experiments altogether, Lakatos staked out a third position, according to which the 'honorific title' of 'crucial experiment' might be 'conferred on certain anomalies, but only *long after the event*, only when one programme had been defeated by another one'. In place of the inductivist and falsificationist idea of 'instant rationality', Lakatos argued that crucial experiments can be identified only 'with hindsight' and scientists 'can be wise only after the event'. Lakatos enjoined historians of science to 'look in history' for rival research programmes, progressive and degenerating problem shifts, the victory of progressive over degenerating programmes, and prolonged wars of attrition between rival programmes which are only later linked 'with some alleged crucial experiment'. Since 'human beings are not *completely* rational animals', Lakatos recognized the need to supplement the 'methodology of research programmes' with an 'external history', or causal explanation of those aspects of the history of science that deviated from his normative ideals.[58] More damaging to his programme of rational reconstruction of the history of science was Lakatos's recognition that, since a degenerating research programme can always make a comeback, even the wisdom of hindsight could never be 'absolutely conclusive'. Despite his concerted efforts to the contrary, Lakatos's historiography of scientific research programmes supported a relativistic sense among postpositivist historians of science of the cognitive parity and uncertain attrition of the diverse and competing research programmes that make up the history of science.

Paul Feyerabend embraced with enthusiasm the relativistic implications of Popperianism that Lakatos strenuously tried to avoid. Claiming that Lakatos was a relativist in rationalist clothing, Feyerabend argued that in the absence of a 'time limit' on the allowable resuscitation of research programmes, Lakatos's standards of rationality were vacuous, arbitrary and dogmatic.[59] Feyerabend's own philosophical career was similarly poised between rationality and relativism, between 'critical rationalism' and 'skeptical fallibilism'.[60] Seeking a radical extension of Popper's critical rationalism, Feyerabend argued that a highly confirmed and entrenched theory can be effectively criticized only by the development of 'strong alternatives', which contradict not only some of the consequences, but also the 'basic principles', of the existing theory, while generating new independently testable predictions. Characterizing strong alternatives as 'incommensurable' with the theories they are designed to criticize, Feyerabend transformed the measured tones of critical discourse into the discordant clash of mutually incomprehensible worldviews. Feyerabend used the history of science to lend plausibility to the resultant methodology of 'anything goes'. Stressing the historicity of the knowledge situation, he used the example of Galileo's defence of the Copernican hypothesis to argue that scientific progress occasionally requires the deployment of 'counter-inductive' strategies and propagandistic machinations' to defend a new theory which, though correct, is 'refuted' by an archaic and inadequate observation language. In this manner, Feyerabend rejected historiographies of science based on the notions of instant rationality and the cumulative progress of science because they overlooked the historical contingencies of the knowledge situation, the importance of 'counter-inductive' procedures in the history of science, and the crucial link between individual liberty and an anarchistic epistemology. Feyerabend deconstructed Popper's critical rationalism in a way that reinforced the relativistic image of the history of science derived from Kuhn's paradigm-based analysis of science.

Research Traditions and the Arationality Assumption

Larry Laudan responded to relativism by linking the rationality of science to its historicity. Critical of the ahistorical character of traditional theories of scientific rationality and progress which defined progress in terms of rationality, Laudan defined rationality in terms of progress. He proposed that *'rationality consists in making the most progressive theory choice,* not that progress consists in accepting successively the most rational theories'.[61] Laudan treated scientific decisions as historical events, extended in time and shaped by the evolution of science; and he developed a 'pragmatic' theory of science as a 'problem solving activity', which identified progress with growth in *'problem solving effectiveness'* and linked rationality to progress-enhancing decisions. Seeking a middle ground between reconstructionist and descriptivist approaches to the problem of scientific rationality, Laudan deployed the distinction between 'internal history' and 'external history' in a way that emphasized 'the evolving character of rationality itself'. He delineated this distinction in terms of the 'arationality assumption', according to which *the sociology of knowledge may step in to explain beliefs if and only if those beliefs cannot be explained in terms of their rational merits*.[62] Laudan had no quarrel with Robert Merton's 'noncognitive sociology of knowledge', which ignored the formation of beliefs and the acceptance of theories in order to focus on the social structures and institutions of science. Nor did he deny that social and psychological forces affect rational decision-making. He merely insisted that these factors are secondary or irrelevant to the primary goal of understanding science as a rational activity. Laudan raised these concerns and considerations in his influential opposition to the 'Strong Programme' in the sociology of knowledge.

Laudan referred to global theories as 'research traditions', which he clearly distinguished from Kuhn's paradigms and Lakatos's research programmes. Although he rejected, as obscure and inapplicable, Popper's objective measure of 'truth' and 'verisimilitude', he strongly opposed Kuhn's claim that global theories are 'implicit' sources of 'tacit knowledge'. Laudan insisted that a research tradition consists in a set of explicit assumptions or propositions which, contrary to Kuhn's notion of normal science, are openly debated and logically evaluated at every stage in their development. Laudan treated research traditions as dynamic entities; he recognized that a research tradition may have at its inception no assumptions at all in common with those at a more advanced stage of its development. While Kuhn and Lakatos thought of global theories as static structures, with 'hard cores' abstracted from the flow of history, Laudan adopted a more historicized sense of the cognitive and evaluational units of scientific development. Laudan measured the scientific progress produced by these developmental entities in terms of their *problem-solving effectiveness*, which he equated with the number and 'weight' of the empirical problems solved by the specific theories of a research tradition minus the number and weight of 'anomalies' and 'conceptual problems' generated by the tradition. Since the problems of a theory are identified and weighed in relation to the problem-solving activity of rival theories, a research tradition is evaluated relative to its historical rivals, and not on the basis of some abstract, absolute standard.[63] Laudan proposed the comparative criteria of 'problem-solving effectiveness' as a way of rationally reconstructing the history of science.

Rejecting the orthodox view – shared by positivists and postpositivists alike – that only empirical considerations are relevant to the *justification* of scientific theories, Laudan, along with Toulmin and Gerd Buchdahl, recognized the role of 'conceptual problems' in the evaluation of scientific theories.[64] Laudan distinguished between 'internal' conceptual problems, which involve logical difficulties with a theory's fundamental postulates, such as vagueness, inconsistency and circularity, and 'external' conceptual problems, which encompass incongruities – including formal inconsistencies, joint implausibilities and mere incompatibilities – between the basic concepts and postulates of a theory and another scientific theory, a broader metaphysical precept, or an extra-scientific world-view, such as theology, philosophy, morality, or politics. Although the orthodox view recognized the role of 'internal' conceptual problems in the context of justification, Laudan (and Toulmin) stressed the importance of 'external' conceptual problems in the evaluation of scientific theories. Usually given the same 'weight' as empirical problems in the evaluation of competing theories, Laudan noted that conceptual problems count more heavily when the empirical adequacies of the theories are virtually equivalent. Laudan's view of the role of conceptual problems and research traditions in the rational reconstruction of the history of science assimilated the history of science to the history of ideas, with its emphasis on 'bodies of doctrine' rather than on specific explanatory theories and the individuals who deploy them.

Continuity and Discontinuity

Laudan shared a growing sense of unease among some postpositivist scholars with the doctrine of incommensurability and its one-sided emphasis on the moment of discontinuity in scientific change. Instead of viewing the history of science as a series of revolutionary upheavals, in which 'entire conceptual systems' give way to new ones, they offered 'evolutionary' accounts of 'transformations' unfolding according to determinate 'invariants' to produce a 'significant dimension of continuity' in the 'problems, concepts, and standards' of science, 'however much they otherwise change.'[65] Laudan linked the doctrine of incommensurability to the 'central cleavage' between 'gradualists' and 'revolutionaries' in the history of science community. Working with a 'consensus' model of science, which related cognitive continuity and cumulativity to the operation of consensus-forming algorithms of decision-making, gradualists – Whigs and positivists – problematized cognitive change within science and treated the Chemical Revolution as a re-enactment of the Scientific Revolution of the seventeenth century, which they regarded as a transition from nonscience to science. In contrast, Kuhnian 'revolutionaries' rejected the idea of a consensus-forming method unique to science, emphasized 'dissensus' and discontinuities in scientific theorizing, and treated cognitive continuity and permanence as in need of explanation. On this model of scientific change, the Chemical Revolution occurred within the domain of science and is simply a prominent feature in a general terrain of cognitive flux. On both models, however, the Chemical Revolution involved a moment of radical discontinuity, in which Lavoisier ushered in the age of modernity in chemistry by making a fundamental break with tradition. Gradualist and revolutionary historians of science uphold the revolutionary nature of Lavoisier's achievement by portraying the phlogiston theory as either an 'inversion' of the oxygen theory or as 'incommensurable' with it. As will be seen in Chapter 3, some postpositivist scholars replaced this Manichean perspective with more nuanced accounts of the moments of continuity and discontinuity in the scientific and philosophical sensibilities of Lavoisier and Priestley.

Laudan encapsulated and systematized these moderate sensibilities in his 'reticulated model' of scientific change, which he used to replace the one-sided sensibilities of gradualists and revolutionaries with 'a single, unified theory' of the moments of consensus and dissensus, continuity and discontinuity, in the development of science. Laudan argued that in order to capture 'both sorts of insights', it is necessary to reassess the twin ideas of the 'hierarchical structure of scientific debates' and the 'holistic picture of scientific change', which, he claimed, were at the heart of Kuhn's 'revolutionary model'.[66] According to the hierarchical theory of scientific justification, factual disagreements among scientists are resolved at a higher level of methodological consensus and methodological disagreements are resolved at a higher level of axiological consensus. Axiological differences, for which there is no higher level of appeal, are either nonexistent or irresolvable on the hierarchical model. The holistic picture of scientific change assumes that 'goals, methods, and factual claims invariable come in covariant clusters', so that when one 'cluster', or 'global theory', clashes with another cluster, there are no objective or shared criteria of evaluation to resolve their factual, methodological and axiological differences. There is a discontinuous break between successive 'clusters', and one theory simply replaces another one. Laudan retrieved the moment of continuity and rationality in scientific change by replacing this 'hierarchical picture' with the 'reticulated model of justification', according to which disagreement at any cognitive level – factual, methodological or axiological – is resolvable by reference to a consensus at any other level.[67] Consensus persists through scientific change because, contrary to the 'covariance fallacy', different and conflicting 'clusters' have one or more cognitive levels in common. Laudan's dynamic model of research traditions further integrated the moments of continuity and discontinuity in scientific change, since it made 'the relative continuity between successive stages' in the evolution of a research tradition consistent with 'many discrepancies between the methodology and ontology of its *earliest* and its *latest* formulations'. According to Laudan, an adequate representation of the moments of continuity and discontinuity in scientific change can be achieved only if the essentialist and formalist model of the history of science as a succession of static, self-contained cognitive structures is replaced by a just sense of the historicity and dynamic complexity of the research traditions that regulate science and its historical development. In this manner, Laudan posited an inextricable tie between the history of science and the philosophy of science. Chapter 7 below incorporates Laudan's reticulated model of scientific change into a more contextualized account of its historicity and temporality.

The French Connection

Epistemological Breaks and Regions of Rationality

The theoreticist sensibilities of postpositivism emerged in France immediately after World War Two in connection with the rise of what Foucault called the 'philosophy of the concept'.[68] Foucault described a conflict in post-war French philosophy between the 'philosophy of experience', associated with existentialism and phenomenology, which grounded truth and meaning in the immediate knowledge of a knowing subject, and the philosophy of the concept, represented by structuralism and the tradition of history and philosophy of science associated with Gaston Bachelard and George Canguilhem, which linked knowledge and meaning to the operation of impersonal rules and theoretical structures. Between 1927 and 1953, Bachelard developed a philosophy of science that explored issues and problems central to the tradition of Anglo-American postpositivism.[69] Like Popper, Bachelard drew a categorical distinction between the 'real object' and the 'thought object', nature and convention, claiming that 'science has no object outside its own activity'. As the domain of reason and theory, science makes an 'epistemological break' with 'reverie', or the 'dreamlike character of ordinary experience'; it brings familiar objects under concepts and theories that are radically different from those of ordinary language and common sense. Like Kuhn and Feyerabend, Bachelard argued that once a science breaks with ordinary experience, it develops through a series of epistemological breaks, in which old conceptions of nature and method give way to new ones. Although Bachelard rejected any notion of the continuous accumulation of scientific truths within a single conceptual framework, he insisted, contrary to Kuhn and Feyerabend, that new conceptual frameworks make progress over the ones they replace by assessing and correcting the range of validity of their predecessors. Emphasizing the historical mutability of scientific standards and norms, Bachelard regarded traditional philosophy as an 'unwitting agent of reverie', presenting as necessary truths the contingent features of particular modes of thought. Striving to escape from the conservative shackles of philosophy, like Kuhn, Bachelard adopted a descriptivist approach to the history of science: knowledge is understood not through abstract philosophical theorizing but by reflecting on successful applications of reason to the comprehension of nature. Like Toulmin, Bachelard believed that the study of the history of science would reveal various 'regions of rationality', but not any overarching principle of rationality. Bachelard's distinction between *l'histoire périmée* (the history of 'outdated' science) and 'l'histoire sanctionné' (the history of science validated by present standards) should not be confused with whiggism. He insisted on the need to understand the past in its own terms, and he rejected the triumphalist assumption of the 'immutable adequacy of present science'. He used present science to illuminate, not evaluate, past science.[70]

Like Popper, Bachelard identified the objectivity of science with the inter-subjective testability of scientific statements. Echoing Nietzsche's view of the disciplinary roots of rationality, Bachelard argued that the social structures and institutions of science inculcate in its members an 'intellectual surveillance of the self', in which the spontaneous play of 'reverie' gives way to the coercive rule of truth and objectivity. Bachelard anchored scientific objectivity in the consensus of the scientific community, which forms not around the objects of ordinary experience, but in relation to the 'thought objects' or theoretical entities of science. Rejecting the metaphysical idea of the mind-independent existence of scientific entities, he also baulked at the suggestion that 'theories uniquely determine the facts to which they are applied'.[71] Bachelard defended the doctrine of 'applied rationalism', according to which the concepts and theories of science are applied to pre-existent objects which are themselves produced by the application of earlier concepts and theories. Truth is neither created by science nor given by the world; it is the result of the correction, or rectification, of the world by science. Applied rationalism provided Bachelard with a middle ground between the idealist view of the mind-dependent objects of science and the realist notion of their pre-existence. Embedded in the institutions and practices of the 'scientific city', the scientific mind rectifies the world with the aid of scientific instruments, understood as 'theories materialized'. Bachelard shifted science from phenomenology, which describes the objects of ordinary experience, to 'phenomeno-technics', which uses scientific instruments to produce scientific objects. According to phenomeno-technics, theory is not distinct from its applications and the concrete objects of science result from the technical application of its theoretical concepts and structures. In a manner characteristic of post-positivism, Bachelard's philosophy of science mediated between rationalism's emphasis on the role of reason in science and empiricism's insistence on the need to subject it to the test of experience.[72] His 'phenemeno-technics' also provided a basis for Bensaude-Vincent's later constructivist analysis of the role of instruments in the formation of Lavoisier's chemistry.

Working within the Bachelardian framework, George Canguilhem wrote histories of scientific concepts rather than histories of scientific theories.[73] Rejecting the holistic theory of meaning, according to which the meaning of a concept derives from its place in a specific theoretical structure, Canguilhem separated the concept used to interpret an observation from the theory used to explain it. Arguing that scientific concepts are 'theoretically polyvalent', Canguilhem muted Bachelard's radical view of the nature and extent of epistemological breaks and obstacles in the development of science. Against Bachelard's view of a sharp break between science and nonscience, reason and reverie, Canguilhem argued that a concept is more or less scientific depending upon its degree of integration into the experimental techniques and explanatory procedures of its day. Stressing the existence of conceptual continuities through even major theoretical changes, he argued that a single scientific episode, such as Black's discovery of specific heat, could be simultaneously (conceptually) 'valid' and (theoretically) 'outdated'. Like their Anglo-American counterparts French postpositivist philosophers of science struggled to develop a balanced sense of the moment of (conceptual) continuity and (theoretical) discontinuity in the development of science.

Theoreticism and Antihumanism

Louis Althusser and Michel Foucault incorporated theoreticism into a broader philosophical concern with the relation between the individual agents and the overarching structures of history. Influenced by the complex intermingling in France in the 1960s of the traditions of phenomenology, structuralism, Critical Theory and Marxism, they linked the historiography of science to a critical assessment of the ontological traditions of Continental philosophy. But while there is little or no consensus among current scholars about the best way to understand Foucault's corpus, it is generally agreed that the major source of tension in Althusser's thought derives from his attempt to connect the diachronic philosophy of Marxism to the synchronic sensibilities of structuralism. The following treatment of Althusser and Foucault will focus on the theoreticist, or postpositivist, strains in their complex philosophies, leaving for later chapters a consideration of those aspects of their thought that challenged the postpositivist view of the autonomy of science and the hegemony of philosophy.

Althusser used the hybrid philosophy of Structural Marxism to combat the historicist interpretations of history associated with the Marxist tradition. He linked the 'humanist' interpretations of history developed by phenomenological and existential Marxists like Maurice Merleau-Ponty and Jean-Paul Sartre to the historicist notion of a teleologically structured subject of history and the essentialist idea of an underlying cognitive complicity, or identity, between the objects of history and the subjects who produce and interpret them. Whereas humanist Marxists viewed history as the journey of the human subject through alienation to self-realization, orthodox Marxists applied the same notion of the realization of an inner potential to the expanding forces of production. But humanism ran afoul of French structuralism which 'decentred' the subject in history. Structuralists maintained that the identity of a social act is determined not by the intentions of the individual actor, but by the impersonal rules and regulations of the society to which the individual belongs. By breaking the constitutive link between the conscious subject and the world it creates, structuralism implies that an objective understanding of the social world is acquired without reference to the consciousness of the individuals participating in it. The social sciences relate to their objects of study in the same objectivizing way that the natural sciences relate to natural objects. In this manner, Althusser treated Marxism as the scientific study of history, conceived as an objective and autonomous 'process without a subject', as a realm of class struggle irreducible to an underlying teleology.[74] He shifted the question of the validity of Marxist theory from the context of historical action and political efficacy to the domain of epistemology and the philosophy of science.

Althusser rejected foundationalist doctrines of immediate knowledge because they postulated a complicity or identity between the known object and the knowing subject. Deploying Bachelard's distinction between the 'real object' and the 'thought object', between reality and theory, Althusser argued that science produces knowledge not by 'reflecting' reality but by transforming already existing conceptualizations of it into new concepts and theories. Science is not a static, formal image of reality but a historical process, undergoing continual transformation. It is an impersonal process, governed by what Bachelard called 'problematics', or 'definite theoretical structures', and punctuated by 'epistemological breaks'. Taking place entirely 'within thought', this process is irreducible to politics or economics. Like other postpositivist philosophers of science, Althusser linked the objectivity of science to its theoretical autonomy, and he ruled out as 'ideological' the search for 'epistemological guarantees' which ground the 'thought object' in the 'real object'. Whereas a theoretical ideology is dominated by the need to answer questions generated by extra-theoretical, social demands and interests, a scientific problematic possesses an internal principle of intelligibility, which operates independently of the external world or society. Althusser construed Marxist philosophy as the theory of theoretical practice, which does not provide external guarantees for Marxist science, but concerns itself with the internal mechanism, or system of proof, that enables a theoretical problematic to produce genuine knowledge of the real world.[75] But the radical interiority of problematics, which provide their own criteria of validity, thwarted Althusser's attempt to identify the 'form of scientificity', the 'discourse of scientific proof', which would have placed problematics under the ideological sway of an external guarantee. Althusser could not sustain a consistent view of the objectivity of science and the scientificity of Marxism. Like Kuhn, he lapsed into relativism.

Epistemes, Discourse and the Historical A Priori

Foucault eschewed Althusser's interest in normativity. He suspended all 'questions of truth and falsity' and he treated knowledge as a historical product with no privileged epistemic status. Using Althusser's theoretical antihumanism 'to eradicate the explanatory role of the subject from the field of history', Foucault incorporated postpositivism's concern with the theory-laden nature of scientific facts and observations into a more general structuralist perspective on the priority of language over experience, the discursive over 'the lived', the symbolic over the natural.[76] As a blend of structuralism and phenomenology, his early studies of the history of madness and psychology yielded 'archaeological' accounts of the 'experiential structures' underlying the theories and institutional practices of psychopathology in different historical epochs.[77] In the early 60s, however, Foucault moved away from his earlier interest in experiential structures and the relation between discursive and non-discursive practices. In *The Birth of the Clinic*, he construed 'archaeology' as a means of elucidating the objective conditions for the occurrence of a linguistic act. Under the sway of structuralism, Foucault rejected subjectivist and hermeneutical theories of meaning, identifying the meaning of a statement with its role in a system of statements, or 'discursive formation', and reducing the knowing subject to a function of autonomous rule-governed systems of discourse. But Foucault resisted the sway of structuralism in one important respect. Whereas structuralism involved the search for cross-cultural, ahistorical, abstract rules that define the conditions of possibility of all discursive formations, Foucualt's archaeologist focused on the variable local rules that establish the conditions of existence of a given discourse. Interested in the 'historical *a priori*', Foucault used the archaeological method to develop a historical explanation and critique of scientific rationality independent of the norms of current scientific practice.[78]

Foucault's philosophy of science came closest to Kuhn's theoreticism in the mid-60s. In *The Order of Things* he characterized the epistemological status of the 'human sciences' – psychology, sociology and ethnology – by identifying them as unstable products of the 'modern episteme', which he also placed against the historical background of the epistemes of the Renaissance and the Classical Age.[79] Understood as a global theory, paradigm or way of viewing the world, the episteme of a historical period or epoch consists in the basic conceptions of order, sign, language and knowledge that define 'the conditions of possibility of all knowledge' for that period or epoch. Influenced by Canguilhem's histories of specific scientific concepts, Foucault examined the role of philosophical concepts in defining the conditions of possibility for specific scientific concepts. *The Order of Things* provided neither a history of the central ideas of the sciences nor a philosophical inquiry into their conceptual foundations, but an account of the rules scientists of a given epoch unwittingly followed in the construction of their disciplines. Although Foucault used the 'archaeological method' to show that the norms and values of the human sciences depend upon contingent historical conditions, unlike his counterparts in the Anglo-American postpositivist community he regarded his relativistic analysis as inapplicable to the 'noble' sciences of physics and chemistry.

In his next book, *The Archaeology of Knowledge,* Foucault provided an explicit and systematic account of the 'archaeological method' of historical inquiry, which, in some respects, carried him beyond the theoreticism of postpositivism.[80] Foucault replaced the humanist notion of history as a process of conscious self-emancipation with the idea of history as an autonomous terrain of discourse which, governed by anonymous rules of formation, is opaque to the consciousness of historical actors but open to the objective scrutiny of the historian. Foucault's notion of the autonomy of history will be discussed more fully in Chapter 7. For now it is significant to note that although Foucault carefully distinguished between a discursive formation and its scientific articulations, between the serial organization of statements to form a discourse and the unitary groupings of statements to form theories, the former, no less than the latter, functioned like the global theories of postpositivism to decentre the subject and articulate an 'epistemology without a knowing subject'. In his subsequent 'genealogical' writings, Foucault used the archaeological method to develop an antihumanist, or decentred, account of discursive and nondiscursive practices in science, focusing less on the rules of discourse than on the configurations of power in which these rules are articulated and deployed. The consequences of Foucault's genealogical inquiries for the sociology of scientific knowledge will be discussed in Chapters 4 and 5 below, while his view of the specificity and autonomy of history will be revisited in Chapter 7.

Idealism and the Spectre of Marx

The rise of Bachelard's philosophy of science in the 1940s eclipsed the idealist philosophy of Emile Meyerson which was influential in France in the 1920s and 30s. Bachelard's view of the discontinuity of scientific development and its materialization in instruments and institutions opposed Meyerson's idealist emphasis on the unity and continuity of thought and its foundation in the human mind. Meyerson regarded the tendency of science to reduce the apparent diversity and multiplicity of nature to a deeper, causal reality of permanence and identity as a 'secret propensity of the human mind', which requires conceptual abstraction and theory construction. He saw in the history of science not a terrain of paradigmatic diversity and theoretical pluralism, but constant and recurring forms of scientific reasoning based on principles of identity and conservation. Part of a broader anti-positivist movement in Europe, which included the neo-Kantianism of Ernst Cassirer and the ethnology of Lucien Lévy-Bruhl, Meyerson's philosophy shaped the historiography of science developed by Alexandre Koyré and had a significant influence on Hélène Metzger, the niece of Lévy-Bruhl and arguably the finest historian of chemistry in the twentieth century.[81] Tragically, the Nazis murdered Metzger in 1944, and it was left to Koyré to disseminate the idealist historiography of science, which he did to great effect with the publication of *Études Galilénnes* and his move to America in the late 1930s. Koyré upheld the idealist thesis that science 'is essentially *theoria,* a search for the truth', which has an 'inherent and autonomous' development, an internal logic of its own.[82] Emphasizing the unity and continuity of the human intellect, Koyré searched for close ties between science, philosophy and theology. Rejecting the positivist notion that the defining characteristic of modern science is an algorithmic empirical methodology, he interpreted the Scientific Revolution not in terms of the acquisition of new observations or additional evidence, but as an intellectual transformation that involved the mathematization of nature. On this view, the Scientific Revolution involved 'transpositions that were taking place inside the minds of the scientists themselves'.[83]

The seeds of Koyré's idealist historiography of science thrived in the fertile soil of America after World War Two. The ground had been well prepared before the war by the idealist historiographies of philosophy developed by E. A. Burtt, A. J. Snow and A. O. Lovejoy, as well as by the ideas of the philosopher A.N. Whitehead and the historian R. G. Collingwood. Koyré's historiography also suited the emergent professionalization of the history of science in post-war America, providing its young practitioners with a coherent set of textual problems and analytical techniques on which to hang their disciplinary identity.[84] They found their model in Koyré's scholarship, which involved close conceptual analyses of the informing ideas and intellectual affiliations of important scientific texts. Koyré's methodology influenced the work of I. Bernard Cohen, Thomas S. Kuhn, Henry Guerlac, A. Rupert Hall and Richard S. Westfall, among others, who published important texts that opened up a vision of the unity, autonomy and cultural significance of scientific thought.

Koyré's idealist historiography of science performed useful ideological and political functions in the anticommunist climate of America and Europe during the Cold War.[85] Koyré was a fervent anti-Marxist, and Hall and Westfall defended his idealist historiography by stressing its opposition to Marxist materialism. Tracing scientific ideas to socioeconomic forces, Marxism rejected the idealist view of theory as an autonomous value-neutral description of reality in favour of the idea that theory is rooted in the practical activities of human beings and oriented towards the improvement of human life. These ideas flourished among British Marxist and left-wing scientists in the 1930s who placed a socialist construal on the thesis of the progressive unity and humanizing function of science associated with Sarton's 'new humanism'. Robert K. Merton, the doyen of the sociology of science in America, developed a watered-down version of the Marxist historiography which focused on the modes of organization and the social structures of science but left its cognitive content untouched.[86] Koyré responded to these ideas by insisting upon the separation of theory and practice in science, claiming that it is only by the study of science's 'own problems, its own history, that it can be understood by historians'.[87] Guerlac endorsed the idea that 'the history of science is fundamentally the history of scientific thought', and Hall rejected 'externalist' attempts to explain the Scientific Revolution in terms of the socioeconomic conditions of seventeenth-century England because they overlooked the 'continuity of scientific thought'.[88] Locating science in 'the human desire to know', idealist historians emphasized its 'transcultural standing'. Westfall concluded that, contrary to the externalist view of science as the product of society, 'much of the modern world appears to me as so many epiphenomena to the growth of science'.[89]

While Koyré's influence on Anglo-American scholars reinforced the general tendency of twentieth-century philosophers of science to ignore chemistry and focus on mathematics and physics, other members of the French school, including Duhem, Meyerson and Bachelard, cut their philosophical teeth on the atomic debates of nineteenth-century chemistry, the fierce Franco–Prussian controversy over the founder of modern chemistry – was it Lavoisier or was it Stahl? – and the nature of the Chemical Revolution – was it more continuous than discontinuous?[90] But it is Hélène Metzger who best represents the importance of chemistry in the French tradition of the philosophy of science. The idealist imperative associated with this tradition informed the impressive and influential body of work that Hélène Metzger compiled before her untimely death. Metzger rejected the presentist orientation of positivist-Whig historians because it ignored the historicity of science and precluded 'a sympathetic understanding of past thought'.[91] An adequate understanding of past science is beyond the reach of an empiricist historiography that relates progress to the crucial experiments of a few men of genius. Distancing herself from Meyerson's view of the unity and universality of human reason, Metzger deployed Lévy-Bruhl's notion of mentalities; she regarded it as the duty of historians of chemistry to unearth the 'orientations de mentalité' that inform and unify the different texts of an author and the diverse theories of a historical period.

Within an idealist interpretive framework, which downplayed the role of experience, agency and discontinuity in the development of science, Metzger produced an interpretation of the Chemical Revolution that prefigured some of the themes and ideas developed by postpositivist historians of chemistry thirty years later. She rejected the positivist-Whig idea of Lavoisier as the experimental founder of modern chemistry, and she shifted the locus of the Chemical

Revolution from Lavoisier's antiphlogistic chemistry to his more fundamental philosophy of matter. Upholding the worth of Stahl and his chemistry, Metzger stressed the continuity that existed between Lavoisier's chemistry and the work of his eighteenth-century predecessors and rivals. She argued that the more important aspects of his chemistry, such as the law of the conservation of matter, the definition of 'simple substances' and the reform of the chemical nomenclature were the outcome of a 'gradual transformation that had been going on throughout the eighteenth century'. Postpositivist scholars endorsed Metzger's suggestion that 'Lavoisier marks the end of an era rather than the beginning of a new age'.[92] Metzger's narrative also prefigured some of the historiographical limitations of postpositivism. In focusing on the conceptual unity and structure of chemical doctrines, she produced an inadequate sense of their specific historical context, agency and sequential development.

Summary and Conclusion

Postpositivism challenged the empiricist sensibilities of positivism and whiggism with a theoreticist conception of the nature of science and its historical development. Replacing the positivist ideal of empirical certainty with a fallibilist view of conjectural knowledge, postpositivism rejected the epistemological distinction between observation statements, grounded in experience, and theoretical statements, based on conjecture. Identifying knowledge with theory, which is ineluctably conjectural, and emphasizing global rather than local theories, postpositivism replaced the positivist-Whig view of the epistemological efficacy of individual subjects and their originary moments of discovery with the theoreticist notion of 'epistemology without a knowing subject' and the structuralist idea of history as a 'process without a subject'. While the concept of history as a 'process without a subject', associated with the French 'philosophy of the concept', only impacted the historiography of the Chemical Revolution after it was assimilated by sociologists of scientific knowledge, who gave it an antitheoreticist formulation, the Popperian notion of 'epistemology without a knowing subject' had a more immediate impact on postpositivist interpretations of the Chemical Revolution. The next chapter will show how the normative interests associated with notions of 'objective knowledge' shaped postpositivist interpretations of the Chemical Revolution, while the following chapters will explore the various ways in which sociologists of scientific knowledge and sociologically minded historians of science formulated ostensibly non-normative histories of science which challenged or deflected traditional concerns with historical agency.

Although Kuhn and Hanson sometimes lapsed into a non-essentialist form of idealism, which assimilated things to thought and reality to theory, the Popperian notion of critical realism represented the dominant tendency among postpositivist scholars, who upheld a metaphysical or categorical distinction between the known object and the real object, between thought and reality, convention and nature. In rejecting the positivist-Whig notion of the unity of man, knowledge and nature, postpositivism rendered unintelligible the essentialist view of knowledge as inscribed in the nature of things and the historicist notion of a logic or *telos* of history. In place of the positivist-Whig vision of the unity, linearity and homogeneity of a single absolute historical time, postpositivism advanced the idea of a succession of distinct and different epochal times, each with its own unity, linearity and homogeneity. In this manner, postpositivism rejected the positivist-Whig problematic of periodization, origins, authorship and priorities, and replaced the positivist-Whig view of the progressive accumulation of knowledge with an image of the history of science as a succession of discrete self-enclosed worldviews, paradigms, research traditions or research programmes.

Postpositivists replaced the positivist-Whig model of science as an inherently consensual and democratic activity with an image of science as a cognitive terrain characterized by a topology of dissensus, discontinuities and revolutionary fissures. While some postpositivist scholars still searched for continuities, as well as discontinuities, in the historical record, which they reconstructed in terms of global principles of rationality, other postpositivists described more localized forms of reason, or abandoned the notion of scientific rationality altogether. The resulting historicized models of science replaced the formalist view of scientific theories as static sets of abstract propositions with a more dynamic emphasis on the role of 'guiding assumptions' in the formation and development of scientific theorizing. But the emphasis these models placed on the conceptual unity and structure of scientific doctrines discouraged the formation of an adequate sense of the specific historical contexts and sequential development of scientific thought. As will be seen in the next chapter, the articulation and application of this interpretive framework, against the backcloth of existing positivist-Whig sensibilities, gave coherence and direction to the resurgence in scholarly interest in the Chemical Revolution that occurred in the 1960s and 70s.

3 POSTPOSOTIVIST INTERPRETATIONS OF THE CHEMICAL REVOLUTION

In accord with the arationality assumption, postpositivist historians of the Chemical Revolution rejected any attempt to locate 'the causative factors in Lavoisier's work' in the socioeconomic conditions of eighteenth-century France.[1] According to Carleton Perrin, for example, the Chemical Revolution involved a '*conceptual* breakthrough' which 'did not follow inversions of polarity in the changing political climate'.[2] Guerlac linked the Chemical Revolution to the unfolding of an 'internal history', claiming that the 'outstanding feature of Lavoisier's Chemical Revolution' involved the emergence of chemistry from its 'industrial and practical background' as 'an autonomous discipline, a body of theoretical knowledge' with 'its own theoretical problems, its own methods of thought and inner logic'.[3] In a similar vein, Marco Beretta emphasized the problems involved in drawing 'any univocal conclusion on the relation between science and politics' in the Chemical Revolution.[4] Treating the 'social context' as 'peripheral' to the unfolding 'logic' of the Chemical Revolution, postpositivist scholars examined specific concepts and theories in relation to particular research programmes, disciplinary research traditions and, in some cases, broader cultural and intellectual themes and movements. The theoreticist orientation of postpositivism also downplayed the role of new observations and experimental evidence in the development of science, leading Freddy Verbruggen to claim that 'the phlogiston controversy, and the disagreement between, for instance, Priestley and Lavoisier, was not a matter of the 'observation', but of the INTERPRETATION of chemical processes'.[5] Underdetermined by man's interests or nature's ways, by the knowing subject or the known object, the Chemical Revolution involved an autonomous transformation in the concepts, theories and categories of eighteenth-century physics and chemistry.

Toulmin gave a clear statement of the theoreticist interpretation of the Chemical Revolution when he claimed that 'the advance which chemistry required when Lavoisier began his work was essentially a theoretical one, which might be suggested by a sufficiently striking experiment, but could not be imposed on one by it'. Toulmin hailed Lavoisier as '"the father of modern chemistry" ... not because by good fortune he was the first to hit on an experiment which *obliged* one to accept the new point of view', but for 'the clear-headedness with which he devised the first really comprehensive and fruitful set of chemical concepts and categories'.[6] From this perspective, the Chemical Revolution involved not only a shift from the phlogiston theory to the oxygen theory of combustion, but also changes in the fundamental concepts and theories of chemical composition and the nature of matter, as well as transformations in the disciplinary identity of chemistry and its cognitive status. In keeping with the idea of history as 'a process without a subject', postpositivist historians showed how these conceptual, epistemological and theoretical changes started prior to Lavoisier's appearance on the scene and continued long after his death. In this manner, postpositivism opened up new perspectives on the discovery of oxygen, the genesis, development and assimilation of Lavoisier's theories, the reasoned opposition of his adversaries, and the nature, degree and extent of the changes that took place during the Chemical Revolution and its overall place in the intellectual landscape of the eighteenth century.

Historiographical Innovation and the Pull of the Past

Postpositivist Re-enactments of Positivism

Postpositivist interpretations of the Chemical Revolution included specialized historical narratives and schematic philosophical models. The philosophical models developed by Kuhn, Musgrave, Toulmin and Philip Kitcher stand out and will be considered in this section. Although they focused on theoretical doctrines and conceptual presuppositions, these accounts of the Chemical Revolution did not break entirely with empiricist and whiggish modes of thought. Denying the logical efficacy of crucial experiments in the debate between Priestley and Lavoisier, Toulmin claimed that the 'superior merit of Lavoisier's theory lay in this: that it gave clear and economical explanations instead of confused ones, not that it gave true instead of false ones'.[7] Musgrave defended the superior rationality of the oxygen theory by questioning the formalist theories of justification associated with Logical Positivism. According to Musgrave, 'philosophers of science who think that evidential support of a theory depends solely upon the *timeless logical relation* between theory and evidence will have to say that 1784 phlogiston theory had as much empirical support as 1784 oxygen theory'. Both theories explained the main facts about combustion, calcination and acidification, and the phlogiston theory successfully accommodated the doctrine of the 'decomposition of water' formulated by Lavoisier in 1784. But, Musgrave insisted that instead of taking a static, formalist view of theory-choice, 'the chemists of the late eighteenth century' noted that the 1784 phlogiston theory merely accommodated in an *ad hoc* and inconsistent way known facts that were predicted by the 'smooth development of the oxygen programme'. Appealing to Lakatos's 'methodology of scientific research programmes', Musgrave argued that eighteenth-century chemists recognized that between 1770 and 1785, the 'oxygen research programme' was 'theoretically and empirically progressive', while its phlogistic rival was 'degenerating'. Musgrave concluded that 'external' accounts of the Chemical Revolution were not only inadequate, but also 'unnecessary ... since the methodology of research programmes explains that revolution *internally*'.[8]

Kuhn offset Musgrave's 'rational reconstruction' of the origin of modern chemistry with the relativist notion of the Chemical Revolution as a 'paradigm shift'. Evoking positivist-Whig sensibilities, Kuhn claimed that prior to the Chemical Revolution, phlogistic chemists adopted a qualitative chemistry which appealed to a 'small number of elementary "principles" ... of which phlogiston was one ... to explain' the instantiation and transformation of such generic qualities of observable substances as acidity, metallicity and inflammability. Although phlogistic chemists achieved some measure of success in this endeavour, Kuhn claimed that they soon found themselves floundering in a morass of diverse paradigmatic responses to the 'anomalies' generated by the rise of pneumatic (gas) chemistry and the emergence of quantitative experimentation.[9] Lavoisier solved these problems, brought order to chaos and pointed chemistry in a new direction when he replaced the qualitative chemistry of 'generic principles' with the quantitative chemistry of 'simple substances' based on isolating and weighing the simplest units of laboratory analysis. Further articulating Kuhn's analysis of the Chemical Revolution, Gerry Doppelt argued that Lavoisier's rejection of the chemistry of principles, which ground observed properties in underlying material elements, was accompanied by explanatory losses as well as explanatory gains; the shift in standards involved in this paradigm shift meant that 'much of the nineteenth century failure to explain the qualities of compounds was no indictment of chemical theory'.[10] Contrary to the revolutionary self-image of the postpositivists, Kuhn's notion of the incommensurability of the phlogiston theory and the oxygen theory constituted a transformation and not a wholesale rejection of the positivist-Whig idea that the Chemical Revolution involved a cognitive 'inversion', or transition from the 'looking glass chemistry' of phlogiston to the real word of oxygen.

Inommensurability

Kuhn's concept of the incommensurability of the phlogiston theory and the oxygen theory gained considerable currency among historians who focused on the social rather than the cognitive aspects of the Chemical Revolution. In *The Formation of the German Chemical Community (1720–1795)*, Karl Hufbauer pointed to the determining role of moral, material and 'manpower' factors in the response of German Phlogistians to the oxygen theory to support the claim that 'when rival theories ... are ultimately incomparable, or "incommensurable", theoretical allegiances are open to extraneous influences'.[11] In *Chemistry Transformed: The Paradigmatic Shift of Chemistry from Phlogiston to Oxygen*, H. Gilman McCann used Kuhn's view of the insufficiency of 'logical factors' in the development of science to support his own view of 'the impact of general social structure, scientific organization, age, nationality, tradition ... and personal contact in the process of revolutionary conversion'.[12] Although Hufbauer and McCann focused on the 'substance of science', their sociological analyses must be distinguished from the 'Strong Programme' in the sociology of scientific knowledge. Whereas the proponents of the 'Strong Programme' rejected the arationality assumption and explained all beliefs, rational and arational, scientific and nonscientific, in terms of their social causes, McCann and Hufbauer focused their sociological analyses on beliefs and ideas which, according to Kuhn's reconstruction of the history of science, could not be explained 'internally'. Functioning within the parameters of the postpositivist distinction between 'internal history' and 'external history', Hufbauer and McCann developed 'external' histories of the nonrational dimensions of the Chemical Revolution.

Viewing the Chemical Revolution as a 'classic case of incommensurability', the historian of chemistry Maurice Crosland noted the similarity between Kuhn's theory of scientific change and the positivist-Whig view that the Chemical Revolution 'completely changed chemistry in the last quarter of the [eighteenth] century'.[13] A similar mingling of old and new historiographical sensibilities shaped Robert E. Schofield's analysis of Priestley's 'natural philosophy', which traced the core polarity of the Chemical Revolution to fundamental theoretical differences between Priestley and Lavoisier. Schofield argued that Priestley's scientific thought was characterized by a quest for 'fundamental physicomathematical explanations(s)' that was at odds with Lavoisier's view of the methodological autonomy of chemistry. According to Schofield, Lavoisier's success as a chemist arose out of a Stahlian rejection of Newtonian 'mechanism' in favour of a 'materialist' search for the 'permutations and combinations' of relatively indestructible 'elements with property-bearing characteristics related to the realm of laboratory experience'. On this analysis, Priestley rejected the shallow triumphs of Lavoisier's materialism and returned to the 'mechanistic' programme of Newtonian dynamic corpuscularity, which denied any permanent identity to the chemical elements and emphasized 'the fundamental significance of determining the ultimate constituents of matter in its mechanistic modes and operations'. Schofield also argued that Priestley's commitment to Boscovich's version of Newtonian corpuscularity – according to which gravity is not an inherent property of matter – accounted for his 'persistent inability to take seriously Lavoisian arguments based on conservation of weight'.[14] Schofield incorporated the positivist-Whig sense of Priestley's methodological deficiencies into the postpositivist view of the incommensurability of the basic categories of Priestley's natural philosophy and Lavoisier's chemistry.

A phalanx of postpositivist scholars incorporated into theoreticism the positivist-Whig view of a Manichean division between Lavoisier's progressive chemistry and the reactionary response to it by Priestley and other phlogistic chemists. While Robert Siegfried lauded the quantitative credentials of Lavoisier's chemistry, J. B. Gough related the rise of 'an objective, rigorous scientific method for determining the relative order of compositional simplicity' to 'Lavoisier's 'gravimetric criterion', which was '*better* than the intuition of his Stahlian rivals and predecessors'.[15] Evoking the antirealist theme in Kuhn's philosophy of science, Crosland claimed that 'Priestley's charming amateurism' placed him 'in a different world from the strictly quantitative and specialized science which was the new chemistry of Lavoisier'.[16] Just as John Brooke portrayed Priestley as 'picking nits in the new French system', so Arthur Donovan dismissed him as one of 'God's fools', who 'airily devalued the whole enterprise' of theoretical chemistry in order 'to glory in the unfathomable wonder of creation'.[17] While Toulmin used his theoreticist reappraisal of the Chemical Revolution not to 'reinstate Priestley's theories', but 'to make it clearer why we rightly prefer Lavoisier's', Kuhn seemed to question the positivist-Whig perspective when he argued that Priestley's opposition to the oxygen theory was not a 'violation of scientific standards', but had its origin in the kind of commitment that characterized normal science and that ultimately facilitated the very paradigmatic change it opposed.[18]

Elsewhere, however, Kuhn evoked positivist-Whig sensibilities when he concluded that a person, such as Priestley, 'who continues to resist after his whole profession has been converted has *ipso facto* ceased to be a scientist'.[19] On Kuhn's analysis, Priestley's opposition to the oxygen theory was not so much 'illogical or unscientific' as cognitively irrelevant and downright unprofessional. In this Kuhnian vein, McCann based his sociological analysis of the Chemical Revolution on a prosopographical incommensurability between Priestley, the aging British amateur, wedded to a qualitative chemistry, and Lavoisier, the young French professional, armed with the new quantitative paradigm.[20] A similar whiggish bias informed Musgrave's claim that the Chemical Revolution came to a rational conclusion with the discovery of water in 1784, almost four years before Priestley engaged his adversaries in earnest debate.[21] The tenacious hold of whiggism on the mind of historians – referred to by Butterfield as 'the historian's pathetic fallacy' - is evident in Schofield's conscious decision to ignore 'injunctions not to read back into history the obviousness of modern paradigms': he traced Priestley's failure to appreciate the 'easy interpretation' his experiments received 'within the frame of the oxidation theory' to his adherence to the 'physicalist' tradition of Newtonian corpuscularity.[22]

Between Realism and Relativism

The clash between historiographical innovation and tradition generated philosophical tensions in Phillip Kitcher's 'realist' model of the Chemical Revolution. Like Kuhn and Feyerabend, Kitcher rejected the standard view of 'theory reduction' (which he called 'Legend'), according to which a theory – regarded as 'nothing more than a set of statements' – is reduced to another theory only if it can be logically derived from that theory. Adopting a 'globalist' perspective, Kitcher treated 'statements and beliefs' as only one component of a scientific 'practice', which also consists in a scientific language, paradigms of experimentation and instrumentation, a set of appropriate problems, and abstract patterns of reasoning instantiated by attempts to solve these problems. Kitcher accepted Laudan's 'multidimensional' model of science and its transformation, and he viewed the history of science not as a monotonic encapsulation of reduced theories by reducing theories, nor in terms of the complete replacement of one paradigm by its totally distinct rival, but as a succession of partially overlapping practices, which afford limited, but real, judgments of the rationality of scientific change.[23]

Kitcher steered his account of the explanatory superiority of the oxygen theory over the phlogiston theory between two extreme positions: the positivist-Whig view that because phlogiston 'failed to refer' to anything in the world, 'the phlogiston theory crumbled under the compelling force of Lavoisier's evidence' and the Kuhnian claim that because phlogiston referred to something in a 'different world ... no objective evidence forced the abandonment of the phlogiston theory'. While Kitcher rejected the positivist-Whig view because it failed to account for Priestley's success as a chemist and his considerable influence on the development of 'Lavoisier's own ideas', he objected to Kuhn's 'many worlds' analysis because it made the notion of any communication, dialogue or influence between Priestley and Lavoisier difficult if not impossible to comprehend. Kitcher traced the inadequacies in these competing perspectives to their shared commitment to a univocal semantics, which he replaced with the claim that a scientific community 'allows a number of different ways of fixing the reference' of a linguistic term. In this manner, Kitcher argued that the language of the oxygen theory *partially* overlapped the language of the phlogiston theory and Lavoisier *partially* understood and, where appropriate, corrected Priestley's language.

Kitcher rejected Kuhn's idea of a 'shift in standards' during the Chemical Revolution; he insisted that phlogistic and antiphlogistic chemists functioned within a shared 'schema' concerning the nature of chemical composition and the 'principle of the balance sheet'. Within a shared, complex set of conceptual and empirical constraints, Phlogistians and Antiphlogistians developed competing quantitative analyses of a wide range of reactions, including the calcination of metals, reduction of calces and metal-acid solutions. Kitcher portrayed a search for plausible and consistent explanations of a 'full body of experiments' that went way beyond 'Legend's tales of single crucial experiments'. The situation was complex and fluid, with plenty of room for honest debate, for the formation of consensus and dissensus among phlogistic and antiphlogistic chemists. But all this changed, according to Kitcher, after the discovery of the composition of water in 1784, when Lavoisier and his followers could not only reveal the empirical inadequacies, *ad hoc* devices and conceptual inconsistencies in the explanatory schema of the phlogiston theory, but could also demonstrate the greater simplicity and coherence of the oxygen schema. Kitcher concluded that well before 1790, Lavoisier and his colleagues 'had shown they could apply their schema to a class of instances well beyond the range of the rival phlogistonian approach'.[24]

Despite the conceptual subtlety and sensitivity of Kitcher's analysis, his comparison of the explanatory merits of the phlogiston theory and the oxygen theory was skewed in favour of the latter. While focusing on the problems that Lavoisier found with the phlogiston theory, he said nothing about the many objections that Priestley made to the oxygen theory, made only passing reference to the problematic nature of Lavoisier's theory of acidity, and downplayed the problems posed for Lavoisier's quantitative methodology by the central role of caloric, an imponderable 'subtle fluid', in his system of chemistry. Like Musgrave, Kitcher slighted Priestley's role in the Chemical Revolution by bringing meaningful debate to an effective conclusion with the discovery of the composition of water in 1784, four years before Priestley joined the fray. The whiggish bias that informed Kitcher's identification of Lavoisier's chemistry and point of view with 'our' chemistry and point of view also shaped his claim that 'even a brief reading of the writings of the Phlogistians reinforces the idea of a true doctrine trying to escape from flawed language'.[25] At least with respect to his treatment of the Chemical Revolution, Kitcher used the innovative interpretive strategies of postpositivism to breathe new life into old prejudices. Compromised and undermined though it was by the rise of postpositivism, the Whig interpretation of history did not entirely vanish from the community of historians and philosophers of science.

Lavoisier's Crucial Year Revisited

Guerlac and the Historiography of the Crucial Year

Guerlac's seminal study of Lavoisier's 'crucial year' shaped many of the more specialized interpretations of the Chemical Revolution developed by postpositivist historians of science. According to the historiography of the crucial year, Lavoisier founded modern chemistry in his quantitative experiments on combustion, recorded in the famous sealed note deposited in the fall of 1772 with the Permanent Secretary of the French Royal Academy of Sciences. The canonical view claimed that Lavoisier performed his experiments on the reduction of lead oxide at the end of October 1772 in order to confirm his earlier discovery that phosphorous and sulphur gained in weight because of the 'prodigious quantity of air' they absorbed during combustion. Rejecting this scenario, Guerlac used two unpublished manuscripts – *The August Memorandum* and *Système sur les Élémens*, written by Lavoisier in the summer of 1772 – to show that Lavoisier's interest in the reduction of metallic oxides was the source, not the outcome, of his interest in 'combustion in general'. Guerlac also traced Lavoisier's interest in the reduction of metallic oxides to his 'curiosity about effervescence' and his encounter in the summer of 1772 with the experimental work of the British pneumatic chemist Stephen Hales, who attributed the effervescence observed during the reduction of metallic oxides to the release of air. Influenced by the work of Guyton de Morveau, which established for the first time the reality and generality of the gain in weight of metals during calcination, Lavoisier linked the phenomenon of effervescence to the 'augmentation effect'.[26]

Although Guerlac challenged the positivist-Whig idea that 'combustion' was the central theme of the Chemical Revolution, he still viewed 1772 as a 'pivotal' year for Lavoisier. He argued that the 'decisive step' that Lavoisier took in this year 'brought him in succeeding years to the discovery of the role of oxygen, to his antiphlogistic theory of combustion, and to a radical refashioning of the science of chemistry'. Guerlac insisted that Lavoisier's decisive breakthrough occurred not in the domain of experimentation and quantitative methodology, but in the realm of conceptualization and theory construction. It occurred not when he performed his experiments on combustion, but when he entertained the idea that air is a chemical agent, an event that occurred well before he discovered oxygen or recognized the existence of a multiplicity of gases. According to Guerlac,

> The crucial event in Lavoisier's career was his realization that air (which nearly everyone believed to be a simple substance defined by its physical, rather than by any chemical, properties) must play a part in chemical transformations – most dramatically those observed in ordinary combustion, the roasting (calcining) of metals, and the reduction of ores or calxes.[27]

The Long Revolution from Stahl to Dalton

While Guerlac continued to focus his attention on the events of 1772, some of his students and critics questioned the 'cruciality' of the 'crucial year', emphasizing the overall theoretical structure and development of Lavoisier's chemistry, or shifting the focus of the Chemical Revolution away from his experimental career entirely. These scholars turned their attention to the development, over a broader range of phenomena and an extended period of time, of eighteenth-century concepts and theories of acidity, matter, the gaseous state, the reform of the chemical nomenclature and the order of chemical composition. Some postpositivist historians presented these developments as part of a 'Long Revolution', stretching from Stahl to Dalton and structured by 'general methodological and ontological guiding principles'.[28] Assimilating the history of science to the history of ideas, postpositivist historians resurrected Metzger's claim that the Chemical Revolution could not be understood solely, or even primarily, in terms of the rivalry between the oxygen and phlogiston theories of combustion, but must be viewed in terms of the more general and fundamental changes that Lavoisier and his contemporaries wrought in the language, ontology, epistemology, methodology and theoretical foundations of chemistry.

Shifting the interpretive focus away from the problem of combustion and 'the augmentation effect', Maurice Finocchiarro argued that Lavoisier performed his experiments on lead oxide in the autumn of 1772 'in order to elaborate his mistaken theory of elements'.[29] In a similar manner, Robert Kohler argued that Lavoisier studied the combustion of phosphorous 'not to prove a theory of combustion', but in order to show that 'the mineral acids and their salts should be included in the list of substances elemented of air'.[30] According to Kohler, this discovery had a profound effect on Lavoisier; it led him to identify 'oxygen' with the principle of acidity, a necessary ingredient in all acids, mineral and vegetable, gaseous and liquid. Like Kohler and Homer LeGrand, Maurice Crosland argued that 'Lavoisier's oxygen theory was from the beginning as much a theory of acidity as a theory of combustion'. Against the positivist-Whig claim that Lavoisier's theory of acidity was a 'major mistake', in which he allowed the metaphysical ontology of Stahlian principles to compromise the scientific logic of 'simple substances', Crosland argued that it provided 'the first chemical theory of acidity' that correlated the 'reactivity of acids' with their 'chemical composition'. Even where the theory was wrong, according to Crosland, it provided 'the next generation with the starting point for a better theory'.[31] LeGrand located this 'starting point' in the 'compositional confusion' generated by the clash between Lavoisier's operational definition of an element, which required undecomposed acids, like muriatic acid (HCl) to be treated as 'simple substances', and the oxygen theory of acidity, which regarded them as compounds of oxygen. This 'confusion' persisted until 1810, when Humphry Davy refuted the oxygen theory of acidity by showing that neither muriatic acid nor oxy-muriatic acid, which he identified with chlorine, contained oxygen. Le Grand noted that Davy's critical discoveries enabled nineteenth-century chemists to reap the full benefit of Lavoisier's operational definition of an element; he concluded that 'the chemical revolution was initiated by Lavoisier, but was concluded by the destruction of his vestiges of chemistry-by-principles, particularly the oxygen theory of acidity.[32]

If Le Grand made the Chemical Revolution a post-Lavoisier affair, J. B. Gough argued that Lavoisier raised the 'central' problem of the Chemical Revolution, which concerned the nature of the 'gaseous state', as early as 1767 in some draft notes to the *Système sur les Élémens*.[33] In these notes, Lavoisier challenged the widespread eighteenth-century doctrine that air is a simple, immutable substance incapable of entering into chemical combination with other substances. This doctrine grew out of Boyle's view of 'elasticity', or 'spring', as the defining property of air and Newton's derivation of this property from a short-range force of repulsion between its particles. Although Newton allowed air particles to exist both in a 'free', 'elastical', or 'repulsive' state and, like other substances, in a 'fixed', 'combined', or 'attractive' state, Boyle treated elasticity as an essential and immutable property of air. Boyle's essentialist view of air shaped the doctrine of its chemical inertness, which linked the phenomena of combustion and respiration to the chemical properties of extraneous particles floating in the air or to the mechanical motion of the air particles themselves. This doctrine treated 'vapors', 'exhalations', or 'factitious airs' as solutions of liquid substances in 'perennial air', which gave them their 'elasticity', or 'expansibility'.[34]

While the British pneumatic chemists were working to isolate and identify a number of chemically distinct airs in the laboratory, Lavoisier began to embrace the theoretical idea that air is a chemical substance. Lavoisier encountered Hales's experiments on the fixation of air in Johann Theodore Eller's thesis on the elements, which he read in 1766, as well as in the chemical lectures of Guillame Francois Rouelle, the leader of the French Stahlians, who lectured at the Jardin du Roi from 1742 to 1768. Impressed with Hales's experimental results, Rouelle insisted that besides functioning as an instrument of chemical reactions in its 'free', or 'elastic', state, air, deprived of its elasticity, enters into a 'fixed' state in which it combines with other substances. Lavoisier rejected Eller's idea of the unity of matter and the transformation of the elements in favour of Rouelle's doctrine of the four elements, earth, air, fire and water, three of which (fire, air and water) could exist in either 'free' or 'fixed' states. Lavoisier's discourse of 1766 also revealed his familiarity with the physical ideas of Hermann Boerhaave, who argued that elasticity is not an essential property of air particles, but the result of their combination with the 'matter of fire'. Influenced by the famous article 'Expansibilité', written by Robert Jacques Turgot in the *Encyclopedie* in 1756, Lavoisier argued in his *Système* that air becomes 'fixed', or enters into chemical combinations, when it is deprived of its 'matter of fire', or 'caloric'. According to Gough, Lavoisier honed these ideas between 1768 and 1773, when he first announced his doctrine that matter exists in three states: solid, liquid, and the 'state of vapor'.[35]

Maurice Fichman explored the complex interaction between the traditions established by Stahl and Boerhaave in the early eighteenth century, and traced Lavoisier's concept of the gaseous state and chemistry to French Stahlians and their interest in the 'chemical status of fire and air'.[36] Pursuing a similar line of inquiry, Robert Siegfried developed a fine-grained analysis of 'Lavoisier's view of the gaseous state' and its application to his work on combustion and calcination in 1772 and 1773. Siegfried argued that the most important consequence of Lavoisier's view of the nature of air was the idea of the three states of matter, which 'allowed him to consider gases as true chemical principles no different from those which occurred naturally in either solids or liquids'.[37] Seymour Mauskopf showed that the phenomenon of 'the detonation of niter' - viewed as 'the release of an "air" or some other enormously expansible material from a solid', - played a significant role in the development of Lavoisier's thermochemistry and its interaction with the phlogiston theory.[38] Robert Morris, on the other hand, focused on the link between Lavoisier's concept of the gaseous state and the idea that the 'matter of fire', or 'caloric', is, like any other chemical constituent, capable of entering into and being released from chemical combinations. This scenario made changes of state, as revealed by changes of temperature, chemical processes for Lavoisier and, according to Morris, placed the caloric theory of heat, rather than the concept of the gaseous state, at the core of Lavoisier's chemistry.[39]

All of these scholars drew attention to the impact of Lavoisier's early ideas on his mature chemistry. While Morris presented 'the caloric theory' as 'the foundation stone upon which Lavoisier erected his new chemistry', Gough claimed that 'Lavoisier's theory of the gaseous state' was 'not only the initial step in the chemical revolution, it was the most important – for it underlay all the major subsequent steps from the early experiments on combustion to the new nomenclature of 1787'.[40] Siegfried argued that the theory of the gaseous state forced Lavoisier not only to reject the phlogiston theory, but also 'to redefine the fundamental concepts of chemical composition – specifically the chemical substance and the chemical element'. Furthermore, according to Siegfried, Lavoisier's 'simple model of "air" as a combination of a ponderable substance with enough imponderable matter of fire (caloric) to sustain a permanent gaseous state is the organizing theme of Part 1 of his *Traité Élémentaire de Chimie*, published in 1789'.[41] John Christie drew attention to some important parallels between Lavoisier's dualistic schema of 'attractive cohesion and repulsive caloric' and William Cullen's programme for linking thermal and chemical properties to changes in the state of matter.[42] Influenced by his work with the physicist Pierre-Simon Laplace in the late 1770s and early 1780s, Lavoisier, like Cullen, envisioned a time when theoretical chemistry, conceived along the lines of celestial mechanics, would provide a quantitative and predictive analysis of the balance between attractive and repulsive forces in nature. According to Christie,

Lavoisier's vision of the mathematization of chemistry placed him in the Newtonian tradition of ethers and imponderable subtle fluids. In shifting the focus of the Chemical Revolution away from the originality of Lavoisier's experiments on combustion and towards the power, fruitfulness and intellectual context of his basic chemical concepts and categories, postpositivist scholars undermined the positivist-Whig notion of Lavoisier's 'crucial year' and any clear 'sense of its cruciality'.[43]

Lavoisier and Eighteenth-Century Chemistry: Reform and Revolution

Carleton Perrin resisted these scholarly developments however, claiming that the 'shift of interest away from the 'crucial year' ... and from combustion' was 'premature'.[44] Perrin claimed that a genuine appreciation of the 'cruciality' of Lavoisier's 'crucial year' required a significant modification in established historiographical perspectives. It needed historians of eighteenth-century chemistry to abandon 'the phlogiston as obstacle thesis' and to realize that Lavoisier's rejection of the phlogiston theory was a 'consequence', not a 'cause', of his revolutionary programme. He also questioned the 'very designation of pre-Lavoisian chemistry as "phlogistic"' because phlogiston was 'only one component' of a broad and diverse set of concepts and practices that constituted chemistry in the early eighteenth century. Finally, he called for a just appreciation of Stahl's genius and Lavoisier's indebtedness to eighteenth-century chemistry rather than seventeenth-century physics. Within this revamped historiographical framework, Perrin traced the major themes of Lavoisier's 'research tradition' to the 1760s, when his young mind was shaped by the complex intermingling of diverse themes and issues drawn from eighteenth-century physics and chemistry.

Perrin noted that, like many of his contemporaries, Lavoisier praised Stahl for using the phlogiston theory to unify the phenomena of combustion, calcination and the artificial formation of sulphur (from sulphuric acid). Exploring the French Stahlian tradition, Perrin described how Rouelle modified Stahl's system to recognize the chemical activity of air, and how George-Francois Venel argued, in his influential article 'Chymie' in 1753, that air played an important role in many chemical reactions in which it followed the same laws as other elements and could be manipulated like other chemical substances. In 1766 Pierre-Joseph Macquer stamped his authority on the idea that air combined chemically, but Johaan Fredercik Meyer accounted for the generation of air in chemical reactions by adopting Eller's view that it was 'factitious, formed by the transformation of water into air'. Clearly, 'the chemical role of air was a matter of controversy, in the small world of Parisian science, at a formative period in Lavoisier's career'.[45]

Perrin drew attention to the changing identity of phlogiston in pre-Lavoisian chemistry. Thus Rouelle rejected Stahl's idea that phlogiston was one of the three chemically active earths and identified it instead with elemental fire, which he incorporated into a broader chemical system based on the ideas of the four elements, elective affinity and generic chemical principles. Perrin argued that the French Stahlians celebrated Stahl not merely because he unified the phenomena of combustion and calcination, but because he *'succeeded in taming the elusive inflammable principle'.*[46] Although Stahl never isolated phlogiston, he showed how to detect and transfer it from one substance to another and his followers sought to do the same for other generic principles, such as acidity and causticity. Contrary to the dominant historiography, Stahl's treatment of phlogiston provided Lavoisier with a model for the study of air.

In keeping with the arationality assumption, Perrin claimed that, as the scion of an affluent family of lawyers, Lavoisier entered the eighteenth-century chemical community free of the immediate practical concerns of medicine, pharmacy, mining and metallurgy. Nurtured by 'a broad exposure to the sciences', the young Lavoisier focused on the 'discrepancies among [the] different authors' he read and the 'gaps in the chemical knowledge' he acquired. Although he was sensitive to the quantitative problems with the phlogiston theory, he showed no evidence that he was unhappy with it, working with it on an operational level while avoiding speculative disputes about its ultimate nature. Perrin argued that Lavoisier acquired his concern with quantification and precise measurement from physics, which also yielded his theory of the 'vapour state'. He used the methods and instruments of physics, such as the thermometer, barometer, hydrometer and balance, in his 'balance-sheet' analysis of chemical reactions. Lavoisier's use of physical techniques in chemistry contravened 'the spirit of chemical methodology' expressed by Venel in the *Encyclopedia*. Venel correlated the quantitative methods of physics with the gross properties of objects, which he contrasted with the more 'indirect, qualitative, and intuitive' methods chemists used to study the 'elusive inner properties of matter'. According to Perrin, the young Lavoisier used physical concepts and methods to 'expand and reform rather than overthrow existing chemical knowledge'.[47]

Distinguishing between 'reform' and 'revolution', Perrin noted that it was only after his discoveries 'in the fall of 1772 that Lavoisier proclaimed a revolution in chemistry'. These discoveries arose out of a 'series of contingencies' associated with Lavoisier's academic responsibilities, which caused him in the summer of 1772 to turn his theoretical interests and strategies to a new set of problems. Appointed to a committee of the Academy to investigate the puzzling loss of diamond when exposed to heat, Lavoisier conducted a futile series of new experiments with the Academy's 'burning lens and a modified version of Hales's pedestal apparatus', during which he came into regular contact with Pierre Mitouard and the phenomenon of the increase in weight of phosphorous during combustion. Having established to his own satisfaction around 20 October that the gain in weight of phosphorous and sulphur during combustion was due to the fixation of air, Lavoisier inverted his earlier view of the relation between metals and their calces, arguing that air came not from metallic lead, but from its calx. He recorded these results in the famous sealed note that he deposited with the Academy on 1 November 1772.

Challenging conventional historiographical wisdom, Perrin claimed that Lavoisier showed no sign in this document of overthrowing Stahl's theory of combustion. On the contrary, Lavoisier thought that his discovery complemented rather than contradicted Stahl's theory of calcination: where Stahl treated 'calcination as only loss of phlogiston', Lavoisier showed that 'there is at the same time loss of phlogiston and absorption of air'. When Lavoisier presented his discovery as one of the most important since Stahl, he sought to emulate, not discredit, his illustrious predecessor. Just as Stahl had tamed the elusive, unisolable phlogiston, by identifying its lawful effects and combinations, so Lavoisier believed in the autumn of 1772 that he had discovered a means to follow fixed air, theoretically and practically, during the course of its chemical combinations and permutations. Perrin described how during the winter of 1772–3, Lavoisier refined his ambitious project into a coherent 'investigative programme', in which he used 'Halesian and other quantitative techniques' to elucidate the role of air in a wide variety of chemical reactions. Lavoisier outlined his research programme in a research memorandum of 20 February 1773, in which he expressed grave doubts about the validity of the phlogiston theory; by the end of the winter of 1773, he was on the verge of abandoning it. Perrin showed how the reduction of the oxides of metals like iron and mercury without the addition of phlogiston in the form of charcoal reinforced Lavoisier's view of the nature of air, according to which the heat and light associated with combustion comes not from any phlogiston in the combustible, but from the matter of heat, or caloric, released by the air when it combined with the burning substance. Although Lavoisier muted his public criticism of Stahl, he made his first reference to a revolution in chemistry in a paper he read to the Academy's Easter public meeting, where he claimed that the further study of air would lead 'to a period of almost complete revolution'.[48]

Perrin showed that throughout the 1770s and early 1780s, Lavoisier applied his theory of heat and the gaseous state to the processes of combustion, calcination, respiration and vegetation; after the discovery of water in 1784, he shifted his focus to the phenomena of fermentation, plant nutrition and organic analysis. In keeping with Laudan's model of the dynamic mutability of research traditions, Perrin argued that the development of Lavoisier's programme involved 'no mere elaboration of the framework conceived in 1772–1773', but the reworking of some of its guiding assumptions. By the time of the *Traité*, Lavoisier had replaced his earlier view of a universal 'igneous fluid', the cause of heat, light and elasticity, with a more differentiated view of '*two* imponderable fluids': caloric and the matter of light. Similarly, Lavoisier replaced his earlier view of the four elements with a tentative list of simple substances. Like Laudan, Perrin pictured a slow, gradual process of conceptual and methodological transformation resulting in a dramatic and discontinuous break with tradition. Lavoisier's *Traité* not only eliminated phlogiston, it also added a new gaseous dimension to chemistry, replaced the combinatorial order of generic principles with the logic of simple substances, and substituted for the traditional qualitative procedures of chemistry a quantitative methodology and instrumentation derived from physics.

Perrin regarded 1772 as Lavoisier's 'crucial year' because the dramatic and revolutionary results of the *Traité* were the outcome of a continuous, if convoluted, chain of events and transformations that had their origin in the autumn of 1772.[49] In this year, Lavoisier broke with conventional views about the order of chemical composition, did some important experiments on combustion and calcination, and linked a new vision of chemistry's future to an investigative programme, or series of experiments designed to elucidate the role of air in chemical processes. Perrin replaced the positivist-Whig view of an 'eureka experience', or a single moment of illumination, with an account of Lavoisier's crucial year as the starting point of an organic, interactive, life-long programme for the investigation of nature, which introduced new concepts, methods and assumptions into chemistry.[50] Other postpositivist scholars, such as Larry Holmes, Arthur Donovan and Evan Melhado, also emphasized 'the programmatic nature of Lavoisier's discovery', which they identified with an unfolding research tradition or programme, rather than with a single act, or moment, of thought or experience. Unlike Perrin, however, these scholars showed little interest in identifying a crucial or nodal point in the evolution of Lavoisier's scientific career, focusing more on the doctrines and disciplines that shaped the evolution of Lavoisier's investigative programme and determined the identity of the Chemical Revolution.

Disciplinary Identity and the Chemical Revolution

Lavoisier's Investigative Programme

Larry Holmes rejected the 'single key', or single problem, model of the Chemical Revolution. He insisted that 'creative thinking' is not an 'isolated act' or a 'small set of great moments', but an organic 'growth process', in which a 'system of thought' slowly changes under the impact of 'intertwined problems' and 'fresh experiments'. Instead of focusing on 'one or another group of subproblems', Holmes emphasized the 'coherent nature of Lavoisier's concerns', which were sustained by 'a comprehensive research programme oriented around the processes that fix and release airs'.[51] Whereas previous historians had treated Lavoisier's work on animal chemistry and physiology as a belated addendum to his great research on combustion, Holmes argued that 'Lavoisier's studies of fermentation, respiration, and the composition of plant and animal matter were intimately connected', as parts of a 'common research programme', with 'the development of his investigations of combustion in general chemistry'. Focusing less on Lavoisier's formal publications and more on his private laboratory notebooks, Holmes described the development of an 'investigative enterprise' that was far from linear and homogeneous. Starting in 1773, Lavoisier passed 'through a long, and sometimes stormy, conceptual passage', in which he experienced false starts, blind alleys, partial solutions and internal inconsistencies in his research programme. Between 1773 and 1774, he inhabited 'an intermediate mental environment, in which he reasoned sometimes in terms of the concept he was seeking to construct and sometimes in terms of the concept of phlogiston that he ultimately left behind'. Holmes concluded that 'despite the local pressures which threatened to divert it along the way', Lavoisier was guided to 'the outline of a new system of chemistry' in 1777 by 'a research programme that appears in retrospect singularly coherent'.[52]

Holmes rejected the myth of Lavoisier as 'the founder of modern chemistry' because it emphasized discontinuities between early eighteenth-century chemistry and Lavoisier's revolutionary achievements. Whereas positivist-Whig historians portrayed the phlogiston theory as a nonscientific intruder that 'postponed the Scientific Revolution in chemistry', postpositivist scholars, like Arthur Donovan and Evan Melhado, claimed that Lavoisier revolutionized chemistry by applying to it models drawn from the physics current in his own time. Both accounts devalued early eighteenth-century chemistry, or presented it in an entirely negative light. Even those historians, such as Arnold Thackray, Robert Multhauf and Maurice Crosland, who paid more attention to this relatively neglected area of the history of chemistry focused on 'those strands ... that can be seen either as first stages in the movement towards the Chemical Revolution, or harbingers of the theoretical framework of nineteenth century chemistry'. Metzger was the exception to this general rule, and Holmes adopted her evolutionary view of the relation between Lavoisier's chemistry and that of his eighteenth-century predecessors. Holmes emphasized the need to disentangle the image of 'Lavoisier as the leader of a great revolution in science' from his image as 'the founder of modern chemistry'.[53]

According to Holmes, Lavoisier saw himself not as constructing 'a whole new system of chemistry', but as adding to the already existing scientific traditions of plant and salt chemistry the chemistry of 'aeriform fluids, combustion and the formation of acids'. If Lavoisier emphasized his break with the traditional chemistry of phlogiston, he also recognized the continuity that linked his research programme to the prevailing chemistry of acids, bases and salts. On this analysis, Lavoisier neither 'overturned the science of chemistry as a whole' nor established it as 'a science for the first time', but 'transformed [and extended] certain extensive areas of a science whose scope exceeded those areas'.[54]

Holmes did not try to elucidate the 'salient features' of the Chemical Revolution, such as the discovery of oxygen, the deployment of the balance, the reform of the nomenclature, the emergence of a new compositional order, or the acceptance of a new paradigm of knowledge. Indeed, he even suggested that the Chemical Revolution was not 'an event *within* chemistry', but was part of a larger and broader revolution, initiated by Hales and Black, in which 'pneumatic chemistry', which encompassed physics, chemistry and medicine, constituted an interdisciplinary challenge to the community of practising chemists.[55] Thus Holmes viewed the Chemical Revolution as 'a complex multidimensional episode', an integrated network of cognitive, experimental, organizational, social and cultural strands, which required for its comprehension the combined and cooperative efforts of historians, philosophers and sociologists of science. Holmes hoped that this cooperative enterprise would avoid both the extremes of idealism, which treats ideas as though they have 'lives of their own', and the sociology of scientific knowledge, according to which 'scientific knowledge is solely the product of social processes'. He called upon historians of science to focus on the 'intrinsic' features of science, which consist in 'those investigative operations, whether they be in the laboratory, the museum, the field, or the lecture hall, where scientists themselves spend the working days of their lives'.[56] Behind the 'literary' veneer of formal publications and abstract ideas, Holmes followed 'the dense trail of interacting theoretical and experimental activity' which, he claimed, constitutes the laborious and sustained process of discovery at the heart of the scientific enterprise.

Although Holmes modified postpositivism to produce a more dynamic and experimentally oriented image of science, he still upheld the basic idea of the significance and priority of guiding assumptions in the development of science. He also worked within the framework of the arationality assumption, restricting sociology to the study of the extrinsic influences of society on a previously constituted scientific or investigative enterprise. As Holmes envisioned it, the historical study of science as an investigative enterprise does not treat ideas 'as having a life of their own', but shows how they 'emerge from sustained explorations of aspects of nature that can be confronted in the laboratory, the observatory, the museum, or the field'. According to Holmes, scientists enter into political, institutional and professional relationships 'in order to facilitate the investigative activity around which they orient their lives'.[57] On Holmes's view, the investigative enterprise of science is driven not by the sociological forces of 'laboratory life' but by the forces of argument, evidence and reason.

Lavoisier's Methodological Revolution

Adopting similar theoreticist and internalist sensibilities, Arthur Donovan projected a more revolutionary image of Lavoisier's science. Like Holmes, Donovan rejected the positivist-Whig idea that the Chemical Revolution was 'the result of a Eureka experience', the main consequence of which was 'the overthrow of the phlogiston theory'. Donovan argued that it took Lavoisier twenty years of 'prolonged, intense, and frequently frustrating work' to establish chemistry as a scientific discipline. Influenced by his work with Larry and Rachel Laudan on the dynamics of scientific change, Donovan focused on the role of conceptual problems and methodological issues in the articulation and development of historically mutable research traditions. Donovan related Lavoisier's scientific career to 'the goals and methods of the science of his era' by exploring 'the methodological significance' of the well-known historical fact, emphasized by Guerlac, that 'Lavoisier saw himself as working in both physics and chemistry'. In order to transform chemistry into a real science without, at the same time, subordinating it to the 'mechanistic and mathematical conceptions of natural philosophy' associated with the dominant Newtonian paradigm, Lavoisier appropriated to chemistry the methodology of experimental physics promulgated by Jean Antoine Nollet, who rejected the reduction of physics to mathematical reasoning and insisted that precise experiment is the key to acquiring reliable *theoretical* knowledge of causal relations. According to Donovan, Lavoisier's 'success as a theorist' is not explicable 'in terms of a single critical discovery', but involved his 'programmatic commitment' to Nollet's model of experimental physics.[58]

Donovan maintained that historians of science could appreciate the way in which Lavoisier's methodological commitment 'carried him from an investigation of fixed and released air to the realization of a scientific revolution' only if they recognized not only the historical mutability of the facts, concepts and theories of science, but also the way in which the more fundamental aims and methods of science vary with time. Donovan claimed that this realization would liberate historians from the traditional notion of the rigid disciplines of science and would enable them to appreciate how the identification and selection of suitable models and domains of inquiry shape the construction and maintenance of a scientific discipline. Donovan used this historiographical perspective not only to illuminate the link between Lavoisier's reform of chemistry and the reform of physics, but also to show how 'his achievement as a chemist' was connected to 'the other pursuits he engaged in as an enlightened and immensely active *philosophe* in the closing decades of the Old Regime'. This does not mean, Donovan hastened to add, that Lavoisier's chemistry was little more than applied politics. On the contrary, Lavoisier's interest in politics and social reform grew out of his Enlightenment belief in the power of the methods of science to solve the problems faced by both administrators and natural philosophers. In accord with the internalist sensibilities of postpositivism, Donovan directed the causal arrow from science to society, thought to reality, and not vice versa, concluding that it was Lavoisier's 'allegiance to a specific but broadly applicable vision of reform that informed and rendered coherent much of his private and public life'.[59]

Donovan placed the Chemical Revolution 'within the larger cultural tradition of the Enlightenment, a tradition that at different stages in its development encompassed both the older and the newer conceptions of science'. Donovan argued that whereas Buffon 'made Newtonianism and philosophical naturalism central to the doctrines of the High Enlightenment', Lavoisier and the British 'improving' chemists adapted the Enlightenment ethos to 'the critical epistemology of the eighteenth century'. Insisting that 'all theories of matter are entirely hypothetical' and explain nothing, Lavoisier and the British chemists explicated the distinctive subject matter of chemistry in terms of the observable properties and regular relations of laboratory substances. Transformed into 'a fully articulated model of what came to be called the positive sciences', chemistry entered 'the nineteenth century as the paradigmatic new science'.[60] No longer regarded as 'simply a philosophical reflection of nature', according to Donovan, science became a collection of autonomous disciplines, each with its own domain and method of inquiry. Donovan maintained that as a manifestation of the Enlightenment, the Chemical Revolution was integral to the Second Scientific Revolution, which occurred between 1780 and 1850 and involved conceptual and institutional transformations that separated early-modern science, or natural philosophy, from the autonomous disciplines of modern science. In an interesting twist of historiographical logic, Donovan used the postpositivist construal of the historicity and mutability of eighteenth century science to support the positivist-Whig sense of the continuity of the Chemical Revolution with the Scientific Revolution.

Lavoisier: Physicist or Chemist?

Evan Melhado accepted Donovan's idea that 'Lavoisier's chemical innovations' issued from physics, but he denied that the Chemical Revolution 'marked the birth of a new science'.[61] Appealing to the 'less prominent works' of Metzger and Kuhn, Melhado developed a 'disciplinary history' of eighteenth-century science, in which he argued that 'the Chemical Revolution was generated externally by physics', which transformed an already existing scientific discipline 'and then withdrew from it'. Claiming that the 'emergence of new specializations have more to do with conceptual *growth* than with conceptual *change*', Melhado argued that in the Chemical Revolution, '[p]hysics did not dominate a still dependent discipline', but provided 'a mature one with a novelty that it could assimilate only by undergoing a revolution'. Melhado sustained this analysis by identifying 'the central motifs' around which the field of prerevolutionary chemistry was closed, thereby establishing the distinction between 'endogenous and exogenous' factors and the role of exogenous factors in fostering 'normal scientific development' or subverting 'disciplinary expectations' and precipitating a 'revolution'.[62]

According to Melhado, when Lavoisier entered chemistry in the early 1760s, the French Stahlians 'were the major agents in narrowing chemistry in eighteenth-century France and in its demarcation from physics'. Focusing on 'calcination, combustion, reduction, and wet-analysis', they emphasized 'the distinction between chemical and aggregative properties' of matter, and they differentiated 'fixed fire' (phlogiston), a chemical principle, from heat ('free fire'), the agent of repulsion. As Fichman had already shown, the French Stahlians responded to the conceptual problem generated by the clash between the Boylean idea of elasticity as an essential, immutable property of 'perennial air' and emergent eighteenth-century concepts of air as a chemical agent by drawing an ontological distinction between the physical and chemical properties of matter. The French Stahlians located the chemical properties of a substance in its microscopic particles and treated its physical properties, which mark and vary with its physical state, as the outcome of the contingent state of aggregation of these particles. Arguing that elasticity is not, as Boyle thought, a property inherent in the individual particles of air, but is, as Newton suggested, a characteristic of their state of aggregation, or rarefaction, the French Stahlians allowed individual particles of air to enter into chemical combinations with other substances. Viewing microscopic 'attraction' as 'the cause of ... reactions and the condensation of heat into phlogistic compounds', the French Stahlians also argued that 'Herman Boerhaave's discussion of fire in its aggregative form (i.e. serving as an instrument capable of, e.g., rarefying bodies) was entirely physics'. The French Stahlians 'placed the phlogiston theory of fixed fire within chemistry and the aggregative theory of free fire within physics'.[63]

Melhado distanced himself from the majority of postpositivist historians, including Fichman, Gough, Perrin and Siegfried, who placed Lavoisier on the chemical side of this disciplinary divide, stressed his indebtedness to Stahlian chemists, and upheld the chemical nature of his problems, interests and concepts. Claiming that Lavoisier 'worked under Boerhaave's banner, not Stahl's', Melhado rejected the claim that 'the revolution in chemistry resulted from endogenous factors' because it 'accounts neither for Lavoisier's unconcern for the phlogiston theory, the supposed source of his chemical view of fire, nor for his distance from affinity theory', an important ingredient in Stahlian chemistry. Melhado argued that whereas the Stahlians focused on phlogiston, the fixed, nonrepulsive form of fire, Lavoisier 'always followed Boerhaave in according the matter of fire the role of expansive agent, even in its fixed form'. For Lavoisier, the fixation of heat occasioned 'disaggregation'; for the Stahlians, by contrast, 'it brought about condensation and combination'. According to Melhado, the young Lavoisier approached the chemical fixation and liberation of air as 'aggregative phenomena', explainable by his 'thermal model of the gaseous state'. On this analysis, Lavoisier exploited chemistry in the service of a physical theory of heat.

According to Melhado, Lavoisier 'stood at the threshold of the Chemical Revolution' in April 1773 when he began to treat calcination, the central problem of Stahlian chemistry, in terms of 'his views about the gaseous state and change of state of aggregation'. Puzzling over the role of charcoal in the production of air by the reduction of metallic calces, Lavoisier wondered whether its phlogiston (fixed fire) was the source not of matallicity, but of the 'ethereal agent of repulsion'. He abandoned this idea in 1775, however, when he realized that charcoal combined with the air in metallic calces to yield 'fixed air' (carbon dioxide). No longer viewing charcoal as a condensed source of repulsive heat, Lavoisier abandoned the supposition of phlogiston, or condensed heat, and remained 'true to the ethereal character of Boerhaave's heat fluid'. According to Melhado, this allegiance explains 'Lavoisier's well-known claim that he had transferred the source of heat in combustion from the combustible (phlogiston) to the air (caloric)'. Lavoisier regarded the Phlogistians as wrong because 'the matter of fire was more abundant in the most disaggregated states of ordinary matter, not the most dense'.[64] Melhado concluded that the idea that 'Lavoisier's chief interest was initially aggregation and not combination' makes sense of this early text, his self-perception as a physicist, his marginal status in the discipline of chemistry in the 1760s, and the participation of other physicists in Lavoisier's 'research network'. In accord with the historiography of the Second Scientific Revolution, Melhado developed his disciplinary analysis not only to clarify 'the historiography of the Chemical Revolution', but also to determine the eighteenth-century 'mechanisms that generated the disciplinary structures of the nineteenth century'.[65]

Revolution 'in' or 'into' Chemistry?

Donovan, Melhado and Perrin clarified their interpretive strategies and historiographical presuppositions in an exchange of papers in *Isis* just before Perrin's untimely death. Perrin defended the idea that Lavoisier 'accomplished a 'revolution *in* chemistry', and not 'a revolution *into* chemistry', as Donovan supposed, by drawing 'a careful distinction between the concepts of *reform* and *revolution* in scientific change, as employed by Lavoisier and his colleagues in the late eighteenth century'. Perrin argued that although 'Lavoisier was reform minded ... he used the word *revolution* ... quite sparingly' when he spoke about scientific change. Although Lavoisier 'relentlessly pursued a programme of [methodological] reform', in which he used physical instruments and measurement to solve 'chemical problems', the revolution he envisioned 'in the fall of 1772 was a conceptual and theoretical one', based on the introduction of the idea of the gaseous state into the pre-existing science of Rouellian chemistry. Perrin insisted that Lavoisier's sparing use of 'phlogiston and chemical affinities' in his early work registered not a distancing from the chemical concepts and theories of his day, but the simple fact that they were not relevant to the 'mineral compositions he was investigating' in the 1760s. According to Perrin, Lavoisier was 'a fully trained Rouellian chemist', who 'had nothing to do with aggregative theory'; he 'was often refereed to as a "physician"' because he used physical methods, not physical theories. Perrin pointed out that although physicists played a prominent role in Lavoisier's research network, Parisian chemists readily embraced his physical reform of the instruments and methods of chemistry; what they 'initially opposed' was 'Lavoisier's vision of a [conceptual] revolution in chemistry'. It took Lavoisier twenty years of hard work and persuasion, aided by the intervention of physicists, to get chemists to accept his revolutionary ideas, but neither Lavoisier nor his collaborator and converts 'expressed an intention to subordinate chemistry to physics'. According to Perrin, Lavoisier's revolution made chemistry the 'peer of physics' and the two sciences 'marched together at an accelerated pace', with physics dealing with 'the gross and general properties of bodies in nature' and chemistry penetrating 'to their inner secrets'.[66]

Donovan responded to Perrin's criticism by stressing the difficulties involved in demarcating between physics and chemistry in the eighteenth century. As a 'well-trained Rouellian chemist', who 'attended Jean-Antoine Nollet's lectures on experimental physics', Lavoisier regarded himself 'as much an experimental physicist as a chemist'. Donovan also challenged Perrin's claim that Lavoisier's colleagues fully appreciated his 'methodological innovations' before 1772, 'the 'crucial year' in which Lavoisier formulated the research programme that led to his antiphlogistic theories'.[67] Rejecting Perrin's idea that the Chemical Revolution was 'an episode in theory displacement alone', Donovan argued that while 'many chemists were familiar with the methodological principles of experimental physics before 1772', they did not regard them as 'central to chemistry until Lavoisier's success with them'. According to Donovan, Lavoisier projected a revolution that was 'both methodological and theoretical': he used the methodological standards of experimental physics to demonstrate the truth of his new chemical theory and he used the success of his theory to establish these standards as 'central to chemistry as well as physics'. Working on the methodological and theoretical levels simultaneously, Lavoisier realized that 'methodological injunctions that do not produce successful theories are sterile, while theories that are not epistemologically well grounded are feeble'. While Perrin viewed the Chemical Revolution as a 'theoretical revolution', Donovan interpreted it in epistemological terms. According to Donovan, Lavoisier believed his achievement was important not simply because it provided a better theory of chemistry, but because it 'provided a set of chemical theories that met the most rigorous standards for determining what qualifies as well-grounded knowledge'. Donovan recognized that 'chemistry before the revolution' was a distinct 'discipline', a 'set of investigative and interpretive practices', but he insisted that it was only after the revolution, after it had been epistemologically reconstructed, that it became a science.

Donovan tempered his appeal to the historical record in this dispute with the claim that his historiographical differences with Perrin rendered an effective resolution of their disagreements unlikely. According to Donovan, Perrin remained attached to the positivist idea that scientific theories consist in deductive sets of general statements and scientific progress involves the replacement of older, less adequate statements with new, more adequate ones. In contrast, Donovan stressed the globalist view that science functions on three hierarchically arranged levels of enquiry: 'the lowest level in terms of generality' contains empirical evidence acquired by observation and experiment; the 'middle level' contains the concepts and theories used to explain this evidence; at 'the highest level science is structured by more general methodological and ontological guiding assumptions'. Since, on this model, a 'revolutionary' change involves 'a significant change in guiding assumptions', Donovan could regard Lavoisier's achievements as 'a revolution' only because 'his successful replacement of the phlogiston theory also involved transforming chemistry into a science by making the methodological principles of experimental physics central to it'.[68] Echoing Donovan's response, Melhado chided Perrin for his positivistic assumption that 'novel [archival] facts' could settle the interpretive differences between them. Stressing the postpositivist thesis of the underdetermination of theories by facts, Melhado criticized Perrin's 'historiographical Baconianism': while Perrin used 'novel facts' to distance himself from 'the idols of the marketplace and the theater', he overlooked the Bacon, who, 'risking error to escape confusion', indulged the understanding and attended to 'the viability or propriety of interpretation on the basis' of available facts.[69] Just as postpositivism made scientific observation dependent on interpretation, so Melhado emphasized the interpretive, hermeneutical basis of the history of science.

Language, Composition and Matter Theory

Condillac, Lavoisier and Nomenclature Reform

Contrary to the historiography of the 'crucial year', which focused on Lavoisier's early training in experimental physics and Rouellian chemistry, the historians of science William Albury, Marco Beretta and Trevor Levere located the core of the Chemical Revolution in the epistemological and linguistic principles underlying the *Méthode de Nomenclature Chimique*, published in 1787, and the *Traité Elementaire de Chimie*, which appeared two years later. Influenced by Crosland's classic study of the history of chemical nomenclature, Beretta related Lavoisier's reform to broader changes in the philosophy of language, natural history, mineralogy and chemistry. While positivist-Whig historians treated the 'striking phenomenon' of the rapid adoption of the French nomenclature by the European scientific community as a byproduct, or side-effect, of the triumph of Lavoisier's antiphlogistic theory, postpositivist historians who focused on Lavoisier's early training as a Rouellian chemist or experimental physicist downplayed the significance of Lavoisier's linguistic reform, treating the *Méthode* as the 'capstone' or culmination, but not the core, of the Chemical Revolution. In contrast, Albury and Beretta insisted that Lavoisier's chemistry was inextricably indebted to Condillac's philosophy. Assimilating the history of science to the history of ideas, these scholars presented Lavoisier as a philosophical chemist, concerned to subject the phenomena of chemistry to the rigours of a new epistemology and a new nomenclature.

According to Albury, Lavoisier used Condillac's view of knowledge as the natural, combinatory product of human needs and sensations to emphasize the necessity of equations and the deployment of a systematic nomenclature in chemistry. Whereas Locke treated language as the passive reflection and communication of the prelinguistic processes and results of reasoning, Lavoisier upheld Condillac's dynamic view of the analytical activity of the knowing mind, in which linguistic reform and theoretical development are inextricably involved in the acquisition of scientific knowledge. According to Albury, Condillac incorporated the traditional methodology of composition and decomposition into a method of quasi-algebraic reasoning, which Lavoisier used to rationalize the science of chemistry by modeling its logical structure on that of mathematical physics. On this view, the art of reasoning is the art of analysis, the terms of which must be clearly identified and defined. Lavoisier insisted that 'all reasoning in scientific matters implicitly contains true equations', and he called upon chemists to solve the problems of chemical composition by 'a true mathematical analysis', in which the unknown composition is identified with a simple recombination of known substances.[70] In a similar manner, Levere linked Lavoisier's algebraic method of reasoning to his quantitative instrumentation, claiming that on Lavoisier's proposal, 'the terms in a chemical statement would have the same invariance as those in an algebraic one; and the balance was to be the instrument demonstrating that invariance'.[71] For Levere, 'language, logic, and instruments' were the 'joint [epistemological] keys' to the Chemical Revolution.

Albury used his analysis of Lavoisier's chemistry to support the Founder Myth, claiming that Lavoisier's 'attempt to devise a formally algebraic mode of reasoning applicable to all chemical problems' made the *Traité* 'the foundation of modern chemistry'.[72] In a similar manner, Beretta insisted on the radical discontinuity between Lavoisier's chemistry and that of his phlogistic and alchemical predecessors. Upholding the positivist view of the unity and progress of science, Beretta linked Lavoisier's revolutionary struggles in chemistry to the Scientific Revolution of the seventeenth century. Claiming that Condillac's philosophy was shaped more by Descartes than by Locke, Beretta argued that Lavoisier was the practitioner of a form of 'Cartesian analysis'.[73] Beretta here took sides in a hoary nationalistic dispute associated with the positivist-Whig historiography of the Chemical Revolution, locating the seeds of modern chemistry not in the English soil of Bacon and Boyle, but in the fertile mind of Descartes and his French progeny.

Mechanism and Materialism in the Chemical Revolution

Postpositivist historians of science interested in the role of matter theory in eighteenth-century chemistry rejected the attempt to assimilate the Chemical Revolution to the Scientific Revolution, preferring instead to emphasize Metzger's view of the continuity between Lavoisier's chemistry and the work of his eighteenth-century predecessors and contemporaries. In this vein, Kuhn emphasized the cognitive dissonance between Boyle's corpuscular chemistry, which denied the existence of chemical elements and reduced chemical qualities to the configuration of the 'neutral corpuscles of base matter', and Lavoisier's anti-reductionist interest in the identification and isolation of chemical elements and the determination of chemical compositions. Kuhn claimed that although after 1670 all chemists, including Boyle's Stahlian opponents, 'employed a particulate theory of matter', they did so in a way that 'correlated qualities with the enduring characteristics (particularly shape) of the indestructible atom of the elements'. Stahlian chemists also 'denied the possibility of the mutual transmutation of the various elements; and, during the eighteenth century ... gradually extended the list of elementary substances'.[74] Kuhn concluded that the 'true progenitors of Lavoisier's chemistry' were among eighteenth-century Stahlian chemists, who bent the reductionist orientation of seventeenth-century corpuscular chemistry to suit their antireductionist interests and sensibilities.

Arnold Thackray arrived at a similar conclusion in his study of the role of Newtonian matter theory in the development of eighteenth-century natural philosophy. Critical of positivist-Whig historians for ignoring the role of 'speculative inquiries into internal structure and elective affinities' in the development of eighteenth-century chemistry, Thackray documented the pervasive influence of the Newtonian idea of the inertial homogeneity and porosity of matter structured by short-range attractive and repulsive forces. While Thackray insisted that a true appreciation of the development of chemistry in the eighteenth century required 'a recovery of the dominating influence of Newtonianism in all its forms', he also emphasized the limitations of the Newtonian tradition and 'the importance of Stahlian thought'. Indeed, Thackray contrasted the 'profound failure of the Newtonian programme' in chemistry with the successes of the French Stahlians, who were 'passionate, literate, and widely informed defender[s] of the autonomy of chemistry against the very real and obvious barrenness of physicalism'. Rejecting the physical analysis of matter into its simple, immutable, unitary atoms, Lavoisier and the French Stahlians insisted upon the specificity and irreducibility of chemical elements and their correlation with specific chemical qualities. Although Lavoisier entertained the future possibility of a successful physicalist programme, John Dalton broke completely with the Newtonian tradition when he formulated an antireductionist chemical atomism which underpinned 'the triumphant nineteenth-century career of chemical science'.[75]

In his extensive studies of eighteenth-century Newtonianism, Robert E. Schofield argued that the claim that Lavoisier freed eighteenth-century chemistry 'from the malign influence of Becher and Stahl', in order to return to 'the correct path for science laid down by Boyle and Newton', distorted the real nature of 'Lavoisier's chemical revolution', which presupposed 'the failure of Boyle's corpuscular philosophy'.[76] As Schofield noted, Boyle's programme for explaining chemical properties and reactions in terms of the 'primary, mechanical qualities' – size, shape and motion – of 'the fundamental particles of an otherwise undifferentiated matter' relegated the elements and principles of chemistry to the status of 'secondary coalitions of primitive corpuscles'. With no independent means of determining these microscopic parameters, however, mechanistic chemists like Nicholas Lemery and Guillaume Homberg succumbed to *ad hoc* speculations; they arbitrarily conjoined operational notions of elements and principles to the mechanistic framework. Schofield further claimed that while French Stahlians reacted to this unsatisfactory state of affairs with 'a growing movement against mechanism in chemistry', British Newtonians, including Priestley, transformed 'kinematic corpuscularity' into 'dynamic corpuscularity', adding interparticulate forces of attraction and repulsion to the microscopic parameters of size, shape and motion. Just as Newton had derived the force of gravitation from macroscopic motions, so his mechanistic followers sought to derive other forces from microscopic motions. Inspired by Query 23 of the *Opticks,* they explained 'deliquescence, composition and decomposition, dissolution, concretion, crystallization, cohesion, and congelation' in terms of short-range attractive forces and 'volatility and evaporation, fermentation and putrefaction, elasticity and disjunction' in terms of 'repulsive forces, which succeed where attractive forces end'. Schofield characterized Hales's enunciation of the chemical role of air, in terms of the movement of particles between a fluid, elastic or repulsive state and a fixed, attractive state, as the 'high point of a purely mechanical chemistry'.[77]

Claiming that dynamic corpuscularity foundered on the same speculative rock 'that sunk kinematic corpuscularity', Schofield showed how Black and

Rouelle incorporated Hales's results into the new philosophical framework of 'empirical materialism', which explained phenomena in terms of 'substances carrying, as distinguishing characteristics, the qualities of causing those phenomena in proportion to their quantities'. According to Schofield, the 'materialist' philosophy of imponderable, repulsive fluids merged in the middle decades of the century 'with the anti-mechanism of Stahlian chemistry to produce a materialistic, antireductionist chemistry'. Schofield further argued that while natural philosophers in Holland and Britain began to explain the phenomena of heat and electricity in terms of the movement and conservation of underlying, imponderable, repulsive fluids, the Stahlians claimed that 'mechanistic descriptions of the figures and motions of primitive particles' had no useful role to play in chemistry, which concerned 'those aggregations of particles which are sensible and combine in the laboratory'. According to the Stahlians, these 'aggregations' give the compounds they compose their characteristic properties; they retain their identities 'in and through the various compounds they compose' and are 'so simple that they cannot, by any known method, be decomposed in the laboratory'. Schofield noted that the Stahlians analysed and classified the compounds formed by the hierarchical aggregations of these irreducible elements, and they used affinity tables not, like the mechanists, as a means of measuring corpuscular forces, but for 'refining the taxonomy of elements and chemical substances, and [as] a way of avoiding dynamics in chemistry'.[78]

Schofield claimed that while Lavoisier identified '*calorique*' as 'the vehicle of repulsion, elasticity, dimensional and state changes', as well as 'the cause of fluidity and the producer of heat and light', he 'then abandoned the imponderable *calorique*' to get on with a chemistry that made 'the measure of weight and its changes' the 'essential feature of chemical analysis'. Against Morris and Siegfried, who placed caloric at the core of Lavoisier's chemistry, Schofield argued that Lavoisier used the 'principle of the conservation of substance, uniquely determined by weight', to transform his definition of an element – as the end-point of chemical analysis – 'from banality to an operational concept'. While Morris and Siegfried identified caloric as a 'genuine and important link' between Lavoisier and the Stahlian chemistry of principles, Schofield looked elsewhere for Lavoisier's Stahlian credentials, noting that, like his Stahlian predecessors, Lavoisier used, but did not discuss, the concept of affinity, and he dismissed references to 'the simple and indivisible atoms of matter' as 'entirely of a metaphysical nature'. Lavoisier also based his new nomenclature on the Stahlian assumption that 'elements are irreducible carriers of qualities' and that 'the nature of the compound is determined by those of its composing elements'. Schofield insisted that Lavoisier did not free chemistry from Stahl, so much as rationalize Stahlian chemistry and change 'the emphasis of future chemists' activities from an overly sophisticated reductionism to the jig-saw puzzle problems of permutations and combinations of elements'. According to Schofield, Lavoisier was the Linnaeus, not the Newton, of chemistry, and the Chemical Revolution was 'the last phase of a counter-reformation in which pre-revolutionary formal qualities were materialized in substances. It was the triumphant stage of an eighteenth-century neo-Aristotelian reaction against mechanism'.[79]

Chemical Composition: From Generic Principles to Simple Substances

Postpositivist historians interested in the origins and development of Lavoisier's concept of a chemical element also questioned Newton's impact on eighteenth-century chemistry. Influenced by Metzger's historiography of continuity, scholars like Perrin, Morris and Maurice Daumas rejected the Kuhnian and positivist-Whig claim that the Chemical Revolution involved an abrupt and radical change in the views eighteenth-century chemists held about the nature, number and distribution of the chemical elements. These historians challenged the idea of a strict demarcation between a pre-Lavoisian chemistry, based on a small number of unisolable and imponderable 'generic principles' abundantly available in nature, and Lavoisier's concept of the specificity of a large number of 'simple substances', isolable in the laboratory and detectable by the balance. They argued that the shift from a qualitative chemistry of generic principles to a quantitative chemistry of simple substances was underway long before Lavoisier appeared on the scene, was the product of intellectual trends common to phlogistic and antiphlogistic chemists alike, and was only completed after Lavoisier died. Indeed Lavoisier treated the core elements of his system - caloric, oxygen and azote – as unisolable principles, present throughout nature as the conveyors of the characteristic properties of the compounds they entered, and subject to the laws of chemical affinity.[80]

While these historians of science assimilated the oxygen theory to the pre-Lavoisian chemistry of imponderables, other scholars argued that pre-Lavoisian chemistry was as much a chemistry of simple substances as the oxygen theory. Thus, although Marie Boas Hall recognized, along with Kuhn, that the reductionist thrust of corpuscular chemistry precluded Boyle from Lavoisier's 'proper understanding of the nature of chemical elements', she insisted that this understanding emerged prior to Lavoisier and was 'intimately associated with discussions of the particulate structure of matter' inspired by Boyle's work. Hall noted that, in the period between Boyle and Lavoisier, most chemists drew a clear distinction between natural philosophy and chemistry, referring the physical properties of bodies, such as solidity, fluidity and density, to the kinematical characteristics of their 'ultimate particles' and explaining 'chemical change' in terms of 'chemical principles and elements'. Mapping this distinction onto the particulate view of matter, Stahlian chemists replaced the 'doubtfully material principles' of traditional chemistry with real chemical entities 'compounded of particles'; they preferred 'discussing simple substances to discussing the simplest conceivable substances'.[81] Some pre-Lavoisian chemists, including Macquer, Venel, de Morveau, Bergman and Scheele, distinguished between 'chemical atoms' or 'molecules', understood as 'the last particle identical with the mass which they composed', and their more fundamental 'constituent particles', which were regarded as objects of inquiry for 'the physicist, not the chemist'. Lavoisier upheld this distinction in his famous pronouncement that 'by the word element', he meant not 'the simple and indivisible molecules which make up bodies', but 'all substances which we have not yet been able to decompose'. According to Hall, the deepening sense among eighteenth-century chemists of the futility of talking about the experimentally undetectable ultimate particles of matter shaped the basic principles of Dalton's atomic theory of matter, which made atoms 'both physical and chemical principles ... at once the last products of chemical analysis, and real physical units which account for the physical properties of bodies'.[82] Hall used her analysis of eighteenth-century theories of the structure of matter and the chemical elements to shift the locus of the Chemical Revolution from Lavoisier to Dalton.

Other postpositivist scholars traced the roots of Lavoisier's concept of 'simple substances' to the development of Affinity Tables and the work of mineralogical chemists in the first part of the eighteenth century. While Hall argued that corpuscular chemists articulated an idea of chemical affinity that was 'inevitably bound up with the acceptance of the existence of an attractive force as an essential property of the particles of matter', A. M. Duncan claimed that early eighteenth-century chemists constructed Affinity Tables – in which substances were arranged 'in an orderly pattern ... which depended on a sequence of experimentally observed properties' – in order to resolve the tension between their desire to make chemistry a science 'like Newtonian physics' and their equally strong desire to do so 'not by theoretical speculation but by codifying the facts'.[83] Starting 'from the older classification of substances according to the principles which were their essences', the 'authors of affinity tables ... adapted these classifications so that substances were arranged according to what they combined with'; they treated the sortal categories of 'principles' as 'classes of specific substances'. According to Duncan, these developments provided the generative context for Lavoisier's *Tabléau des Substances Simples,* 'the famous list of elements, which reads very like a list of the headings of the columns of affinity tables drawn up according to Lavoisier's system of chemistry'.[84]

James Llana and David Oldroyd also detected an affinity between Lavoisier's notion of 'simple substances' and the views of the mid-century mineralogists J. H. Pott, Axel Cronstedt and Torbern Bergman, working in Germany and Sweden. Operating within the tradition of natural history, which upheld the continuity and multiplicity of the Great Chain of Being, these chemists replaced the older classification of minerals in terms of 'congeries of properties', generated by a handful of generic principles, with a system of classification that emphasized 'the chemical distinctions ... among substances' composed of analytically determined (specific) 'simple substances'. In this manner, phlogistic mineralogical chemists converted the concept of the earthy principle into the concept of the class of specific earthy substances and argued for the reasonableness of basing nomenclature on composition rather than properties. According to Llana, 'the table of simple substances in the *Méthode* was the natural outcome of the natural history tradition in mineralogy and chemistry'. Porter arrived at a similar conclusion, tracing Lavoisier's analytical definition of simple substance and compositional nomenclature to the combined efforts of 'chemically-literate mineralogists', familiar with the practical activities and techniques of metallurgists and mineral essayers in the mining industry, and 'enlightened rulers in Sweden and Germany' seeking to use chemistry to serve the interests 'of mining – and thus of the various state treasuries'.[85] Oldroyd linked these developments in chemical theory to Lavoisier's linguistic nominalism. Deploying philosophical strategies developed by Locke and Condillac, Lavoisier replaced the notion of underlying 'real essences' not with the radical nominalist claim that 'species are wholly figments of the human imagination', but with the idea that simple substances are 'nominal essences', understood as classes of isolable objects identified by the analytical techniques of the day. Emphasizing the link between science and metaphysics, Oldroyd concluded that,

> the major part of the chemical revolution was the change in matter theory, with its accompanying methodological shift. By comparison, the change in the theory of combustion was perhaps, in the long run, a matter of subsidiary importance.[86]

Integrating these partial considerations into a coherent view of eighteenth-century theories of chemical composition, Robert Siegfried and Betty Jo Dobbs related Lavoisier's pragmatic definition of an element as the end product of analysis to a 'century-long transition away from the metaphysical towards the operational concept of the element'.[87] The elements of early eighteenth-century chemistry, such as earth, air, fire and water, were property-conferring principles which designated kinds of being or types of qualities. Initially thought to be unisolable in the free state and to be apprehensible only by their effects, as the century wore on they 'came to be thought of as obtainable *in impure form* as the end product of analysis' and were eventually transformed into classes of specific substances, isolable in the laboratory and detectable by the balance.[88] Just as mid-century phlogistic mineralogical chemists transformed the concept of the earthy principle into classes of specific isolable substances, so the pneumatic chemists transformed the concept of the element air into the idea of the 'aerial' state, which encompassed a multiplicity of specific airs. A little later, in the 1780s, phlogistic and antiphlogistic chemists cooperated in a research venture that replaced the concept of elementary water with the idea of ordinary water. Although this conceptual transformation did not undermine belief in the existence of phlogiston, it did shape the desire of late eighteenth-century phlogistic chemists to identify phlogiston with a specific substance, isolable in the laboratory and detectable by the balance. Thus, in 1783–4, Priestley endorsed Richard Kirwan's suggestion that phlogiston was identical with inflammable air, or hydrogen. In 1785, however, Priestley rejected this hypothesis in favour of the suggestion that inflammable air is a compound of elemental water and the unisolable phlogiston. Priestley's final reversal to a chemical ontology of 'generic principles', devoid of any reference to 'simple substances', resonated with Lavoisier's followers, who called for a 'simplifying reduction of the number of simple bodies' and upheld the traditional view of chemical elements as the ultimate universal constituents of matter, a notion Lavoisier also defended in his unpublished *Cours de chimie experimentale*.[89]

But Siegfried and Dobbs did not allow their sense of the continuity of eighteenth-century chemistry to underestimate the break with tradition involved in the practical deployment of Lavoisier's pragmatic definition of an element. According to Siegfried, 'it was Lavoisier who first published an actual table of simple bodies and organized a system of chemistry around it'. Insisting on the incompleteness of Lavoisier's 'compositional revolution', they argued that it was only after Davy's refutation of the 'misleading oxygen theory' and the formulation of Dalton's atomic theory that metaphysical questions of 'ultimate simplicity' in chemistry gave way to compositional questions based on 'relative simplicity as determined by weight changes'.[90] In a similar vein, Ferdinando Abbri stressed Davy's 'almost ironical destiny'. Motivated by the traditional desire to discover the 'true elements of nature', Davy ended up, through his refutation of the oxygen principle and discovery of 'further simple substances', giving credence to Lavoisier's chemistry of simple substances.[91] While Siegfried and Dobbs located the culmination of the Chemical Revolution in the work of Davy and Dalton, and Abbri stressed the inability of 'some British chemists' to comprehend the revolutionary changes that were consuming them, other postpositivist historians replaced metaphors of inversion and incommensurability with more open-ended accounts of the Chemical Revolution, viewed as a complex process of shifting perspectives and changing emphases, an intricate pattern of intermingling continuities and discontinuities, in which the traditional ontology of chemistry was transformed rather than replaced.

Continuity and Discontinuity

Manichean sensibilities associated with the positivist-Whig eureka conception of a scientific discovery combined with Kuhnian notions of the incommensurability of the oxygen theory and the phlogiston theory to reinforce the idea that the Chemical Revolution marked a moment of abrupt and discontinuous change in the development of science. These sentiments prevailed among post-positivist historians of science who examined the dissemination and assimilation of Lavoisier's chemistry in chemical communities outside France and Britain. Thus while Hufbauer emphasized the role of non-cognitive factors in the German community, Anders Lundgren, Román Gago and H. A. M. Smelders found that the overthrow of the phlogiston theory in Sweden, Spain and the Netherlands 'took place without much debate and without much opposition'. In a similar vein, Abbri showed how Italian chemists embraced Lavoisier's work as 'a really revolutionary event in science'.[92]

French and Scottish Responses to Lavoisier's Chemistry

In contrast, postpositivist accounts of the response of French and British chemists to the oxygen theory stressed the developmental nature of theoretical doctrines and research traditions and argued that the changes that took place during the Chemical Revolution were gradual and sequential, with the 'successive adaptation of new components into a modified traditional framework'.[93] Registering the contextualizing tendencies of postpositivism, these historians interpreted the spread of Lavoisier's chemistry not in terms of the subsumption of the peripheries of the European scientific community under the hegemony of its Parisian centre, but in terms of the dynamic interaction between relatively distinct and autonomous chemical cultures, with their own principles, procedures and assumptions. As Douglas Allchin noted, these interactions did not result in the wholesale replacement of phlogiston by oxygen so much as the formation of 'complementary schemes aimed at different phenomena or different aspects of the same phenomenon'.[94] Considering French responses to Lavoisier's chemistry, Perrin struck a similar chord, rejecting the traditional assumption that 'the development of chemistry between 1772 and 1790' involved a 'simple confrontation' between 'two conceptual packages labeled phlogiston and oxygen'. Perrin refuted this dualistic perspective by noting the lack of unanimity among eighteenth-century chemists on 'the *nature* of phlogiston', a pre-Lavoisian 'undercurrent of doubt and controversy concerning the doctrine', and the gradual sequential unfolding of Lavoisier's revolutionary ideas, which were perceived as a threat to 'traditional chemistry' only after the publication of his 'Reflections on Phlogiston' in 1784. Perrin characterized a 'spectrum of responses', ranging from wholesale acceptance to outright rejection, by identifying when, to what degree and with what disciplinary affiliations Lavoisier's French interlocutors accepted the 'diverse components' of his system, which included, in the 'order of their unveiling' and 'ascending degree of difficulty of adoption', the absorption of air in combustion and calcination, the analysis of the atmosphere into two distinct gases, the oxygen theory of acids, the caloric theory of heat and the vapour state, the composition of water, the rejection of phlogiston and the new nomenclature. Perrin identified the years 1784–89 as the key period of transition in France, which witnessed the almost complete acceptance of Lavoisier's system by 1790. Stressing the conversion of senior chemists, like Claude-Louis Berthollet, Antoine-Francois Fourcroy and Guyton de Morveau, to the oxygen theory, Perrin rejected 'Planck's principle', which linked the adoption of new ideas in science to the emergence of new generations of scientists. Whatever their age, 'scientists can and do convert to novel and controversial theories, though the process is a gradual and complex one'.[95]

Postpositivist historians of science also found plenty of food for thought in British responses to Lavoisier's chemistry, which were laden with theoretical, philosophical, linguistic and cultural overtones. While Perrin explored Black's slow cautious conversion from the phlogiston theory to the oxygen theory, which occurred in 1789 and signalled the end of widespread Scottish resistance to the French chemistry, Christie linked this opposition to the perceived explanatory power of the phlogiston theory in Scotland, where Alisdair Crawford and P. D. Leslie used it to explain animal heat and James Hutton deployed it in his geological studies.[96] Meanwhile, Donovan traced Scottish resistance to the Scottish philosophical tradition, which 'emphasized skepticism and caution' in the evaluation of scientific claims, as well as to propriety concerns about Lavoisier's indebtedness to Black's earlier findings concerning fixed air and latent heat. Evoking wider cultural considerations, which emphasized differences between the individualistic empiricism of the British Enlightenment and the corporate rationalism of the French Enlightenment, Donovan argued that prominent Scots objected to the new chemical nomenclature because they associated it with the French model of centralized authority in science and culture, preferring instead to let 'the language of chemistry evolve naturally through the accumulation of incremental changes in the marketplace of ideas'.[97] Donovan further claimed that the intervention of events associated with the French Revolution thwarted the prospect for a rational 'accommodation' between the Scottish critics and French champions of the new chemistry, bringing in their wake John Robison's conservative crusade against the Jacobins, which linked Lavoisier's chemistry, the metric system and the Revolutionary calendar to the obliteration of the past inherent in the 'unrestrained advocacy of rational analysis'. As editor of Black's posthumously published lectures, Robison 'attempted to portray his master as an ally against the French', reviving the erroneous claim that Lavoisier failed to acknowledge his debt to Black and inflating Black's reservations about the oxygen theory into a 'profound opposition' to it.[98] If, as Donovan claimed, Robison's polemic hastened the collapse of the Scottish Enlightenment by forcing a division between science and religion, his edition of Black's lectures provided the historian with an important but 'imperfect key' with which to disentangle the historical reality of Scottish chemistry from 'Robison's complex mythography'.[99]

Priestley Responds to Lavoisier

Postpositivist accounts of the reception of Lavoisier's chemistry in England lay to rest the positivist-Whig image of Priestley's lonely inconsequential defence of an archaic, useless and patently erroneous phlogiston theory. Although Priestley was eventually left to 'stand alone' in his defence of the phlogiston theory, he was not isolated by inferior knowledge or inferior methods from his fellow countrymen who, contrary to the positivist-Whig scenario, did not rush to embrace Lavoisier's view as the true theory of chemistry. Challenging the positivist-Whig image of the phlogiston theory as a useless and confused metaphysical construction, postpositivist scholars showed that in England, as in Scotland, the phlogiston theory provided natural philosophers with a powerful research tool and explanatory strategies. Indeed, the phlogiston theory proved indispensable to the development of British pneumatic chemistry, enabling Cavendish and Priestley to isolate and examine over a dozen new gases and allowing Priestley to incorporate these results into a systematic view of the workings of nature. Postpositivist scholars also challenged the positivist-Whig distinction between the quantitative methodology of the oxygen theory and the qualitative sensibilities associated with the phlogiston theory by distinguishing between Continental phlogistic theorizers, who treated phlogiston as unisolable in the laboratory and undetectable by the balance, and Cavendish, Priestley and Richard Kirwan, who tried, in the 1780s, to identify phlogiston with inflammable air (hydrogen). Even when Priestley reverted to the earlier view of phlogiston as an unisolable and imponderable generic principle, he registered his quantitative sensibilities when he critically noted that the presence of caloric and the acidifying principle in the oxygen theory made it as much a chemistry of 'imponderables' as the phlogiston theory. On this analysis, Priestley not only appreciated the principle of the conservation of weight, he applied it more consistently than Lavoisier and his followers, who conveniently overlooked the role of imponderables in their own supposedly quantitative theories.[100]

Despite their shared empiricist ontology and methodology, British chemists were divided by fundamental compositional differences over the nature and identity of phlogiston, water and the acids. The absence of other than a minimal set of shared beliefs explains why Priestley could continue rationally to defend the phlogiston theory long after Cavendish and Kirwan had made their separate peaces with the Antiphlogistians in the late 1780s.[101] David Knight, Golinski and I linked British responses to the new chemistry to a wider framework of epistemological, linguistic, cultural and political values and interests. These values and interests depicted the 'imposition' of the French theory and language as an authoritarian threat to the 'liberty' enjoyed by the British community of individual experimentalists, who used a shared observational language to communicate experimental facts and to preserve the autonomy of individual observations and judgments.[102] In a similar vein, Michael Conlin appealed to the contrast between Priestley's 'empiricist methodology' and the 'rationalist methodology' of the Antiphlogistians to explain 'Priestley's American Defense of Phlogiston'.[103]

Interpretive tensions produced at the intersection of postpositivism and the positivist-Whig historiography shaped my overall interpretation of Priestley's science. While I shared Schofield's sense of the gross inadequacies of the positivist-Whig account of Priestley's science, I was unconvinced by his attempt to assimilate Priestley's thought to the Newtonian tradition of natural philosophy. Besides being at odds with the textual evidence, the imposition of a Newtonian framework on Priestley's thought constituted a distortion of his philosophical sensibilities, which were decidedly anti-Newtonian.[104] In contrast to the dualism and voluntarism of the Newtonian doctrine of passive matter, Priestley used the rationalist principles of a monistic metaphysics to argue for the intrinsic activity and sentience of matter. This anti-Newtonian stance was reinforced by Priestley's empiricist sensibilities, which excluded from science any reference to imperceptible microscopic forces and particles. It seemed clear to me that a just sense of Priestley's science required a just sense of its place in the totality of this thought.

Influenced by the postpositivist view of the cognitive unity of science and metaphysics and the crucial role of non-empirical, or conceptual, issues in the development and evaluation of scientific theories, Ted McGuire and I interpreted Priestley's science in relation to his synoptic vision of reality, generated by the subtle interplay of the doctrines of necessity, materialism, Socinianism and associationism. I used this interpretive framework to render intelligible Priestley's work in electricity and pneumatic chemistry, and to throw new light on his role in the Chemical Revolution. In order to counter the view, shared by positivist-Whig and Kuhnian historians of science, that the Chemical Revolution involved a radical, discontinuous break with the past, I explored the continuities as well as the discontinuities that existed between Priestley's 'aerial philosophy' and Lavoisier's chemistry on a variety of cognitive levels. I replaced monomial accounts of scientific change, which allowed positivists to identify *empirical* continuity with *scientific* continuity and postpositivists to identify *theoretical* discontinuity with *scientific* discontinuity, with a multidimensional, contextualist interpretation of the Chemical Revolution, which posited the intermingling and interconnection of contrary moments, such as the continuous and the discontinuous, the gradual and the sudden, across a spectrum of cognitive dimensions that encompassed empirical, theoretical, methodological, epistemological, ontological, axiological and institutional factors. I used this interpretive framework to show how the Enlightenment concept of the self-defining subject established a unitary framework of regulative principles, dealing with the relation between science and metaphysics, the method of analysis, and the relation between thought and language, which were differently interpreted in the competing views that Priestley and Lavoisier developed about the ontology of chemistry, the nature of experimentation, the reform of the chemical nomenclature and the institutional organization of science.[105] My view of the Chemical Revolution as an integral part of the Enlightenment, understood as a unitary movement of thought with diverse and conflicting manifestations, reinforced Holmes's cautionary reminder of 'the truism that any historical transition – whether scientific, political, or cultural – involves elements of change and continuity'.[106] This interpretation of the Chemical Revolution will be discussed more fully in Chapter 7, where it will be related to a broader interpretive model of the history of science.

Summary and Conclusion

Identifying knowledge with theory, postpositivist historians of chemistry shifted the interpretive focus of the Chemical Revolution from the domain of empirical foundations and experimental methodology to the realm of theoretical doctrines and research traditions. They formulated general accounts of the Chemical Revolution that transformed the positivist-Whig conception of a cognitive inversion, in which the metaphysical illusions of phlogiston gave way to the real world of oxygen, into the idea of the incommensurability of the phlogiston theory and the oxygen theory. This interpretive shift replaced the positivist-Whig view of Lavoisier as the experimental 'founder' of modern chemistry with a focus on Lavoisier the theoretical innovator, who challenged tradition with new theories of acidity, heat, the gaseous state and chemical composition. In place of the positivist-Whig view of a continuous experimental and quantitative tradition that linked Lavoisier to the seventeenth-century luminaries of the Scientific Revolution, postpositivist historians described Lavoisier's extensive indebtedness to eighteenth-century natural philosophy and chemistry, emphasizing the importance of Boerhaave's physics and the positive role of the Stahlian tradition in the formation and development of Lavoisier's chemistry. Breaking with the positivist-Whig historiography of the 'crucial year', postpositivist historians rejected the idea that the Chemical Revolution was the result of Lavoisier's 'eureka experience' and saw it instead as an extended conceptual process, which started before Lavoisier appeared on the scientific scene and was completed only after he left it. Emphasizing the role of comprehensive research traditions in the evolution of Lavoisier's theorizing and the structural dynamics of the Chemical Revolution, postpositivist historians of science focused on the disciplinary identity and scope of the Chemical Revolution; they explored the complex and ambiguous relation of Lavoisier's research to the traditions of eighteenth-century physics and chemistry.

Assimilating the history of science to the history of ideas, postpositivist scholars explored a variety of links between the conceptual upheavals of the Chemical Revolution and eighteenth-century theories of knowledge, matter, language and composition. These close-grained analyses qualified the popular postpositivist image of discontinuous changes and incommensurable transitions in the development of science with a more balanced sense of the moments of continuity and discontinuity in the Chemical Revolution. These interpretive sensibilities sensitized postpositivist scholars to the complex and uncertain nature of the assimilation of French chemistry by the European scientific communities; it enabled them to rescue Priestley from the Manichean denigrations of the positivist-Whig historiography. Emphasizing the transcultural autonomy of science, postpositivist scholars developed 'internal' accounts of the Chemical Revolution, which recognized its entanglements with Enlightenment ideas and culture, but upheld its causal independence of the socioeconomic conditions of eighteenth-century Europe. However, sociological criticisms of the arationality assumption generated, in the 1980s and 90s, an approach to the Chemical Revolution which challenged or supplemented postpositivist accounts of the disembodied theories, concepts and methods of science with an appreciation for the experimental, discursive and institutional practices of individuals or small groups of scientists, functioning in specific sociocultural contexts.

4 FROM MODERNISM TO POSTMODERNISM: CHANGING PHILOSOPHICAL IMAGES OF SCIENCE

Opposition to postpositivist interpretations of the Chemical Revolution emerged in the 1980s and 1990s among historians of science influenced by the burgeoning discipline of the sociology of scientific knowledge. These historians stressed interpretive rather than evidential inadequacies with postpositivist models of the Chemical Revolution; they faulted these models not so much for errors or inaccuracies in their specific accounts of the Chemical Revolution as for their questionable assumptions concerning the nature of science and its historical development. Treating science as a social activity, social historians replaced the postpositivist focus on what scientists believed and thought with a concern for what they did and why they did it. They focused on concrete processes and specific agents rather than on abstract structures and global paradigms; and they replaced idealist notions of the unity and autonomy of scientific thought and theory with an awareness of the particularity and materiality of experimental and discursive practices. Critical of the essentialist tendencies and normative interests of postpositivism, sociologically minded historians of science developed nominalist descriptions and naturalist accounts of the specific historical and social practices that shaped science and its historical development. Unhappy with the inability of relatively 'static' structures, like paradigms and research programmes, to do justice to the specificity and diversity of history, they called for 'temporalized and contextualized' accounts of the history of science.

John Christie and Jan Golinski brought these sociological considerations to the attention of historians of chemistry in 1982. In their seminal article 'The Spreading of the Word', they proposed a new historiography for eighteenth-century chemistry predicated on the rejection of the arationality assumption. Christie and Golinski criticized 'the old internalism' because it identified science with 'the processes and products of the intellect'. Insisting that factors falling on both sides of the divide of 'the old internal/external dichotomy' constituted eighteenth-century chemistry as a specific 'historical practice', they replaced the normative distinction between internal and external, rational and arational, beliefs with a 'firmly contextualized' account of the texts and practices of eighteenth-century chemists.[1] Influenced by the discipline of the sociology of scientific knowledge, they adopted a naturalist approach to the history of science, which broke the link between explaining and justifying scientific beliefs. Within the framework of naturalism, other social historians of science generated case studies that denigrated whiggism, rejected 'philosophical appropriations of history' and articulated fully contextualized models of the relation between science and society.

This historiography has all but eclipsed positivism and postpositivism in the minds of contemporary scholars, but it is not without its problems, many of which are rooted in its reactive formation. This disciplinary amnesia is exacerbated by an underlying cultural break of epochal proportions associated with the transition from modernism to postmodernism. A proper comprehension of sociological naturalism and its implications for the historiography of science calls for an exploration of the broader cultural contours associated with this epochal transition. This chapter will focus on the philosophical and cultural sources of this cognitive upheaval by linking positivism and postpositivism with modernism, and by showing how the transition from modernism to postmodernism resulted in a radically new interpretive framework within which sociologists of scientific knowledge articulated the interpretive strategies and explanatory principles of their new discipline. The next two chapters will scrutinize the paradigmatic articulations of this interpretive framework in the sociological community and their effects on the historiography of the Chemical Revolution.

Naturalism and the Strong Programme

The 'Strong Programme', enunciated by David Bloor in 1976, founded the modern discipline of the sociology of scientific knowledge. Bloor criticized traditional sociologists of science, led by Robert K. Merton, and sociologists of knowledge, like Karl Mannheim, for acquiescing in the arationality assumption. Examining the institutional and social factors relevant to the 'rate of growth and direction of science', these sociologists limited the scope of their explanatory inquiries to the domain of rationally unfounded ideas, leaving 'untouched the nature of the knowledge thus created'. Challenging this traditional division of labour, Bloor rejected the normative distinction between internal, rational beliefs and external, arational beliefs in the name of 'naturalism', which explains the formation of scientific beliefs and ideas without assessing their epistemic status or validity. Bloor claimed for the Strong Programme the same 'kind of moral neutrality' traditionally associated with the empirical sciences. He challenged the hegemony of postpositivism in the history of science by drawing a categorical distinction between the normative, reconstructionist interests of the philosophy of science and the 'neutral', descriptive and explanatory, goals of the sociology of scientific knowledge.[2]

Evoking the traditional positivist image of the unity of science, according to which '[t]he search for laws and theories in the social sciences is absolutely identical in its procedure with that of any other science', Bloor justified the Strong Programme by claiming that in contrast to philosophical models of the progress and rationality of science, it embodied 'a genuinely scientific attitude towards science'. In place of the longstanding philosophical definition of knowledge as justified belief, which contrasts custom and consensus with the individual virtues of competent observation and rational inference, Bloor and his colleague Barry Barnes treated scientific knowledge as a social phenomenon, equating it with whatever a cognitive community collectively regards as knowledge. Bloor outlined the method of study of this social object in the Strong Programme, which is not so much a sociological theory, with detailed social mechanisms and laws, as a 'meta-sociological manifesto', specifying the necessary conditions for an adequate sociology of scientific knowledge. Bloor linked the scientificity of the Strong Programme to the principles of causality, impartiality, symmetry and reflexivity. On this model, an adequate sociology of scientific knowledge must focus on the social causes of scientific beliefs (causality), and posit the same kinds of causes (symmetry) for all beliefs – including those of the sociologist of scientific knowledge (reflexivity) – whatever their epistemic status (impartiality). By thus challenging the normative interests of the philosophy of science, sociologists of scientific knowledge opened up the domain of past and present science to the empirical, 'neutral' scrutiny of science itself.[3] Whereas the 'historical school' of postpositivist philosophers of science identified 'epistemology with the philosophy of history', sociologically minded historians of science looked to the social sciences to provide an adequate explanation of the origins and development of scientific knowledge.

Under the umbrella of the Strong Programme, a range of diverse and frequently antagonistic philosophical, sociological and historiographical strategies competed for the allegiance of sociologically minded historians of science. A main fault line in the emerging sociological tectonic appeared in response to the accusation of philosophers of science that all sociologists of scientific knowledge are 'constructivists', who reduce science to a cultural artifact, with no cognitive relation to, or input from, the external world. Many sociologists of scientific knowledge embraced this designation enthusiastically, construing it as either an ontological thesis, according to which scientific representations shape rather than reflect the objects of the world, or as a methodological ruse, in which input from reality is ignored in order better to appreciate the social constitution of scientific knowledge.[4] But Bloor and his associates in the 'Edinburgh School' eschewed constructivism in all its forms, favouring materialism and naturalism over idealism. Bloor argued that if sociologists of scientific knowledge treat science as a synthesis of sensory inputs and cultural responses, they have no need to deny the independent existence of the external world and the role of observation in the rational evaluation of scientific knowledge. Still Bloor remained at one with his constructivist colleagues when he denied the relevance of such normative considerations to a naturalist study of scientific knowledge; he treated empirical evidence and reasoning as one among many contingent causes of the formation of scientific beliefs. The Edinburgh School viewed 'social contingencies not as disturbances to "reason", but as among the necessary causes and conditions of all specific acts of reasoning'; they studied the contingencies of the social settings that lead reason to different conclusions in different social contexts.[5]

A proper appreciation of the implications of the sociology of scientific knowledge for the historiography of science in general and the historiography of the Chemical Revolution in particular must accommodate the motley collection of ontological, epistemological and methodological perspectives gathered together under the broad canopy of (sociological) naturalism. Members of this diffuse collective adopted labels such as 'constructivism', 'laboratory studies', 'actor-network model' and 'science studies' to register their escape from the confines of 'orthodox SSK' championed by the Edinburgh School. Accordingly, while the acronym SSK will be used in this study to refer to the Edinburgh Schools and its allies, the term 'sociology of scientific knowledge' will be deployed in its broader sense when considerations of denominational differences among sociological naturalists are irrelevant. The point to be emphasized at this juncture is that practitioners of the sociology of scientific knowledge did not necessarily champion a relativistic view of science so much as side-step or simply ignore the normative issue of scientific rationality. They insisted that 'all forms of knowledge should be understood in the same manner', not that they 'are to be judged as equally valid'. Instead of the self-refuting thesis of 'philosophical relativism' that 'every belief is as good as every other', sociologists of scientific knowledge embraced the ethnocentric idea that 'there is nothing to be said about either truth or rationality apart from the descriptions of the familiar procedures of justification which a given society ... uses in one or another area of inquiry'.[6] But they denied that this constituted a devaluation of scientific knowledge. Adopting Wittgenstein's view of the finitude of human reason, sociologists of scientific knowledge rejected as false the dichotomy between absolutism and nihilism. They insisted that a state of cognitive anarchy, or 'anything goes', is not the only alternative to the unrealizable goal of a universal, context-free knowledge. They argued that as social products, all knowledge-claims are grounded in variable 'language games', or 'forms of life', which render judgements intelligible and justifiable.[7] So construed, naturalism provided sociologists of scientific knowledge with an intelligible philosophical framework within which to explore how and to what extent scientific knowledge is 'rooted in social life'. Abandoning the modernist problematic of normativity, sociologists of scientific knowledge succumbed to postmodernist suspicions of 'narratives of the legitimation of knowledge'.

Modernism and the Philosophy of Science

Modernist Normativity versus Postmodernist Naturalism

The Strong Programme was a methodological, or regulative, proposal; it said nothing substantive about the nature of society and its relation to science. It was only in conjunction with extrinsic philosophical and historiographical principles and tenets that its methodological strictures yielded substantive models of society and historiographies of science. These principles and tenets projected an image of science that was radically different from those developed by postpositivism and the positivist-Whig historiography of science. Fundamentally opposed to essentialism, historicism, realism and the associated problematic of reference and rationality, this image conveyed the postmodernist sense of 'the fragmentary, heterogeneous, and plural character' of thought, language, reality and the self.[8] Most immediately, the antinormative problematic of the sociology of scientific knowledge resonated with the emergent sensibilities of postmodernism, which eschewed all 'narratives of the legitimation of knowledge'. Rejecting the metanarratives of modernity, which sought to unify and legitimate the separate disciplines of science by grounding them in the 'grand narratives' of philosophy, Jean-Francois Lyotard, in his seminal study *The Postmodern Condition*, argued that the autonomous disciplines of science are the sole source of whatever limited, local and provisional form of legitimation they possess. Encouraging pluralism, relativism and antirealism in the philosophy of science, postmodernism – reflexively attuned to its postmodern identity – encompassed a plurality of philosophical arguments and strategies designed to uphold a nominalist image of science and its fragmented historical identity.[9]

Although the sociology of scientific knowledge rejected the modernist problematic of normativity, it cannot be completely assimilated to postmodernism. It is more enlightening to link the emergence and dynamic development of the sociology of scientific knowledge to an account of the transition from modernism to postmodernism, which connects changing attitudes towards normativity to shifts in the philosophical zeitgeist. But some terminological clarification is in order before this cultural upheaval can be discussed with any clarity. The vocabulary of 'modernism', 'postmodernism' and their cognates is problematic, acquiring different connotations from the perspective of art, architecture, science, social theory, history and philosophy. With an eye on the development of the philosophy and sociology of science over the last 200 years, the term 'modernism' and its cognates will be used in this chapter to identify and characterize the philosophical and cultural sensibilities ushered in by the socioeconomic transition from feudalism to capitalism and usually associated with the principles and practices of the Enlightenment. The term 'postmodernism' and its cognates will be used to characterize the challenge to the philosophical sensibilities of modernism associated with the rise of poststructuralism and the 'failed revolt' of May 1968 in the streets of Paris. A full appreciation of the transition from modernism to postmodernism, and its impact on the emergence of the sociology of scientific knowledge, will be facilitated by first considering the emergence of the modern from the premodern world and the effect of modernism on positivist and post-positivist philosophies of science.

Autonomy and the Origins of Modernism

If, as Frederic Jameson argued, postmodernism is 'the cultural logic of late capitalism', then modernism may be regarded as the cultural matrix of the earlier, commercial and industrial, phases of capitalism.[10] As will be discussed more fully in Chapter 7, the emergence of capitalism involved a 'long drawn-out transition from feudalism', marked by the Renaissance, Reformation, Scientific Revolution and Enlightenment, in which the modern institutions and cultures of individuality and autonomy replaced the premodern world of communal integration and the unity of being. In place of the unitary world of feudalism, in which the individual was constituted by a web of relations to nature, society and God, modernism developed techniques and terminologies of purification and mediation between the distinct realms of nature, society and the self. In this manner, the Enlightenment mind replaced the constraining dialogue between madness and reason that existed during the Renaissance with the order of exclusive reason, which sundered spirit and nature, faith and knowledge, freedom and necessity, feeling and reason.[11]

Self-assertion and self-definition were the guiding principles of modernity, accounting for its focus on the present and its sense of superiority to the past. While the Enlightenment mind rejected the Medieval and Renaissance idea of an intrinsic link between man, nature and society, Kant, pressed by the modernist urge for autonomy, developed his 'transcendental philosophy' as a 'way of demonstrating the autonomy, the self-grounding authority of modern thought and therewith modern culture', an impulse that shaped nineteenth-century German Idealism and the Hegelian notion of 'self-positing spirit'. While Enlightenment thinkers from Descartes to Kant identified modernity with the principle of subjectivity, an all-consuming activity of self-knowledge, self-disclosure and self-production, Hegel's doctrine of 'absolute spirit' moved beyond the level of isolated subjects to the autonomous domain of objective reason.[12] In a Kantian vein, Max Weber identified the spirit of modernity with the differentiation of the traditional unified worldview of religion and metaphysics into the autonomous cultural- and value-spheres of science (truth), morality (normativity) and art (taste), a perspective Jurgen Habermas revived in the twentieth century when he insisted upon the irreducibility of 'the objective, subjective, and the intersubjective dimensions of rationality'. The same logic of purification and autonomization shaped Marxist programmes for the overcoming of man's embeddedness in nature and domination by the past; it also fostered twentieth-century aesthetic modernism, which called for 'the differentiation of the purely aesthetic from other realms of human endeavor, such as ethics, politics, religion, or economics'.[13]

Modernism consisted in 'the relentless development of the objectivating sciences, of the universalistic foundations of morality and law, and of autonomous art, all in accord with their own immanent logic'.[14] Sundering the unified cosmos of the premoderns into distinct spheres of cognition and action, defined, validated and regulated by their own 'inner logic', modernism populated reality with an array of normative distinctions and dualisms, such as those between subject and object, society and nature, human and nonhuman, reality and language, knowledge and faith and science and nonscience. An early twentieth-century opponent of modernism summed-up its basic push towards separation and autonomy in the following terms:

> Instead of the whole intellectual and social order being subordinated to spiritual principles, every activity has declared its independence, and we see politics, economics, science, and art organizing themselves as autonomous kingdoms, which owe no allegiance to any higher power.[15]

The secularized world of modernism made the autonomous individual or community the subject of knowledge, the source of values, the possessor of rights and the focus of utility.

Reason, Language and History

The modernist philosophy of autonomy involved an analytical conception of formal reason, a representational theory of thought and language, and a secular and progressivist view of history. These accounts of reason, language and history shaped the development of positivist and postpositivist philosophies of science and provided the critical foil for the emergence of postmodernist sensibilities associated with the sociology of scientific knowledge.

The transition from feudal unity and integration to capitalist autonomy and differentiation gave modernism its critical, normative and formal orientation. Breaking with scholasticism, religious authority and feudal power, modernism upheld the self-sufficiency of reason and grounded knowledge in rational processes independent of accepted tradition, dogma or authority and oriented towards the instrumental service of humanity. Bereft of specific metaphysical and empirical content, modernist reason was formal and procedural, concerned with abstract forms of knowledge and methods of inquiry.[16] The emergence of a sense of epistemological autonomy also coincided with a shift in the activity of the knowing mind, in which the associative and analogical understanding of the Middle Ages and the Renaissance gave way to analytical reason. Instead of 'drawing things together' in a web of resemblances, the modern mind discriminated between things by 'establishing their identities' and thereby their 'differences' from other things.[17] Positing 'centred' subjects and autonomous structures, the analytical and formal logic of identity and difference shaped positivist and postpositivist accounts of the disciplinary unity, autonomy and internal development of science.

Rejecting traditional sources of legitimation and justification, modernism had 'to create its own normativity out of itself'. Propounding universal principles, rather than local traditions and arbitrary conventions, the autonomous spheres of modernism did not 'simply coexist uncompetitively'. Criticism and denunciation were the order of the day. The modernist urge to extend the boundaries and stretch the limits of rationality generated 'an endless series of reductionist and antireductionist moves', in which the reductionists tried 'to make everything scientific, or political (Lenin), or aesthetic (Baudelaire, Nietzsche)' and the antireductionists showed what such attempts left out.[18] In this competitive, expansionist cauldron, modernism generated reductionist models of scientific knowledge which emphasized the unity of its theoretical structure, the universality of its concepts and methods and the necessity of its instantiations. The modernist problematic of legitimation also issued in the famous distinction between the context of discovery and the context of justification, which positivist and postpositivist scholars used to distinguish the historian's task of describing how scientific discoveries are, in fact, made from the more important job of the philosopher who develops abstract models of knowledge for 'the criticism as well as the reform of what exists'.[19]

The modernist dualism of subject and object assimilated knowledge and language to the category of 'representation'. Upholding Descartes's view of the dualism of mind and body and Locke's concept of knowledge as a 'mental process', modernism replaced the premodern 'hylomorphic conception of knowledge', which involved 'the subject's becoming identical with the object', with the concept of knowledge as the possession by the mind of 'accurate *representations* of an object'. When the modernist problematic of representation broke with the Renaissance doctrine of *signatures*, which treated language as a system of signs enmeshed in the resemblances and similitude of things, it dissolved the premodern 'fusion of semantic and causal relations', which ascribed to names the ability 'to possess the power of things to which they refer'.[20] By separating the causal from the rational, the descriptive from the normative, modernism tied rationality to referentiality, viewing justification, or legitimation, as a referential relation between ideas or statements and the world they seek to represent. Bequeathing to positivist and postpositivist philosophers of science the problematic of realism, the modernist problematic of truth and reference became an ineluctable target of postmodernist deconstructions predicated on the claim that 'there is no reality which is not itself already image, spectacle, simulacrum, gratuitous fiction'.[21]

Modernism's self-legitimating break with the past made time an integral component of reality and produced a tensile sense of history as simultaneously revolutionary and teleological. Looking forward to a future that would shatter the past, modernists maintained that events 'took place not only in history but through history'. Inherently dynamic and revolutionary, modernism conquered space with time, making it 'a locus through which to get from A to B, rather than a place to live in'.[22] Modernists interpreted the passage of time not as a cycle, decadence or a falling, but as an abolition of the past. They saw themselves as separated from the past not by 'a certain number of centuries', but by revolutions, breaks and ruptures 'so radical that nothing of the past survived in them'. The modernist identity was inextricably linked to the notion of time as the sure arrow of progress that separated them from the premoderns. But the modernist sense of future novelty collided with the logocentric requirements of self-legitimation and produced 'teleological narratives' of progressive emergence and justification, in which history unfolds according to a predetermined plan. The resulting historiography of the victors celebrated a single revolutionary break with the past, such as *the* Scientific Revolution, which requires no more revolutions 'to fulfill the promise of the new'.[23] The historicist notion of the unity and homogeneity of modern time shaped the positivist historiography of science and influenced in a more ambiguous way the historical sensibilities of postpositivism.

Philosophy of Science

Philosophical modernism provided a dynamic template for the development of positivist and postpositivist strategies of legitimation, which involved 'grand narratives' of the hierarchical unity of the different cognitive levels of science – observation, theory, method and instrumentation – grounded in general theories of meaning, knowledge and language. More specifically, the modernist philosophy of autonomy and separation shaped the positivist view of the unity of science and its demarcation from nonscience. Modernism's self-legitimating search for universal principles linked the unity of science to the deployment of a single, international method, language or theoretical structure grounded in a shared domain of unitary experience. The positivist preoccupation with the unity and demarcation of science further harmonized with the modernist vision of the formation of autonomous spheres of thought and action, developing within strict disciplinary boundaries according to their own inner logics. But there was a tension in the narrative of modernity between the celebration of 'the epistemic autonomy of increasingly specialized domains of knowledge ... and their unification by formal procedures of reasoning and justification'. While positivists 'subordinated the claims of disciplinary autonomy within the sciences to formal method and the unity of science', postpositivists emphasized 'the autonomy of disciplines and research programmes within disciplines'.[24] Theoreticism substituted the autonomy of reason for the autonomy and unity of the subject.

Tensions between modernist themes of unity and autonomy also informed positivist and postpositivist construals of the arationality assumption. While positivists and postpositivists alike upheld the arationality assumption, which emphasized the social autonomy of reason, theoreticism also maintained an absolute distinction between thought and reality, thereby rendering unintelligible positivist attempts to ground thought in experience, language in the world. Similar tensions shaped postpositivist critiques of the thesis of the formal unity of science. Although the historical school of postpositivists rejected the abstract, formal, static models of explanation and justification developed by Logical Positivism, they still distinguished between the abstract, generalizable content of theories and their specific historical contexts. They also envisioned universal structures, or patterns, for the internal development of science, as is evident in Kuhn's 'timeless cycle of epochs: normal science, crisis science, revolutionary science, and the return to normal science'.[25] If the positivist doctrines of empirical certainty and eureka discoveries in science represented classical modernism's attempt to ground knowledge in the 'self-certainty of subjectivity', then postpositivism's notion of 'knowledge without a knowing subject' marked 'late' modernism's fallibilistic construction of the rationality, universality and autonomy of scientific thought and action.

Epistemological and semantic differences between positivism's vision of a unitary experience for science and postpositivism's insistence on theoretical autonomy and pluralism in science surfaced in their divergent articulations of the modernist problematic of realism. While the 'naïve realism' of positivism restricted the ontological and epistemological significance of science to its observational terms and statements, the main thrust of postpositivism was in the direction of 'critical realism', which anchored the ontology and epistemology of science to its theoretical terms and principles. The antirealist themes of instrumentalism – associated with a minority of positivists – and idealism – a fleeting theme in some of Kuhn's writings, for example – should be viewed as minor deviations from the modernist problematic of realism. While instrumentalism reiterated premodern concerns with 'saving the phenomena', idealism prefigured postmodernist themes associated with sociological constructivism. Modernist realism was an ineluctable target of postmodernist hyperrealism, which identified reality with its 'simulacrum'.

Modernism's valorization of time as simultaneously revolutionary and teleological shaped the positivist-Whig vision of a single Scientific Revolution followed by the synchronized, homogeneous and continuous development of the different disciplines of science along the same line of historical progress. Registering classical modernism's attempt to ground knowledge in the 'self-certainty of subjectivity', the positivist-Whig historiography of science located the motive force of this development in the eureka discoveries, crucial experiments and creative insights of individual men of genius at one with nature. Although the postpositivist doctrine of 'knowledge without a knowing subject' replaced the positivist-Whig vision of the unity, linearity and homogeneity of a single, absolute historical time with the idea of a succession of distinct epochal times, each with its own unity, linearity and homogeneity, it still upheld modernism's valorization of time, emphasizing the growth of knowledge, the link between rationality and progress and the developmental nature of paradigms, programmes and traditions. While modernism upheld 'monarchical' metaphors of 'exclusive time', which depicted diachronic relations of 'subordination and succession' between the past and the present, postmodernism favoured images of the 'liberalism of space', which emphasized 'coordination and coexistence' and spoke of synchronic juxtaposition rather than diachronic development.[26] Replacing modernist metaphors of temporal progress with postmodernist images of spatial dispersion, sociologists of scientific knowledge called for more contextualized accounts of the *places* or *sites* of the production and proliferation of scientific knowledge.[27]

Postmodernsim

The term 'postmodernism' was used sparingly before World War Two, usually in connection with a negative assessment of twentieth-century culture.[28] Postmodernism as a coherent philosophical and cultural movement emerged in the social and cultural upheavals of the late 1960s and early 1970s, though it reached maturity only in the reactionary atmosphere of Reagan-era economics and Cold War strategy. Reflecting these divergent political contexts, the postmodernist movement split into two broad countervailing currents of thought. One current of thought, emanating from Continental philosophy, delineated a radical critique of power and a sceptical view of knowledge, expressed in the slogan 'anything goes', while the other current of postmodernist thought adopted programmes of social engineering 'in which society as a whole asserts itself without bothering to ground itself', and was prepared to affirm one set of beliefs over another, but in a fallible and relativistic manner.[29] While the antinormative orientation of sociological naturalism was closer to the affirmative relativization of knowledge to context and circumstances, many of its analytical categories came from the sceptical cauldron of Continental philosophy. The mingling of the traditions of Anglo-American analytical philosophy and Continental speculative philosophy was a characteristic feature of the sociology of scientific knowledge.

The initial oppositional formation of postmodernism gave to all subsequent forms of postmodernism its characteristic strategy of 'deconstruction'. Postmodernism's original break with the modernist problematic of legitimation occurred with the reaction of French left-wing intellectuals, such as Lyotard, Foucault and Jean Baudrillard, to the refusal of the French Communist Party to support the student revolution in Paris in May 1968. Viewing the 'official left' as a 'double and pseudo-rival of the Gaullist regime', these thinkers concluded that trying to replace one system with another system would end up with a system that resembled the one to be replaced. The 'disillusioned children of 1968' turned their back on class struggle and the global patterns of legitimation associated with it; they looked elsewhere, in the struggles of local communities and minorities, for the revolutionary spirit of 68. But they used these local struggles and the ideas they spawned not to overthrow modernism and replace it with a new system, but to rethink or reform it from the inside, to 'deconstruct' or go 'beyond' it, to struggle 'against and within it'.[30]

The Deconstruction of Reason, Language, and History

Postmodernists linked the predilection for totalizing and systematizing thought and practice to the 'rational demands for unity, purity, universality, and ultimacy' inherent in the modernist problematic of autonomy and legitimation.[31] Concerned with the totalitarian implications of modernism's universalistic aspirations, they replaced the idea of a unique and consensus-forming method in science with the image of a multiplicity of conflicting scenarios and outlooks. While 'sceptical' postmodernists replaced the global order of modernism with a sense of the fragmentation, disintegration and meaninglessness of life, 'affirmative' postmodernists adopted more localized, less dogmatic modes of legitimation, contextually ranking values and building domain-specific models of rationality. Postmodernism replaced modernist notions of 'centred' subjects, autonomous structures and pre-existent essences determining individual events and actions with a view of the role of local agents and specific practices in the temporary formation of fleeting constellations of meanings, values and knowledge-claims. The relativization of truth and meaning to contexts and circumstances involved the replacement of representational and realist models of thought and language with antirealist views of a language-laden reality and relativist doctrines of the rhetorical constitution of scientific knowledge. The terminological move from (modernist) 'language' to (postmodernist) 'discourse' represented a shift in interest from representing to doing things with words, from the representational function and universalizing aspirations of theory to the role of concrete practices and implicit skills designed to simulate and intervene in the world, rather than represent it. Rejecting modernist valorizations of unity, universality and autonomy, postmodernism insisted that, embedded in local contexts of control and domination, science is neither value-free nor necessarily progressive. In this manner, postmodernism blurred the boundaries between reality and rhetoric, science and literature, truth and belief, and knowledge and power inherent in the modernist demarcation between science and nonscience.

Postmodernists questioned the very possibility of history as generally understood. They replaced the legitimating tyranny of modernist time with the reign of 'pastiche', in which an ontology of 'undifferentiated immediacy', the 'singularity of singular events', exists without reference to underlying meanings, determining structures, originary sources or intended goals. Replacing the critical orientation of modernism with a 'flat and affirmative universe of discourse', postmodernism effaced the past as 'referent', leaving behind a fragmented heap of images and texts for which no original ever existed, or was any longer accessible.[32] Unable to organize past, present and future into coherent narratives, sceptical postmodernists settled for a canvas of images, spectacles and simulacra, while their affirmative counterparts constructed narratives of discontinuity and marginality. For postmodernism, imagery and memory have priority over what is remembered and historical writing is necessarily presentist, constructivist, provisional, political and contingent

Upholding a vision of the 'overwhelming spatiality' of local forms of life that 'suffocate the temporal dimension', postmodernists organized life, language and experience 'by categories of space, rather than by categories of time, as in the preceding period of high modernism proper'.[33] The postmodernist valorization of space took two forms: a jubilant and celebratory form and a melancholic and critical one. While the architectural manifesto of postmodernism, *Learning from Las Vegas*, celebrated the escape from 'the planned monotony of modernist megastructures' in the 'vigor and heterogeneity of spontaneous urban sprawl', Foucault put postmodernism's valorization of space over time to analytical and critical ends, showing how techniques of surveillance and control inscribed in the contours of physical and social spaces – prisons, factories, schools, barracks and hospitals – articulated the disciplinary power inherent in (post)modern society. It was the disciplinary space of Foucault rather than the playful space of Vegas that sociologists of scientific knowledge had in mind when they inscribed knowledge, power and identity 'in place and not in time', delineating not the historical development of knowledge, but the dispersed (physical and social) sites, disciplinary structures and global networks involved in its production and dissemination. As a 'time of nominalism', postmodernism emphasized the transience and instability of the 'particular "contextual" orders' of research and inquiry.[34] The denial of temporality, of the diachronic course of history, registered nominalist sensibilities, naturalist inclinations and antirealist tendencies, which regulated the sociology of scientific knowledge and had a profound impact on the historiography of science.

Postmodern Philosophical Frameworks

Sociologists of scientific knowledge deployed a variety of postmodernist philosophical arguments and strategies to break with idealism, essentialism, realism and historicism; in the process, they developed a disciplinary identity fraught with philosophical tensions generated by the transition from modernism to postmodernism. These arguments and strategies comprised a motley collection of previously distinct Continental and Anglo-American philosophical traditions, appropriated to the nominalist and deconstructionist needs of postmodernism and the explanatory and interpretive interests of the sociology of scientific knowledge. This collection of traditions was articulated by an underlying developmental dynamic, in which the modernist idea of science as a form of knowledge, or representation of the world, gave way to postmodernist, or 'performative', models of science as a mode of practice, or intervention in the world. Although they are valued by historians more for their narrative usefulness than their philosophical consistency or coherence, these traditions and the interpretive strategies they evolved will be considered here in terms of their philosophical origins, coherence and systematic implications for the historiography of science. These strategies encompassed Wittgensteinian criticisms of essentialism developed by the Edinburgh School, antirealist and constructivist themes associated with the Continental philosophies of poststructuralism, phenomenology, hermeneutics and genealogy, and a general 'turn against theory' that united a broad sweep of Anglo-American philosophers, historians, literary critics and philosophers of science. The general upshot of these developments involved a terminological transformation in which a sociological and historical vocabulary of 'practice', 'power', 'discourse', 'text' and 'audience' replaced earlier philosophical categories of 'fact', 'observation', 'theory', 'experiment' and 'method' in the explanation of science and its historical development.

The Edinburgh School's nominalist semantics of 'finitism' was the basis of antiessentialist arguments developed by sociologists of scientific knowledge opposed to theoreticism and scientific ('critical') realism. The Edinburgh School treated scientific realism as a form of essentialism, in which theoretical objects, such as electrons, protons, and neutrons, are treated as '*essentially identical* entities', with essential properties 'manifested identically in every particle'.[35] These entities and properties provide the concepts and theories that describe them with inherent meanings and rules of usage that determine their future application and development. Denying the essentialist view of the reality of universals and the efficacy of rules, sociologists of scientific knowledge adopted a form of moderate nominalism, which based the use of words on the resemblances between things. Promulgated by Hobbes, Locke and Hume, the linguistic doctrine of resemblances was championed in the twentieth century by the later Wittgenstein, who had a profound influence on the Edinburgh School. Since, according to finitism, concepts and theories are applied to objects and things on the basis of the analogy rather than the identity of appearances, their future usage is open-ended and uncertain. It follows that the application of concepts and the development of theories is a matter of innovative judgements, or local decisions, which are to be explained in empirical psycho-sociological terms and not reconstructed in normative philosophical terms.

Postmodernist developments in the Continental traditions of structuralism, phenomenology and hermeneutics provided an alternative philosophical channel for the development of these nominalist tendencies, as can be seen in the antiessentialist and deconstructionist themes associated with poststructuralism, ethnomethodology and Reception Theory. These developments replaced the notion of a realm of predetermined meaning, truth or significance, whether in the form of synchronic structures, transcendental egos, authorial intentions or interpretive traditions, with the concept of dispersed and contingent networks of the material and discursive practices of local agents in situated contexts. The philosophical movement of poststructuralism gave the clearest and most immediate expression to the general cultural and aesthetic themes and tendencies of postmodernism. Associated with the names of Jacques Derrida, Foucault and Jacques Lacan, poststructuralism involved a radicalization of the decentring and antifoundationalist strategies and tendencies of structuralism. Structuralism decentred the constitutive subject of existentialism and phenomenology – the 'metaphysics of presence' – only by transferring its foundational function to the fixed and centred principles of abstract, anonymous structures. Treating languages and texts as instantiations, or embodiments, of abstract, synchronic, rule-governed structures, structuralism 'bracketed' the real world of material objects and human subjects and detached consciousness from the historical practices of individual agents in specific contexts. While poststructuralism endorsed structuralism's 'demotion of the subject from constitutive to constituted status', it also dissolved the stable, totalizing texts of linguistic structuralism into indeterminate and diffuse sources of heterogeneous meanings and significations.[36] This change in philosophical sensibilities ushered in a terminological transformation, in which 'language', understood as 'a chain of signs without a subject', gave way to 'discourse', which related issues of meaning to the production and use of 'utterances' by networks of speakers and listeners, authors and audiences. In accord with the interventionist and constructivist sense of the intertwining of words and things, the rules of discourse define not how words represent objects, 'but the formulation and ordering of the objects themselves'.[37] Viewing languages and texts as social practices, rather than closed systems or isolated objects, poststructuralism replaced the abstract formalism of structuralism with considerations of concrete signifiers, situated practices and specific technologies of power. Whereas structuralism made the text a window onto the universal human mind, poststructuralism emphasized the 'materiality' of texts, understood as concrete instruments in the embodied interactions of local agents and their material practices. In this manner, poststructuralism disentangled 'the disorderly manifold' of individual desire and artistic creativity from the coercive uniformity of unified subjects and centred structures, whether in the form of metaphysics (Derrida) or power (Foucault).[38]

Interpretive Strategies of Continental Philosophy

Derrida and Textualism

The transition from structuralism to poststructuralism started in the domain of aesthetic theory and literary criticism, which accounts in part for its overriding concern with the philosophy of language. Like theoreticism in the philosophy of science, literary structuralism rejected essentialist theories of meaning inherent in earlier Romantic and humanist interpretations of literature. Rejecting the idea that languages and texts derive their signification from the representational ideas and intentions of their speakers and authors, structuralists viewed literature as an autonomous linguistic structure to be understood without reference to the world beyond or the author behind its texts. Viewed as 'systems of signs without subjects', texts have meaning in virtue of the structuring relations between the terms of their narratives, and not because of any relation, semantic or otherwise, to the agents that produce them or the historical contexts that surround them. As a species of modernism akin to theoreticism, structuralism substituted the autonomy of objective reason for the certainty of subjective meaning.

Derrida criticized structuralism for failing to break completely with the metaphysics of presence and the totalizing ambitions of modernism. The idea that meaning is immediately present to a sign, whether in the form of a privileged experience or a moment of transcendental illumination, lingered on in the notion of centred structures, with solid foundations and fixed meanings.[39] Since, according to structuralism, the meaning of a sign is dependent on its relation to other signs in its system, then according to Derrida, all signs and the meanings they signify are embroiled in an open-ended play of signification. Although a given interpretation of a text, based on a particular set of relations among its signifiers, becomes privileged because of the intervention of social conventions or ideological predilections, other interpretations, based on other relations within the system, are epistemologically equivalent to the privileged one. '*Deconstruction*' dislodges privileged interpretations by revealing in a text's open-ended patterns of signification epistemologically equivalent alternative readings of it. Articulating the poststructuralist notion of discourse, Derrida's concept of '*writing*' constituted a direct challenge to structuralism, replacing the idea of a predetermined, fixed, hierarchical and synchronic set of ideal meanings with the notion of an ephemeral and egalitarian array of diverse discursive practices of reading and writing.

Derrida's concept of 'writing' challenged the claim to priority of the spoken over the written word associated with the perennial ethos of Western metaphysics, with its commitment to the immediate 'presence' of 'some ultimate "word" ... essence, truth, or reality' as the foundation of meaning and knowledge. Within this dominant tradition of Western philosophy, writing is a derivative or secondary means of communication, dependent upon the primary reality of speech, which directly symbolizes the meaning or idea present to the speaking subject. Viewing language as a differential system of meaning, Derrida's doctrine of '*différance*' argued that the meaning of a sign is never fully present to the sign itself or the occasion of its utterance. Derrida's doctrine of '*différance*' also stressed the materiality of the sign and the exteriority of meaning. No longer viewed as a mere supplement to speech, a material simulation of a pre-existing idea or meaning, the activity or practice of writing is constitutive of the meanings and representations it conveys, of the texts it constructs and deconstructs. Shifting from the modernist notion of representational language to the postmodernist conception of performative discourse, Derrida insisted that 'deconstructive readings and writings' are not just 'analyses' of texts: 'They are also effective or active (as one says) interventions that transform context without limiting themselves to theoretical or constative utterances even though they must also produce such utterances'.[40] Materializing the idealist doctrine of theory-laden observation and experience, Derrida linked the intelligibility of a written text to the activity of writing it. To write, or assert a proposition, is at the same time to uphold or undermine the material, institutional and political context in which the proposition is entertained and the text produced. In this extended sense of 'text', Derrida's doctrine of 'textualism' maintained that '[t]here is nothing outside the text'.[41]

The doctrine of textualism encouraged some sociologists of scientific knowledge to identify a scientific field with its 'literary inscriptions'. Latour and Steve Woolgar, in particular, used Derrida's view of the constitutive role of writing in the representation of facts to link the generation and acceptance of facts in science to the deployment and dissemination of 'inscription devices', or machines such as chart recorders, oscilloscopes, scales or even NMR spectrometers that 'transform pieces of matter into written documents'. Tied to networks of textuality and techniques of inscription subject to the logic of *differance*, factual disputes in science are open-ended and ultimately irresolvable. Furthermore, since there is nothing beyond the text, information in science is to be analysed not in terms of the semantics of truth and meaning, but in relation to communication technologies that constitute, mediate and stabilize factual representations or experimental practices. In keeping with textualism's 'performative' model of language, constructivist historians of science emphasized the rhetorical and illocutionary functions of literary styles, textual illustrations and iconography in the construction of scientific arguments and the making of scientific facts.

But Latour was quick to recognize that the rhetorical forms and functions of science depend upon the deployment of laboratory instruments and the manipulation of nondiscursive materials in the outside world. Derrida himself intimated at the need to attend to 'the social, political, and economic institutions' sustained by 'the material practices of inscriptions'. Most sociologically minded historians of science also chaffed at the interpretive limitations inherent in textualism which, by excluding reference to anything outside the media to determine the meaning and content of the message, favoured a form of technological determinism. Emphasizing the historical situatedness and contested nature of scientific representations, they maintained that 'every literary form of fact-making is linked to local complexes of technical and social practice and that stabilizing any representation is always at the same time a problem of political order and moral discipline'.[42] Golinski also argued that since the rhetorical aims of authors and speakers cannot be fully discerned without comprehending how 'actual audiences in specific contexts' interpret what they read or hear, then 'rhetorical analysis should be framed within a more comprehensive process of *hermeneutics*'.[43] The philosophical rationale for linking the texts of science to their social and instrumental contexts lay outside the scope of Derrida's textualism; it was provided by the hermeneutics of Reception Theory, Foucault's genealogy of the power-knowledge nexus, Heidegger's existential phenomenology, and ethnomethodolgy's focus on the situated rationality of everyday life.

Hermeneutics and Reception Theory

Reception Theory is a postmodernist branch of hermeneutics which came into prominence in the 1970s. Shifting the focus of attention from the rhetorical function of securing assent, hermeneutics treated language and texts as vehicles of meaning. Instead of dealing with the intentional or technological production of 'inscriptions', Reception Theory focused on the interpretation of texts, which involves readers as well as writers, audiences as well as speakers. The modern discipline of hermeneutics first emerged in response to the concerns of Renaissance scholars and Reformation divines with the translation and interpretation of ancient and sacred texts. Taking on the mantle of philosophy in the eighteenth century, hermeneutics went on to serve a plethora of philosophical conceptions of human history, language and life. Developing in tandem with phenomenology in the twentieth century, it subsequently reflected the influences of Husserl's 'transcendental phenomenology', Heidegger's 'hermeneutical phenomenology' and the nominalist sensibilities associated with the emergence of ethnomethodology and the rise of postmodernism. The influence of hermeneutics on the historiography of science reflects its variegated sources and heterogeneous development.

While Husserl's vision of the 'transcendental ego' and its 'ideal meanings' encouraged hermeneuticists like E. D. Hirsch Jr to identify the meaning of a text with an ideal authorial intention, or 'type', Hans-George Gadamer shifted the philosophical centre of gravity of hermeneutics from the prelinguistic constitutive subject to the determination of autonomous global traditions. Influenced by Heidegger's opposition to Husserl, which made meaning a matter of historical interpretation and not transcendental consciousness, Gadamer grounded the multiple, but delimited and related, meanings of a text in the continuity of an interpretive 'tradition' which allows the reader's 'horizon of historical meaning and assumption' to 'fuse' with the 'horizon' of the author of the text.[44] But Gadamer's postpositivist sense of the unity of interpretive and cultural 'traditions' soon gave way to postmodernist celebrations of the multiplicity and diversity of the 'narratives' produced by these traditions.

By making the history of a text's interpretations relevant to the interpretation of the text itself, Gadamer and his followers (such as A.C. Danto) undermined the orthodox idea of a fixed and objective knowledge of the past. According to hermeneutics, historical statements are 'narrative statements', which reconstruct historical events 'within the frame of reference of a story'. Since a story has a plot, with a beginning and an end, it follows that historical events 'cannot be represented without being related to other events that follow them in time'.[45] Instead of being privileged, the original participant's interpretation of an event is incomplete and open to modification by subsequent commentators with richer, more complete points of view or perspectives. But, as Bernadette Bensaude-Vincent noted, the mutability of meaning and perspectives associated with the narrative view of history does not necessarily imply that historical events have no objective reality. Rather, polysemy and interpretive flexibility indicate the importance of an event. On this view, a foundational event such as the Chemical Revolution involves not 'a mythical gesture creating something out of nothing and predetermining the future', but 'a set of events and circumstances' that is open to a variety of 'interpretations and revisions'. Since future generations of historians will continue to judge and rejudge the past, the completed history of a completed event can never be told. It follows that historians of science should abandon the traditional and unrealizable goal of a final, objective knowledge of the 'actual past' and, instead, 'undertake the reconstruction of historical realities by displaying the wide variety of their potential meanings'.[46] In this manner, Mi Gyung Kim and Christopher Mienel argued that the identity of the Chemical Revolution and Lavoisier's role therein was constructed in the nineteenth century by 'chemists and chemist historians' in order to further their personal, professional, political and cultural interests.[47] Reiterating the transition from modernism's representational to postmodernism's performative model of knowledge, these scholars maintained that the job of the historian is not to represent an independently existing past, but to show how it was constructed and to participate in its reconstruction.

Reception Theory, which emerged in the 1970s, was less compromising in undermining the notion of an objective historical reality, shifting the locus of meaning from authors' intentions and interpretive traditions to the readers of texts.[48] According to Reception Theory, knowledge and meaning do not inhere in texts or traditions, but in the interactions between texts and readers, which are different, temporary and unique. The postmodernist text is a flimsy thing, 'no more than a chain of organized black marks on a page', which the reader constructs into a meaningful document. The meaning of a scientific text is determined neither by its author's intentions nor by its objective content, but by the linguistic communities and audiences in which it circulates. Reinforced by the constructivist analysis of science and Wittgenstein's idea that linguistic meaning depends upon localized 'forms of life', Reception Theory encouraged historians of science to link the interpretation of scientific texts to 'local cultures of theoretical and experimental work'. In practice, sociologically minded historians of science mingled rhetorical and hermeneutical considerations in an interpretive stance that,

> spurned questions of authorial intention, focusing rather on the interaction of words with audiences – on the affective powers of rhetorical, narrative, and theoretical devices, on the ways in which recipients assimilate texts and images into their own expectations and appropriate them to their own purposes.[49]

Unfortunately, Reception Theory provided little guidance in the identification and characterization of these extra-textual 'expectations' and 'purposes'. Phenomenology, ethnomethodology and Foucault's genealogical inquiries were more forthcoming in this respect, drawing attention to the dimensions of practice and power involved in textualist meaning and discourse.

Foucault and the Genealogical Method

After May 1968, Foucault shifted the focus of his inquiries from discourse to power. He transformed his 'archaeological' analysis of the anonymous rules of discursive practices into a 'genealogical' account of the relations between these rules and the equally anonymous configurations of power which sustain and constrain them. Abandoning his earlier interest in the autonomous rule-governed grounds of discursive practices for an inquiry into the close relation between knowledge and power, he replaced epistemological questions with questions about the historical origins and political effects of knowledge. Giving a nominalist and constructivist twist to Nietzsche's perspectival view of knowledge as a form of power, Foucault disclosed not the principles and sources of a centralized 'repressive' power, but the 'microphysics' of the localized and dispersed mechanisms of a 'productive' form of power. Foucault emphasized the constitutive role of the disciplines and sites of knowledge production and dissemination in the formation of scientific beliefs and the regulation of scientific behaviour.

Foucault replaced his earlier archaeological preoccupation 'with systematicity, the theoretical form, or something like a paradigm', with genealogical accounts of 'whatever is singular, contingent, and the product of arbitrary constraints'.[50] Foucault's genealogical inquiries were profoundly influenced by Nietzsche's genealogy of morals, according to which the philosophical notion of the constitutive subject of knowledge, truth, morality and freedom is a fabrication inherent in the need for human mastery over a capricious and chaotic world through the creation of an artificial internal realm of order and permanence (identity). According to Nietzsche, truth is the product of the 'adaptations and falsifications of daily experience', and its emergence can be traced in the history of the techniques of punishment and torture designed to control human behaviour and inculcate an internal sense of self-control, order and responsibility. Foucault combined his Nietzschean sense of the disciplinary function of knowledge with a historicized version of Merleau-Ponty's view of the invariant structures – 'nascent logos' – of bodily and perceptual awareness to produce an influential account of the 'spatialization of reason', which related the production of knowledge and experience to historically specific configurations of bodily techniques, disciplinary routines and architectural structures.[51] Foucault differed from Nietzsche insofar as he eschewed all concepts of human agency or intention, treating 'psychological motivation as the result of strategies without strategists'. The genealogist studies not the conflicting strategies and goals of pre-existing antagonists in the struggle for power that is life and history, but the matrix of struggle, the agonistic space created by anonymous techniques, practices and rituals of power within which antagonists acquire their roles and identities. These rituals and practices require and generate knowledge of the objects they control, so that knowledge and power are internally related. Foucault studied the interconnection of 'power/knowledge ... that determines the form and possible domains of knowledge' integral to the kind of 'disciplinary power' that, he maintained, pervades and shapes the modern world.[52]

Foucault distinguished, historically and functionally, modern 'disciplinary power' from premodern 'sovereign power'. Arising in conjunction with the emergence of European absolute monarchies in the seventeenth century, the familiar form of political sovereignty emanates from a unitary source, represents legitimacy through the impartiality of the law, and intervenes in discontinuous, sometimes spectacular, ways to repress opposition and deviant behaviour. But sovereign power's apparatus of control (courts, prisons, police and the military) proved inadequate to deal with the growing population and changes in the relations of production associated with the emergence of capitalism in the eighteenth and early nineteenth centuries. This situation called for more rigorous techniques of surveillance and control made available by 'disciplinary power'. Instead of emanating from a single source, disciplinary power permeates extensive social relations and, instead of operating through discontinuous acts of intervention, functions through the continuous 'production or enhancement of various "goods", such as knowledge, health, wealth, or social cohesion'.[53] Knowledge is the most important effect of disciplinary power and, in turn, shapes and facilitates that power.

Foucault traced the emergence of disciplinary power to the penal reforms of the early nineteenth century. He identified Jeremy Bentham's 'Panopticon' – a design for a circular prison in which the illuminated inmates are observed from an unobservable, because unilluminated, central tower – as the 'ideal form' of the mechanisms of surveillance and control inherent in disciplinary power. Foucault did not identify disciplinary power with any particular institution, but viewed it instead as a heterogeneous set of 'micro practices' – instruments, techniques, procedures – designed to produce 'docile bodies' in a range of institutions, such as prisons, schools, hospitals, factories and the military. Permeating these complex social networks, rather than anchored in dominant individuals or systems of rules, disciplinary power is insidious, heterogeneous and dynamic. Its structures are the outcome of infinitesimal mechanisms of power deployed in the endless local struggles to sustain and undermine networks of domination. The modern 'carceral society' is the result not of the imposition of an overarching economic, social, or political structure, but of the 'swarming' of disciplinary practices, wherein the circulation of specific 'enclosed disciplines' constitutes 'an indefinitely generalizable mechanism of "panopticism"'.[54]

Disciplinary power achieves the goal of total and continuous surveillance by treating human beings as 'docile bodies', produced, analysed, compartmentalized and objectified by the constant and regular configuration and manipulation of their motion in space and time. The control of space was particularly important to this new form of power, as can be seen in the deployment of orderly grids in hospitals, schools, factories and the military, which enclosed and partitioned space 'to be able at each moment to supervise the conduct of each individual'. This '[d]isciplinary space ... was a procedure, therefore, aimed at knowing, mastery, and using'.[55] Within architectures of surveillance, these disciplinary institutions developed their own techniques and procedures of examination and normalization for the generation of the kind of (scientific) knowledge needed to control the individuals involved. Thus, the social sciences of psychology, criminology, demography and social hygiene were first situated in particular institutions of power, including hospitals, prisons, schools and the military. The 'subjectivity' of the knowable human beings in these institutions was the product of the disciplinary technologies to which they were 'subjected'.

Foucault and his followers rejected the modernist dualism of liberating knowledge and oppressive power. Maintaining that knowledge is empty without power and power is blind without knowledge, they treated the discursive practices of science as embedded in technological processes, institutional organizations and laboratory instrument. Knowledge and power, the scientific and the social, are simultaneously constituted by a myriad of disciplinary practices. Disciplinary power operates *through*, not *on*, these practices. It does not act on the vaunted neutrality of science from the outside, but functions through the depoliticization of science, through its so-called neutrality. Viewed as the genealogy of disciplines, the business of the history of science is to trace the 'specific forms of descent, emergence, and transformation of specific discursive practices'.[56] The history of science and the history of society are one and the same.

Foucault's genealogical model had a profound and pervasive influence on the sociology of scientific knowledge and sociologically minded historians of science. Although Foucault himself exempted the natural sciences from his genealogical sense of the inextricability of power and knowledge, a phalanx of historians and sociologists threw his caution to the wind and readily deployed genealogical interpretive devices in their histories of both the natural and the social sciences.[57] Some of these scholars blended hermeneutics and genealogy, treating texts as the means whereby scientific disciplines and practices constitute systems of cognitive and social control in the production of scientific discourse. More generally, they appealed to the 'micro physics' of the power-knowledge nexus to articulate and reinforce the more pervasive nominalism of postmodernism, according to which scientific knowledge is inherently private and local, not public and universal. Upholding the more general constructivist idea that experimental phenomena are not simply given to the scientist directly by nature, but are constructed by the machinations of the power-knowledge nexus, they deployed phenomenological views of the epistemological and ontological priority of practice over theory, treating scientific knowledge not as static, internally structured systems of propositions, separable from external considerations of application and use, but as dynamic, complex, heterogeneous and contested fields of skills, practices, equipments and procedures operating in sites 'of the production of health, wealth, military force, etc.'[58]

Foucault's concept of the 'swarming' of disciplinary mechanisms provided a nominalist reduction of the apparently public and universal nature of scientific knowledge and institutions to the circulation through society of the linguistic and material technologies of local disciplinary power and knowledge. Instantiating postmodernism's 'spatialization of reason', Foucault's model of the architecture and spatial organization of institutions of surveillance encouraged historians of science to examine the partitioning of space, the techniques of compartmentalization, standardization, normalization, and the disciplining of the inhabitants of places of the production and dissemination of scientific knowledge, such as laboratories, classrooms and museums. The idea that the practitioners are as disciplined as the objects of science resulted in a (renewed) historiographical interest in the disciplines of science, which focused attention on the mechanisms of communal unity and uniformity in the training of scientists and the practice of science. Merging the traditional view of 'disciplines' as branches of knowledge with the idea of them as modes of social practice, Foucauldian historians treated scientific disciplines as forms of disciplinary power.[59] Incorporated into phenomenological conceptions of knowledge as a form of practice, Foucault's analysis of the power-knowledge nexus exerted a powerful influence on sociologists of scientific knowledge and sociologically minded historians of science.

Phenomenology and Ethnomethodology

In practice, sociologists of scientific knowledge and sociologically minded historians of science ignored the ontological strictures of Foucault's genealogical method, which eschewed any reference to the meaning or intentions of human actions. Thus many of them incorporated Foucault's antihumanist account of the relation between knowledge and power into phenomenological analyses of the meaning and significance of scientific terms and practices. These analyses took two characteristic forms, one concerned with cognition and the other focused on practice in science. The cognitivist form – evolving through Edmund Husserl's 'transcendental phenomenology', Alfred Schutz's 'phenomenological sociology' and Harold Garfinkel's (early) 'ethnomethodology' – traced the cognitive, or conceptual, foundations of knowledge to the operations of the transcendental ego, the social life-world, the presuppositions of common sense, or the situated logics of commonplace activities. The other line of phenomenological inquiry – inspired by Heidegger's 'practical hermeneutics' and Merleau-Ponty's concept of 'embodied knowledge' – located the foundations of knowledge not in the realm of abstract cognition, but in the concrete ensembles of tacit knowledge and skills associated with the embodied practices of human beings. Sociologists of scientific knowledge incorporated these phenomenological influences into their inquiries in ways that marked a transition from phenomenology's essentialist interest in the universal conditions of thought and being to the nominalist framework of ethnomethodology, which used 'the adjective *local*' to emphasize 'the heterogeneous grammars of activity through which familiar social objects are constituted'. Ethnomethodology envisioned 'a patchwork of "orderliness" without assuming that any single orderly arrangement reflects or exemplifies a determinate' structure, norm or paradigm.[60]

Husserl designed his 'transcendental phenomenology' to meet the threat of scepticism posed by speculative metaphysics and the burgeoning empirical sciences in the late nineteenth century. As a form of conceptual analysis, transcendental phenomenology sought *a priori* knowledge of the essential cognitive structures – 'ideal meanings' – of all possible experiences generated by a purified form of consciousness, or the 'transcendental ego'. In the 1920s, Schutz moved away from Husserl's egological premises and transcendental interests in an attempt to delineate the phenomenological foundations of Weber's sociology of *verstehen*, according to which the aim of the social sciences is to understand, or interpret, social actions, not (causally) explain social behaviour. Schutz charged the phenomenological sociologist with the task of understanding the social world by illuminating the intersubjective rules governing 'the subjective meaning of the actions of human beings from which social reality originates'.[61] Wedding Husserl's phenomenology to the positivist vision of the unity of science, Schutz called upon sociologists of scientific knowledge to render explicit the implicit unitary set of norms, rules, procedures and values that prestructure the objects and practices of the scientific community. Whereas positivism valorized the detached abstractions and objectivity of science, phenomenologists stressed, on ontological, epistemological and ethical grounds, the incompleteness of scientific concepts and their constitutive dependence on the concrete realities of the life-world.[62]

In 'What is Ethnomethodology', published in 1967, Garfinkel gave a nominalist and constructivist construal to Schutz's identification of the life-world with the general rules of meaning and interpretation used by members of a community to define themselves and their everyday actions. Ethnomethodology is predicated on the concept of 'indexicality', according to which the meaning or intelligibility of linguistic expressions and practical activities is not due to the instantiation of pregiven rules of action and interpretation, but is a contingent ongoing accomplishment of local agents in specific contexts. Echoing Wittgenstein's idea that linguistic usages and forms of life go hand-in-hand, Garfinkel maintained that procedures for making actions 'accountable' are constitutively embedded in the social organization of the actions themselves. It is the business of ethnomethodologists neither to explain nor to interpret these procedures, but to describe them in all their particularity, whether in the form of implicit features of the organized activities or of more explicit accounts – such as coding and record keeping – designed to identify what has been done and to show others how to do it. While ethnomethodologists emphasized the importance of 'ad hoc' considerations in accounting practices designed to render explicit the implicit identity and reproducibility of social actions, they did not lapse into subjectivity or a concern with agency. For ethnomethodology, the social world is constructed not by the voluntary choices of its members; rather, it is a necessary presupposition of members' participation in organized productive activity. Speakers and actors 'achieve sensible and precise communication by placing apparently vague lexical particulars within logically coordinated sequences of activities'.[63] In this manner, ethnomethodology located the rationality and intelligibility of science not in any methodological guidelines about 'how science is done *in general*', but in 'the ways in which various scientific practices compose themselves through vernacular conversations and the ordinariness of embodied disciplinary activities'.[64]

Extending ethnomethodology to an analysis of the natural sciences, sociologists of scientific knowledge like Karin Knorr-Cetina developed '*an empirical, constructivist epistemology*', designed to show that understood as sites of contentious 'sense making', laboratories are ruled by the same kind of indexical, occasional and locally accomplished logics 'operative in the social world at large'.[65] Knorr-Cetina portrayed laboratory researchers not as systematic thinkers or observers, but as opportunistic 'tinkerers' who exploit, in a competitive cauldron, local resources, procedures, ideas and materials to produce 'patch-work' epistemic products constituted by the occasional character of local 'know-how' and untrammelled by explicit methodological rules. But other ethnomethodologists and phenomenologists of science, such as Michael Lynch and Joseph Rouse, faulted Knorr-Cetina, and the more general phenomenological project associated with Husserl, Schutz and Garfinkel, for their excessive 'cognitivism'. They argued that the cognitivist view of science as a 'sense-making' activity, though rightfully sensitive to the particularity and heterogeneity of that activity, paid insufficient attention to the embodiment of scientific consciousness in the material environment and practical activities of the laboratory and was, thereby, perfectly compatible with the modernist view of knowledge as propositional and representational.[66] Breaking with cognitivism and adopting a contextualized version of Merleau-Ponty's concept of embodied reason and Heidegger's hermeneutics of practice, Lynch and Rouse replaced the modernist view of science as a unified or dispersed system of cognitive representations of reality with the postmodernist idea of a patchwork of diverse and disparate technical and discursive interventions in the world, dependent on tacit knowledge, implicit skills, local practices and embodied techniques.

Heidegger located the intentionality, or meaning, of experience not in its cognitive structure, but in the totality of man's perceptual awareness and practical involvement with the world. Heidegger's phenomenology was hermeneutical and existential: it interpreted the concrete features of man's multifarious historical situations (the 'ontic' realm) in terms of the universal ontological structures, or 'existentiala' (care, anxiety, being-onto-death, estrangement, guilt), that define man's finitude and make possible and intelligible his concrete existence. Heidegger's philosophy was decidedly not perspectival. It treated human history not as a discontinuous succession of lived experiences, but as grounded in ontological structures that provide the conditions for human freedom. Merleau-Ponty linked these ontological structures to man's bodily existence, arguing that the movement of the lived body, 'with its perceptual and motile capacities', establishes the modes of man's spatial awareness, so that 'objective' phenomena are intertwined with the many ways in which things present themselves in accordance with our practical activities and bodily comportment.[67]

But Merleau-Ponty's concept of embodied spatiality was insufficiently nominalist and contextualist for postmodernism. By focusing on the 'primordial possibilities inherent in "naked perception"', Merleau-Ponty failed to appreciate what Foucault later demonstrated so forcefully: the 'various historical-material transformations' of the naked body. Ignoring Foucault's hostility towards phenomenological intentionality, Lynch called for the integration of his anti-humanist analysis of the disciplinary structures and functions of power into ethnomethodological studies of the 'sense making' activities of laboratory life. Lynch used this hybrid method to show how the sense-making collectives of organized phenomena, spaces, tools and techniques constitutive of laboratory work are situated in 'machinic assemblages and disciplinary labor processes'.

Joseph Rouse used Heidegger's 'practical hermeneutics' to integrate Lynch's blend of genealogy and ethnomethodology into a robust practice-model of science. Ethnomethodologists elucidated the 'work-specific competencies' through which scientists make science, but they never argued, and nor did their view imply, that these competencies are inherently practical in nature. In contrast, when Rouse identified scientific research as a form of 'craft knowledge', he used Heidegger's 'practical hermeneutics' to argue that all decisions in science, no matter how theoretical some of them may seem, are ultimately practical decisions, dealing with the problem of how to cope with the world, not how to represent it.[68] Rouse also linked the Heideggerean view of the social contextuality of practice to Foucault's genealogical analysis of the power-knowledge nexus in an analysis that integrated ethnomethodology's focus on the situated internal dynamics of laboratory life into a broader vision of science as practice, power and culture.

Heidegger's concept of man's 'being-in-the-world' linked knowledge and understanding not to adequate conceptual representations of the world, but to 'performative' interventions in it. Man's being-in-the-world is practical and hermeneutical. Human beings interpret themselves and their world in what they do and how they do it. They construct and render intelligible their spatio-temporal existence by the ways in which they comport themselves in systems of practices, rules and equipment. They render their actions intelligible not by internalizing an explicit set of rules or beliefs about them, but by acquiring the implicit and embodied skills, or 'know how', involved in performing them. Whereas rules and beliefs can be rendered explicit and the object of conscious acceptance or rejection, skills are inherently implicit or tacit, and involve forms of life or ways of being-in-the-world which can be inhabited but not consciously chosen. Heidegger's 'practical hermeneutics' avoids the dualism and subjectivity associated with modernism by insisting that our understanding of the world is not based on the imposition on it of a 'conceptual scheme' or paradigmatic structure, but is the result of our determinate being-in-the-world, which 'may change over time, but not as the result of deliberate choice or action'.[69]

Rouse utilized the views of Kuhn and Ian Hacking, which will be considered in the next section, to develop a nominalist version of Heidegger's practical hermeneutics in which he rebuffed Heidegger's own proposal that science is exempt from the contextualizing hermeneutics of practice. Rouse argued that everyday scientific research is governed not by decontextualized considerations of the coherence, autonomy and systematicity of theoretical knowledge, which relegates experiments to mere instances of theory and treats the places of experimentation as incidental to scientific practice, but by practical – material, technical and social – consideration integral to the local sites of knowledge production. Intimately bound to a local research context, with its focus on the manipulation of materials, the construction of experiments and the organization of particular networks of social relations, theories function not as systematic webs of belief, but as 'loosely joined groups of overlapping, not always consistent extendable models', deployed opportunistically and circumspectly. Stressing the crucial role of experiment in the 'development of science as theory', Rouse claimed that the 'empirical character of science is constructed through the experimenter's local, practical know-how'.

Taking his cue from Foucault's notion of the 'swarming' of disciplinary mechanism in the carceral society, Rouse argued that this inherently local and private knowledge is extended beyond the laboratory, and takes on the highly valued demeanour of public decontextualized knowledge, not by 'generalization to universal laws instantiable elsewhere, but by the adaptation of locally situated practices to new local contexts'. Viewing laboratories as world-transforming sites of a 'Foucauldian microphysics of power', Rouse argued that the 'decontextualized objects' of science result from 'the standardization of highly specific theories, facts, institutions, and procedures into more general-purpose equipment'. The same is true for 'scientific discourse and its evaluation', which expands its range and significance through negotiations within and between 'behaviorally self-adjusting communities' and the 'practical interests that govern them'. Science is to be 'understood in its use', and the job of the Heideggerean historian of science is to abandon any notion of the decontextualized, global autonomy of theory in science and show how '[t]his use involves a local, existential knowledge located in a circumspective grasp of the configuration of institutions, social roles, equipment, and practices that makes science an intelligible activity in our world'.[70] As will be seen in the next section, the nominalist themes of 'use', 'discourse', 'practice' and the 'specificity' of knowledge generated by poststructuralism, hermeneutics, genealogy and phenomenology were further articulated by Anglo-American sociologists of science influenced by Wittgenstein, Kuhn and the 'new empiricism'; these themes also provided the interpretive matrix of sociological accounts of the Chemical Revolution to be considered in the next two chapters.

The Anglo-American Turn Against Theory

The Continental turn against theory, which valorized the concrete over the abstract, the practical over the theoretical, and the experiential over the conceptual resonated in the corridors of Anglo-American academia, though with different philosophical overtones and implications for the sociology of scientific knowledge. While sharing the nominalist and deconstructionist sensibilities of their Continental cohorts, a motley collection of cultural historians, literary critics, social scientists and philosophers rebuffed the 'antihumanism', or formalism, of the French 'philosophy of the concept', which informed poststructuralism as much as structuralism. While eschewing the constitutive subject of knowledge dramatically decentred by the French tradition, Anglo-American scholars still emphasized the role of agency, experience and context in human action and understanding. They favoured a mode of historical inquiry that shifted the philosophical centre of science from the objective determinations of impersonal structures and specific practices to the experimental activities of local agents, intervening in the world in goal-directed ways in accord with their aims, interests and intentions.

Contexts and Agents

The movement 'New Historicism' in American literary studies emerged in the 1980s as a dialectical rejoinder to Continental structuralism and poststructuralism.[71] Scholars like Stephen Greenblatt, Louis Montrose and Frank Lentrichia appealed to postmodernist and poststructuralist notions of the multiplicity and variability of historical texts and actions to reject the earlier historicist view, associated with Marxist literary criticism, that a literary text mirrors 'a unified and coherent world-view' held by a whole population or by a specific social class. They insisted instead that literary texts could be understood only in relation to

specific terrains of antagonistic institutions, conflicting beliefs, contrasting practices and authorial responses to situated sources of cultural authority. Blending
textualism and Reception Theory, the New Historicists treated past events as
texts to be deciphered in terms of systems of signification used by social agents
to persuade others and make sense of the world. Stressing the hermeneutical
view of the dynamic retrospective malleability of historical inquiry, they used
Foucault's notion of the productivity of power to show how disciplinary devices,
such as surveillance and confession, produced and formed cultures and their
agents, rather than repressed human nature and its potentials. At the same time,
they rejected 'various structuralist and poststructuralist formalisms' associated
with these interpretive devices. Reflecting the political activism of the 1960s
and the progressive tradition of historiography in the United States, the New
Historicists expressed disdain for any notion of the autonomy of literature or
discursive practices detached from the concrete struggles of history. They faulted
the formalism of structuralism and poststructuralism because it opposed the
'freely self-creating and world-creating Individual of the so-called bourgeois
humanism' – the 'centred subject' – with 'the spectres of structural determinism
and poststructural contingency', with 'on the one hand, the implacable code, and
on the other, the slippery signifier'. According to Montrose, it was necessary to
replace the 'antimony of objectivist determinism and subjective free-play', which
allowed 'no possibility for historical agency', with the concept of a dynamic interplay between active agents and flexible structures. Montrose offered in place of
formalism the concept of history as an 'interdependent process of subjectifica-
tion and structuration', in which the social structures and systems produced by
the purposive actions and practices of individual or collective agents enable and
constrain those agents, while 'ultimately exceeding their comprehension and
control'.[72]

The 'socialist-humanist history' developed by British 'culturalist Marxism' in
the 1970s prefigured New Historicism's view of the importance of agency and
the flexibility of structures in the development of history. Expressing traditional
British hostility towards theory – associated with 'methodological individualism' and its sense of the 'incipiently totalitarian' nature of the idea of laws of
historical development – New Left historians rejected Soviet economism and
all forms of historical determinism. They stressed instead the role of culture,
experience and ideology in the communal action and revolutionary struggles
of oppressed classes and groups of people.[73] They viewed historical agency as
the precondition, not the mechanical result, of revolutionary change, and they
reduced theory to the level of informed commonsensical guesses, or 'models'
of this agency, which are grounded in close-grained case studies of the past. E.
P. Thompson articulated these interpretive sensibilities in a famous broadside
against the theoreticism of Althusser in particular and the 'decentred subject' of

structuralism in general. Thompson argued that as a philosophy of stasis, which restricts history to 'changes in the permutation of fixed variables' within determinate structures, structuralism placed severe restrictions on the extent and conceptualization of historical change. In contrast, Thompson claimed, 'history as process' is a diachronic affair of 'open-ended and indeterminate eventuation', subject to the 'determining *pressures*' of protean, rather than fixed, structures. It involves neither the imposition of determining structures on passive subjects nor the juxtaposition of fixed structures and individual wills, but the interaction of '*grouped* "wills"'. Facilitated by 'human practice', this interaction makes lived experience, not abstract theory, the cornerstone of knowledge.[74] In order to comprehend the phenomena of the protean past, historical concepts have to 'display extreme elasticity and allow for great irregularity'. They function not as rules, or 'static representations', but as 'expectations' that 'facilitate the interrogation of evidence'. Thompson's historiography of process and agency harmonized with New Historicism's opposition to formalism and reinforced the tendency of Anglophone historians of science to link local experimental practices to the experiences and intentions of human agents actively involved in shaping and reshaping their world.

As will be seen more fully in the next chapter, the conflict between Continental philosophies of form and structure and Anglo-American historiographies of process and agency bequeathed to the sociology of scientific knowledge an interpretive tension that left its mark on sociological accounts of the history of science. Siding with the antihumanist formalism of Continental philosophy, Knorr-Cetina described 'scientists' as '"methods" of going about inquiry' and 'technical device[s] in the production of knowledge'; Lissa Roberts focused on 'the network of practices within which human actors are constructively enmeshed'; while Latour located the meaning and value of a scientist's activities in the networks of power and knowledge that make them possible. In response, Anglo-American social historians of scientific knowledge, like Shapin and Golinski, argued that 'the notion of agent – taken as the volitional human actor – is central to the sociologist's vocabulary', and they insisted on the need 'to identify the interests that actors have apart from the networks in which they might enroll'. Bowker and Latour summarized this dialectic when they noted that the 'Anglo-Saxon's' focus on 'the analysis of "intent" and "interest"' blocked access to the 'French world of discourse', which fulfilled 'the promise of structuralism in general and semiotics in particular' by treating discourse not as an expression of authorial intent or interest, but 'as ways of disciplining, as agents of social order'. In this context, Peter Galison and Andrew Pickering sought a balanced assessment of the active agency of individual scientists, as revealed in their will, intentions, skills and techniques, and the objective constraints placed on that agency by the 'mangle' of experimental practices in which they were involved.[75]

But the concern of Anglo-American sociologists and historians with the role of agency in history did not involve a wholesale resuscitation of the pre-structuralist concept of the centred or constitutive subject of knowledge and history. While they regained an earlier appreciation for the contribution of individual scientists to the creation and maintenance of their scientific identity and roles, they did not conceive of these individuals as exceptional men of genius, as agents of great discoveries elevated above other individuals by their contributions to the unfolding *telos* of history. Rather, they viewed them as local agents actively involved in the translation, interpretation, appropriation and replication of the products and activities circulating in the scientific community to which they belong. They treated scientific texts not as portals into the realm of pre-existing truths and meaning, but as sets of 'specific rhetorical, dialectical and aesthetic devices that enable scientists and technologists to recruit allies, control delegates, neutralize opponents and isolate dissidents'.[76] Anglo-American historians and sociologists of scientific knowledge portrayed individual scientists not as 'universal' agents of objective knowledge and global meaning, but as 'specific' agents operating in local contexts of power, discourse and practice.

Wittgenstein, Pragmatism and Local Knowledge

Representing the Edinburgh School, Barry Barnes and David Bloor used a version of Wittgensteinian nominalism to replace formal theories of abstract meaning and essential truths with sociological accounts of the individual use of concepts and the application of rules. They faulted the arationality assumption, with its notion of ('internal') epistemic factors determining the development of science independently of ('external') social factors, because it presupposed the 'logical compulsion', or 'logical necessity', of deductive inference forms, such as *modus ponens* which, according to formalism, all rational beings find compelling irrespective of the social circumstances or causes that otherwise shape their beliefs. Barnes and Bloor detected the same formalist sensibilities in the rational concept of the determination of actions by rules and in the standard 'deductivist' view of scientific theories, according to which the correct use of concepts is the logical consequence of predetermined meanings and rules, so that individuals function as 'rational automaton[s]' in the use and applications they make of them. Finding the implications of formalism and deductivism stultifying for social inquiry, Barnes and Bloor argued that the formalist analysis of rules and rule-governed behaviour foundered on the problem of the identity and application of rules.[77] The formalist and deductivist solution to this problem, which assumed the reality of universals and their identical instantiations, was undermined by the nominalist claim that different instances or individuals resemble but are not identical with one another. The question of the similarity and difference of particular cases arises in every instance of the use of concepts and the application of rules, thereby undermining any notion of a relation of 'logical implication', or 'sufficient determination', between, on the one hand, general rules, concepts, values and theories and, on the other hand, particular actions, beliefs, strategies and inferences. Not only do rules not contain rules for their application, they cannot be supplemented with such rules without generating an infinite regress of justifying rules. Barnes and Bloor, thus, cited formalism for failing to explain or justify the pervasive and important reality of the widespread acceptance and successful deployment of deductive inference forms, concepts, rules, values and norms.[78]

Satisfied that logic alone could not bridge the gap between general rules and their particular applications, members of the Edinburgh School 'turn[ed] to causes for an answer to the question' of the widespread use and successful application of rules and regulations. But they rejected naturalist or nativist construals of these causes, claiming that any reference to the shared cognitive proclivities of the human species is insufficient to explain the strange and elaborate forms and fluctuations of cognitive systems with time and place. Arguing that the cognitive mutability and variability of the history of science results from causes of 'an entirely local character', amenable to sociological rather than biological analysis, the Edinburgh School made every instance of the use and application of axioms, concepts and rules a matter of innovative judgement at the individual level, shaped and reinforced by agreement at the social level. Underdetermined by nature, language or past usage, each instance of the use or application of a concept or rule must ultimately be explained separately by reference to specific, local and contingent social factors, such as interests, negotiations, practices, habits or 'taken-for-granted' interpretations.[79] Finitism shifted the focus of analytical attention away from formal theories of the abstract meaning of concepts and rules and towards sociological accounts of their particular uses and applications.

Bloor summarized the semantics of finitism and its more general impact on the sociology of scientific knowledge thus: Use determines meaning; meaning does not determine use. Finitism's valorization of use over meaning and, by implication, the nonverbal over the verbal and the particular or the concrete over the general or the abstract, permeated the analytical sensibilities of the Edinburgh School, shaping its account of scientific learning and reasoning, its appropriation of Kuhn's influential doctrine of normal science, its identification of natural philosophy as a form of work, and its antihistoricist historiography of science. Thus Barnes argued that children and novices acquire knowledge not by being told in the abstract what a term or concept means, but by having their attempts to apply it to particular instances monitored by figures of parental or cultural authority. Within this framework, Shapin argued that '[f]or there to be solutions to the problem of knowledge there have to be practical solutions to problems of trust, authority, and moral order'. Barnes used this view of learning and knowledge to integrate Kuhn's notions of 'tacit knowledge' and the priority of paradigms into an account of normal science that replaced the traditional view of scientific theories as formal systems of statements with the idea of 'clusters', or 'constellations', of accepted problem-solutions sufficiently similar to enable inferences from particular instance to particular instance on the basis of analogy and direct modelling. According to the nominalist principles of finitism, scientific inference moves not downwards from general to particular statements, but from particular to particular instances, 'or upwards from the particular to the general'.[80]

Finitism's broad philosophical valorization of the nonverbal over the verbal, of action and deeds over thought and words, surfaced in Shapin's and Schaffer's influential proposal that historians of science treat eighteenth-century natural philosophy as 'a form of work', in which verbal and formal methodological pronouncements functioned as 'rhetorical tools for positioning practices in the culture' or 'claims about the proper audience for natural philosophy and the proper behavior of experimentalists'.[81] Treating the connections between methodological rules and political interests as the 'historically contingent' results of the ways in which local agents use available cognitive resources to further particular social interests, the 'social use' model of science recognized no intrinsic, lawlike connection between kinds of social factors and kinds of knowledge – no proletarian science, no bourgeois science, no German science and no Jewish science.[82] Finitism was ineluctable antihistoricist.

The resurgence of pragmatist sensibilities among American philosophers, such as Richard Rorty, Richard Bernstein and Joseph Rouse, endorsed finitism's nominalist critique of formalism and positivism and reinforced its valorization of concrete action over abstract thought and verbal representation. At the core of pragmatism is the idea of human beings as social agents seeking to change as much as to know the world. Highlighting the dimensions of social conduct and moral action in their understanding of human practice and the pursuit of knowledge, pragmatists treated objectivity as a means not of escaping into reality from the 'limitations of ones community', but as 'the desire for as much intersubjective agreement as possible'. Like finitism, pragmatism treated rationality as a 'criterionless muddling through', or as 'conversations' within and between local communities: the epithets 'truth' and 'knowledge' are used to signal or communicate intersubjective agreement and achieve communal 'solidarity'.[83]

Anglo-American finitism and pragmatism mingled with and reinforced the nominalist and deconstructionist strategies of Continental philosophy in a way that encouraged historians of science to proliferate case-studies 'concerned to show the ways in which the making, maintaining, and modification of scientific knowledge is a local and mundane affair'.[84] Passionately opposed to the idealist denigration of situated knowledge, these case studies sought to dispel the 'enchantment' of science as a form of 'knowledge from nowhere' by 'displaying the contingency, informality, and situatedness of scientific knowledge-making'. To this end, the history of science witnessed the deployment and proliferation of a broad spectrum of interpretive strategies that rebuffed the 'grand narrative of inherent scientific universality' with descriptions of the discernible marks left on it by the sites of its production,

> whether [these] sites be conceived as the personal cognitive spaces of creativity, the relatively private spaces of the research laboratory, the physical constraints posed by natural or built geography for conditions of visibility or access, the local spaces of municipality, region, nation, or the 'topical contextures' of practice, equipment, and phenomenal field.[85]

As will be seen more fully in Chapter 6, these localist sensibilities encouraged historians of science to emphasize the specificity and diversity of eighteenth-century natural philosophy and chemistry and to concern themselves with the discursive and material means whereby Lavoisier and his interlocutors in the Chemical Revolution 'inappropriately universalized' the results of their inherently local and specific inquires and practices.[86]

Kuhn, Experimental Realism and the New Empiricism

Anglo-American philosophers of science in the 1980s furthered the turn against theory that characterized sociological naturalism by developing alternative empiricist understandings of science, which stressed the empirical adequacy rather than the theoretical truth of science. Known as the 'new empiricists', Nancy Cartwright, Bas van Fraassen and Arthur Fine adopted the finitist concept of scientific theories as loose collections of 'mutually inconsistent models', or simulations of local phenomena, which facilitate a dialogue between theory and experiment that may be retained while the overarching theory is replaced. Working within the theoreticist framework of theory-laden observation and the discontinuous theoretical development of science, they argued that science has nevertheless exhibited empirical progress or increased problem-solving effectiveness across its revolutionary transitions. Antirealists about the theoretical terms of science, they insisted that the progressiveness or empirical adequacy of a scientific theory is entirely independent of questions of its theoretical truth or formal adequacy: it is a relative and comparative question dependent for its answer on the judgement of the scientific community.[87]

Ian Hacking's philosophy of 'experimental realism' integrated the epistemological sensibilities of the new empiricism into the practice-model of science by severing the theoreticist link between the truth of theories and a realist commitment to the existence of unobservable entities. As a form of 'entity realism', experimental realism allows experimentalists to be antirealists or instrumentalists about theories, but realists about the entities they purportedly described. Belief in the existence of the entities being investigated is, according to experimental realism, justified only by measuring and manipulating them, or by 'intervening' in their causal processes. Instantiating the shift from the modernist problematic of representation to the postmodernist problematic of intervention, experimental realism linked scientific beliefs and knowledge not to the formation of adequate representations of the world, but to successful interventions in its causal processes. Endorsing Habermas's view that technical control of nature is the 'raison d'être' of science, Hacking eschewed not only theoreticism, but also the 'linguistic idealism' of Latour and Woolgar, who treated the laboratory production of facts as matters of rhetoric and persuasion, rather than 'material intervention'. Hacking contrasted the 'dynamic nominalism' of the social sciences, according to which the (human) objects of social inquiry are constituted by the social circumstances in which they are produced, with his own hard-headed 'experimental realism', which maintained that though generated out of the chaos of natural events only in the laboratory, the phenomena of the natural sciences persist beyond the circumstances of their production. While the social sciences 'make up people', the natural sciences reshape the world. According to Hacking, the laboratory does not involve 'the construction of a scientific fact', so much as the assimilation into 'our stream of knowledge' of one among many of the facts that might have been discovered.[88]

Opting for a narrower interpretive focus than the practice-model of science developed by phenomenologists, ethnomethodologists and genealogists, Hacking's experimental realism concentrated on the link between the formation of concepts and experimental intervention, but ignored the role of wider social practices in the dialectic between theory and experiment. Hacking wished to show that experimentation in science does not aim at solving theoretical problems, so much as utilizing the available hard-won resources in equipment, techniques, skills and accumulated results. Experimentation does more than articulate preconceived theory; rather, it opens up new domains of inquiry and prepares them for subsequent theorizing.[89] Hacking's experimental realism here meshed with the notion of 'generative logics of experiment', also emerging in the works of Thomas Nickles, Diderik Batens and J. P. van Bendegem. Challenging the consequentialist logic of justification associated with the positivist and postpositivist distinction between the context of discovery and the context of justification, Nickles argued that despite the fulminations of generations of philosophers, working scientists never abandoned the idea of a generative logic of discovery, associated most notably with Bacon and Newton, according to which theories are evaluated not in terms of their ability to explain and predict phenomena, but in terms of their method of production, or derivation from, phenomena.[90] For

Nickles and the new empiricists, experimentation is the primary means of generating, not a secondary way of evaluating, scientific concepts and theories.

Hacking's experimental realism and the practice-model of science encouraged historians of science to pay attention not only to the constitutive and productive role of experiments in the development of scientific theories, but also to the substantive contribution of instruments and instrumentation to that process. Rejecting the traditional view of scientific instruments as unproblematic devices bereft of inherent cognitive or cultural significance and used only to illustrate or evaluate conclusions reached by theoretical reasoning, historians of science began to see them as important determinants in scientific thought and action. Numerous historical studies emerged of the role of instruments and experiments not only in the construction of scientific authority, audiences, culture and politics, but also in the formation of scientific knowledge itself.[91] Given the heavy dependence of the science of chemistry on instruments and instrumentation, it is not surprising that historians of chemistry like Bensaude-Vincent, Levere, Holmes, Golinski and Roberts found this approach productive in their analyses of the role of Lavoisier's instruments and apparatus in the controversies associated with the Chemical Revolution. Besides rehearsing the localist sensibilities of finitism, these studies upheld – in the suggestion that not only the facts and theories, but also the instruments of science, follow independent lines of development – postmodernism's nominalist sense of the complex heterogeneity of science.

But it was Kuhn's towering influence in the philosophy of science that provided the primary source of Anglo-American inspiration for the development of the sociology of scientific knowledge. The ambiguous ethos of Kuhn's epoch-making *Structure* pervaded almost every nook and cranny of the research programme of the sociology of scientific knowledge, but three broad areas of influence can be pinpointed here. Firstly, Kuhn's work put sociologists of scientific knowledge in touch with Ludwik Fleck's earlier constructivist analysis of science *Genesis and Development of a Scientific Fact*, which was originally published in German in 1935, but languished in obscurity until Kuhn referenced it in *Structure* and it was published in English in 1979. Fleck argued that even the 'hardest' experimental facts of science are the social products of local 'thought collectives' characterized by specific 'thought styles'. Neither a purely objective feature of the natural world, nor an arbitrary product of a collective imagination, a 'fact', for Fleck, registers the experience of passivity of a 'thought style' in the face of material resistance: 'the *fact* thus represents a *stylized signal of resistance in thinking*'.[92] Fleck also offered a solution to the 'problem of construction', linking the construction of the public, universal knowledge of science out of these moments of local 'resistance' to technologies of linguistic communication and the circulation in the wider community of the specific experimental practices and phenomena of local 'collectives'. In this manner, Fleck pre-empted by half a century the principles, practices and problems of contemporary laboratory studies and constructivism.

Secondly, *Structure* provided an authoritative narrative for the articulation and justification of the principles of finitism. Whereas postpositivist scholars identified paradigms with gestalts or embodied worldviews, sociologists of scientific knowledge used the Kuhnian idea of 'tacit knowledge' to support their claim that science is a practical activity, or 'craft', subject to social control and determination. In accord with the Wittgensteinian semantics of use, finitists emphasized Kuhn's notion of the priority of paradigms over rules and reiterated his claim that science is learned by doing it, rather than by 'acquiring rules for doing it'. Kuhn here utilized Michael Polanyi's notion of 'tacit knowledge', according to which science is an activity subject to the social control and determination of 'inarticulable' practices and skills, which are learned in the manner of a craft rather than a proposition. Just as an apprentice learns a craft by submitting to the 'authority' of the master craftsman, so, Barnes argued, learning a science involves a rigorous process of 'training and specialization', which is maintained and sustained 'by a developed system of social control'.[93] While Kuhn himself endorsed Merton's view that social institutions can help or hinder the pace and tempo of scientific activity, but not affect its cognitive content, Barnes and Bloor used Kuhn's views on the role of education and authority in the articulation of a paradigm to support their assault on the arationality assumption.

Finally, Kuhn's idea that 'paradigms are integral to the definition of scientific communities' shaped not only finitism's view of scientific knowledge as 'local knowledge'; it also influenced finitism's valorization of microsociological analyses. Stressing Kuhn's distinction between the social cohesion of a group that flows from its acceptance of a paradigm and the place of that group in the broader institutional and social context, the Edinburgh School believed that the goal of showing how 'social relations penetrate to the core' of scientific beliefs would be realized not by referring them to their wider social context, but by showing how groups of scientists formed and functioned around a particular way of practising or doing science. If paradigms are understood as specific forms of life, the Kuhnian clash between incommensurable paradigms is, in effect, a clash 'between incommensurable modes of community life'. While Kuhn treated these conflicts as relatively infrequent revolutionary upheavals, sociologists of scientific knowledge viewed them as evanescent ripples in the everyday ebb and flow of local controversies, which they treated as disagreements not about the order of nature, but about the organization of scientific communities. Reversing the familiar order of sociological historiography, which moved from the wider social context to the internal cognitive content of science, the Kuhnian model called for the production of empirical studies of the inherent or internal social profile of scientific communities, before moving outward to consider how these communities intersect with the wider social context in which they flourish and intervene. As will be seen in the next chapter, disagreements about how to characterize the microsociological worlds of science and their relation to the wider social world in which they exist soon surfaced among sociologists of scientific knowledge and constituted an important axis of creative division and interaction between them.[94]

Summary and Conclusion

The discipline of the sociology of scientific knowledge emerged in the 1970s and 1980s as part of a fundamental break with positivism and postpositivism rooted in the philosophical upheavals associated with the cultural transition from modernism to postmodernism. Initially formulated within the philosophical parameters of modernism, the sociology of scientific knowledge soon developed in a direction that broke with the modernist problematic of autonomy and legitimation. Postmodernism jettisoned the normative dualisms of modernism, abandoning the notions of formal reason, representational thought, and progressivist history which underpinned positivist and postpositivist accounts of the internal development of science. Postmodernism stressed the oppressive and totalitarian implications of the modernist problematic of autonomy and legitimation, replacing all talk of universal truth, objective reason and essential structures with nominalist accounts of the role of local agents, specific practices and individual events in the formation of temporary constellations of meanings, values and beliefs. The positivist vision of the 'unity' of science gave way to accounts of its irreducible 'disunity'.[95] Against the modernist problematic of representation, postmodernists argued for relativist and antirealist accounts of the role of discourse and practice in the constitution of scientific knowledge. Elevating the concrete over the abstract, the fragmented over the integrated and the dependent over the autonomous, postmodernism insisted that embedded in local contexts of power and control, science is value-laden and not necessarily progressive. Postmodernism's nominalist ontology of undifferentiated particularity also undermined the historicist notion of the unity and homogeneity of historical time, replacing the progressive movement of modernism with an overwhelmingly spatial vision of reality that downplayed the temporal dimension.

Sociologists of scientific knowledge deployed an intricate if not entirely consistent array of Continental and Anglo-American philosophical strategies to articulate their nominalist analyses of scientific knowledge and practice. These analyses rejected essentialism, idealism, realism and historicism in favour of finitism, which makes knowledge a local instrumental, contingent affair. Thus, while the Edinburgh School replaced theoreticism's essentialist theory of meaning with a form of Wittgensteinian nominalism that relativized meaning to local usage, poststructuralists challenged the metaphysics of presence with anti-realist theories of meaning that denied the existence of extra-discursive objects. Against the idealist doctrine of the determining role of 'guiding assumptions', or 'global theories', in the development of science, they marshalled Wittgensteinian and poststructuralist notions of the underdetermination of actions by rules, eth-nomethodological accounts of the indexicality of thought and action, Kuhnian ideas of the role of concrete examples and tacit knowledge in the development of science, deconstructionist assaults on the totalizing texts of structuralism, and Foucauldian concerns with the microphysics of the power-knowledge nexus. They opposed idealist and formalist assimilations of scientific reasoning to theorizing with nominalist semantics of use, designed to undermine deductiv-ist interpretations of the structure and generalizing function of theories and to support sociological accounts of their local use and applications. They argued for a 'tinkering' view of experimentation and model-building in science, empha-sizing the occasional character of local 'know-how' and the generative logics of discovery championed by the new empiricists. Influenced by the earlier work of Ludwik Fleck and buttressed by Hacking's 'experimental realism', constructiv-ists emphasized the constructedness of scientific facts and the integral role of scientific instruments and institutions in their creation or production. They also utilized Heidegger's 'practical hermeneutics' to rebuff theoreticism more gener-ally and to make a categorical identification of science with practice.

A core principle of the sociology of scientific knowledge was the identifica-tion of science with practice, but just what that practice consisted in was a bone of contention. Viewing science as a form of discursive practice, some sociolo-gists of scientific knowledge shifted the locus of science away from the context of the discovery and justification of scientific facts and theories to the context of their dissemination and reception in the scientific community. Variously influ-enced by Derrida's textualism, Reception Theory and Foucault's genealogy of discursive practices, they located the core of science in its texts and inscription devices, which they viewed not as unproblematic means for the communication and transmission of prior knowledge of an independent reality, but as rhetori-cal, hermeneutical and political devices involved in the simultaneous formation of knowledge, nature and the community of knowers. Other scholars – operat-ing under the influence of Heidegger's 'practical hermeneutics', Merleau-Ponty's doctrine of embodied knowledge, Foucault's conception of the disciplinary structure and function of scientific knowledge, Hacking's experimental realism, and the general valorization of the nonverbal over the verbal associated with the semantics of use – identified science as a form of 'work', or 'craft knowledge', and sought to describe the disciplinary structure, tacit knowledge, implicit skills, embodied techniques and spatial organization of the sites of the production, rather than the discovery, of scientific knowledge.

Finitism's nominalist identification of science with practice saddled sociologists of scientific knowledge with the crucial 'problem of construction', with the problem of explaining how the lauded universality of public science is produced by the inherently local and situated practices of the particular sites of the production of scientific knowledge. Pragmatism's valorization of concrete practice over abstract theory and verbal representation reinforced this concern, as did the stress placed on the specificity of historical contexts by New Historicists and British Cultural Marxists. These Anglo-American historians challenged the antihumanist focus of Continental philosophers on impersonal practices, situated logics, semiotic actants and networks of power with a concern for the role of the experiences and intentions of human actors in the 'process' of shaping science and the world. This challenge marked an interesting tension among sociologically minded historians of science between those who focused on the constitutive constraints of anonymous practices and those who took notice of human volition and action in the making of scientific knowledge. The pervasive nominalism shared by both parties to this dispute also informed their opposition to historicism. Appropriating Gadamer's view of the flexibility and variability of historical meaning, finitism replaced the historicist notion of a logic, or driving force, of history with a sense of the unalloyed contingency of historical events and the multiplicity of historical contexts. Integrated into postmodernism's vision of the dispersed spatiality of human existence, finitism involved 'the randomization of history'.[96]

5 THE SOCIOLOGY OF SCIENTIFIC KNOWLEDGE AND THE HISTORY OF SCIENCE

Sociologists of scientific knowledge responded to the naturalist call of the Strong Programme with arguments and considerations that straddled the traditional boundaries and 'levels' between philosophy and sociology. Their concerns and interests were as much philosophical, and especially epistemological, as they were sociological. By showing how the social dimension of knowledge shaped or determined what count as facts, discoveries, inferences, objectivity and credibility in science, finitism upheld the claim that 'social factors' constitute, rather than merely influence, the content and development of scientific knowledge. Squarely opposed to epistemological individualism, which grounded knowledge in individual experience or cognition, finitism maintained that the 'fundamental and irreducible point' of 'the sociological study of scientific knowledge' is to show 'in what ways that knowledge has to be understood as a collective good and its application as a collective process'.[1] This chapter traces the development of the discipline of the sociology of scientific knowledge in the 1970s, 80s and 90s, showing how the cultural transition from modernism to postmodernism shaped its evolving conception of the social constitution of science. It also relates these broad philosophical and disciplinary developments to the historiography of science promulgated by Steven Shapin and Simon Schaffer in *Leviathan and the Air-Pump* and other essays, which were important points of mediation between sociologists and historians of science.

Rejecting 'the idealist tendency to refer directly to concepts as "scientific", "symbolic", "ideological" and so on', sociologists of scientific knowledge insisted that the conflict between ideas is really a confrontation between the people who hold and use them.[2] But just how people and their social interactions were to be identified and understood proved to be a considerable bone of contention among these sociologists and their successors in the community of science studies, who adopted a range of frequently discordant methodological perspectives. The content and trajectory of these perspectives on the constitution of science by society was undergirded by the cultural transition from modernism to postmodernism. While the Edinburgh School and the practitioners of SSK upheld modernist dualisms of subject and object, society and nature, associated with the representational model of science-as-knowledge and truth as correspondence, constructivists like Knorr-Cetina, Latour, Lynch and Pickering favoured the postmodernist view of science as a form of practical intervention in the world and truth as a measure of its success. While constructivists integrated their view of science-as-practice into a 'performative' model of society, according to which society is 'something achieved in practice by all actors', human and nonhuman, scientific and social, the Edinburgh School used the 'ostensive' model of society to explain agents' theoretical beliefs and cognitive representations of the world in terms of their pre-existing social interests. Within this scheme of things, Foucault pursued a third way: instead of identifying knowledge with power, or delineating the social determinations of science, he 'articulated', in a nonreductive manner, 'discursive on nondiscursive practices'.[3] Steven Shapin and Simon Schaffer blended these models of science and society in their influential analyses of seventeenth- and eighteenth-century natural philosophy, according to which science sustains itself by 'cultivating', rather than by constructing or representing, social interests. These interpretive strategies had a significant influence on sociologically minded historians of the Chemical Revolution.

The Interest Model and the Empirical-Relativist Programme

The Edinburgh School looked for strategies to develop the Strong Programme into a fully-fledged sociological account of the nature and development of scientific knowledge in models of explanation that focused on the social causation of beliefs and concepts. Adopting the 'Duhem-Quine underdetermination thesis', and treating science as a unitary system of conceptual representations ('Hesse nets'), the Edinburgh School argued that since the future extension and application of these 'nets' is underdetermined by logical reasoning and experimental evidence, the actual direction of theoretical development in science is the causal effect of social factors. When Bloor challenged the arationality assumption, associated with the work of Mannheim and Merton in the sociology of knowledge and the sociology of science respectively, he reverted to Durkheim's stronger sociological view ('social realism') that all knowledge claims are collective forms of representation rooted in the social organization of groups of individuals.[4]

The Edinburgh School deployed two strategies – 'grid-group' analysis and the Interest Model – to develop the Durkheimian doctrine of 'social realism' into a coherent sociology of scientific knowledge. Bloor used the grid-group typology of communities developed by the anthropologist Mary Douglas to challenge and modify Kuhn's account of scientific change. He used the grid-group analysis to replace epistemological accounts of scientific change in terms of formal anomalies and inconsistencies with sociological accounts of how the relative stratification and openness of scientific communities condition members' response to anomalies and paradigm shifts. Crucial to SSK's reading of Kuhn on this issue was the claim that 'anomalies' do not act independently, but 'only as tools in the hands of human actors who use then to advance specific ends'.[5]

Grid-group analysis appealed to only a few historians of science in the early 1980s, but the Interest Model had a greater and more pervasive impact on members of this scholarly community.[6] Proponents of the Interest Model referred to the 'interests' of individuals or small groups of scientists to explain their beliefs and judgements and to changes in their interests to explain changes in their beliefs and judgements. Early users of the Interest Model, such as MacKenzie and Barnes, identified these interests with 'external', macrosociological interests, appealing for example to conflicting class interests to explain the dispute between 'biometricians' and Mendelians in early twentieth-century British genetics. But other practitioners, like Pickering and Shapin, soon modified this stance, allowing 'internal', microsociological skills and competencies to 'represent a set of vested interests *within* the scientific community'.[7] SSK individuated interests not according to their intrinsic social identity or domain, whether economic, political, cultural, technical or professional, but in terms of their general ontological status and epistemological function. At the core of the Interest Model was the demand that whatever their social nature or domain, the interests referred to in the model are material, historical, local and contingent entities.

The Interest Model rejected any notion of knowledge as a passive reflection or picture of reality in the minds of isolated individuals; rather it depicted knowledge as an interest-laden cognitive activity rooted in the biologically based need of human beings to control their environment. However, unlike Habermas who treated these technical 'knowledge-constituting interests' as immutable and transcendental features of human nature, the Edinburgh School argued that the 'technical' is always mediated by the 'cultural', the 'instrumental' by the 'ideological' and the 'biological' by the 'social'. Human beings may be possessed of 'inherent capabilities for reasoning', but how they 'actually reason in any specific case' depends upon a matrix of sociocultural and psychological causes and conditions.[8] Lissa Roberts put this point nicely when she noted that the term 'utility' did not function in the eighteenth century 'as a transcendent and universal ideal that substantively informed all practice', but as 'a multifaceted set of discourses that were deployed in keeping with the variety of meanings and purposes with which actors at the time charged them'.[9] Rooted in the material needs of a given society, the social interests that constitute the scientific beliefs and knowledge of that society are oriented towards predictive success and control of reality which, in turn, is shaped and evaluated according to the sociopolitical sanctions and interests that inform them. Knowledge is not 'contemplatively produced by isolated individuals; it is produced and judged to further particular collectively sustained goals'.[10]

The Edinburgh School distinguished its form of 'new contextualism' from earlier, Marxist models of the determination of science by society by emphasizing its antihistoricist conception of the contingency and specificity of social interests and their relation to knowledge and cognition. Barnes and Shapin contrasted the 'coercive model', or deterministic view of the relation between social interests and scientific knowledge, with their 'social use', or 'instrumental', model, which treated knowledge as an active instrument used to further 'particular social goals'. Instead of establishing an intrinsic, deterministic relation between scientific judgements and social circumstances, the Interest Model claimed only that any social interest '*may* have a bearing on judgment and that contingent social factors of some kind *must* have'.[11] Responding to the objection that theories are as much underdetermined by interests as they are by data, Bloor argued that while the initial choice of which of the many available theories will be used to express a given interest in a particular situation is a matter of chance, a pure 'historical contingency', subsequent feedback and positive reinforcement of the initial choice renders the theory 'the favored vehicle of expression of interest'. Emphasizing SSK's valorization of agency, Shapin noted that while the 'coercive model' reduces actors to 'judgmental dopes' compelled by reason or interests, the 'instrumental model' emphasizes the element of choice in the instrumental rationality of all agents, who produce and evaluate knowledge in order to further their interests: 'The role of the social, in this view, is to prestructure choice, not to preclude choice'.[12] The Edinburgh School generated and supported historical case studies designed to show how particular actors in specific social situations deployed available concepts and theories of nature to further a variety of locally integrated social and technical interests.

But the Edinburgh School denied that their enterprise was a purely descriptive one. They enjoined historians of science not only to describe how, in fact, scientific knowledge had developed or was constructed, but also to explain why it took the course it did rather than some other equally feasible and historically available path. The job of sociologists and historians of scientific knowledge is not simply to follow scientists around, describing their thoughts, deeds and practices, but to explain what they thought and did in terms of social causes that may be outside their field of consciousness. Assimilating the aims and methods of the social sciences to those of the natural sciences, the practitioners of SSK stressed the search for 'laws' in this sociological enterprise. While Bloor's naturalism rejected the modernist problematic of normativity, it upheld its valorization of the unity of science and the problematic of representation and explanation.

At the same time, Bloor and his colleagues emphasized the complexity of the causal situation, noting the operation of other 'natural causes', such as narrow professional interests or psychological and sensory inputs, besides broad social ones, but they insisted that the social component was the ubiquitous and overriding cause of the formation of the collective representations of science.[13] Viewing scientific knowledge as the synthesis of social interests and sensory inputs from the external world, they regarded their commitment to a materialist ontology, empiricist methodology and realist epistemology as the hallmark of scientificity. As a form of 'social realism', which reconstructed concepts of nature in terms of social accounts of the production of those concepts, SSK emulated the aims and methods of the natural sciences, while valorizing the content of the social sciences. But SSK, like all forms of social realism, soon ran into the problem of 'reflexivity': constituted by social interests in the same manner as the natural sciences, SSK has no privileged claim to credibility or acceptability. As will be seen below, subsequent developments in the sociology of scientific knowledge highlighted the problem of reflexivity and questioned the modernist aspirations of SSK to the status of an objective and explanatory social science.

In the early 1980s, Andrew Pickering and the Bath Relativists, Harry Collins and Trevor Pinch, developed a constructivist version of SSK. Within the framework of finitism, they challenged SSK's claim to scientificity, as well as its adherence to the modernist problematic of representation and the doctrine of the unity of science. They faulted SSK's conception of 'science as a single conceptual network' because it drew an impoverished picture of the diversity and complexity of the practices of science.[14] They shifted the epistemological focus of finitism from the problem of the application and extension of concepts and theories to that of the replication of experiments in science. Stressing finitism's nominalist view of the particularity of experimental instances and the radical underdetermination and open-ended nature of their negotiated replications, Collins traced the speedy resolution of actual scientific controversies to the relations of authority, credit and trust that structure the scientific communities in which they occur. Viewing the social context of science in terms of informal networks of trust and authority between 'core sets' of researchers, communicating with one another about specific problems and issues, Collins, Pickering and Pinch focused their inquiries on the internal, microsociological interests of scientific communities – their theoretical skills, experimental techniques and professional identities – rather than on the wider social interests emphasized by the Edinburgh School. Collins's Empirical Relativist Programme (ERP) was 'empirical' in the sense that it used 'testimonies by members of the core set', rather than prescriptions by philosophers of science, 'to document rational configurations of theoretical commitment and experimental practice'.[15]

ERP radicalized Kuhn's philosophy of science with an infusion of interpretive strategies derived from phenomenology, hermeneutics and the philosophy of the later Wittgenstein. According to Collins, the job of the historian or sociologist of scientific knowledge is to examine scientific controversies in order to render explicit the implicit, and therefore seemingly 'natural', ways of seeing and acting associated with a scientific paradigm. Though short and infrequent episodes in the development of science, controversies expose through dialectical attrition relations of trust and authority concealed in widely accepted and firmly established instances of scientific knowledge and practice. Collins labelled as 'relativism' the radical state of uncertainty that resulted from the need of the historian or sociologist to suspend their own 'natural' way of seeing and interpreting the world in order to understand, as opposed to merely explain, other ways of seeing and interpreting it. Ultimately, the only way to discern the taken-for-granted rules of another discipline or culture is through 'face-to-face' contact, or 'participation with the area under study'.[16]

Collins is here influenced by Peter Winch's Wittgensteinian version of *verstehen*, which upheld a strict epistemological and ontological distinction between the social sciences and the natural sciences. Winch argued that whereas the natural sciences explain events, including human behaviour, by establishing, through observation and experimentation, relationships of regularity with events extrinsic to the ones being explained, social scientists interpret human actions by establishing, through communication and participation with performers of the action, the concepts, meanings and rules intrinsic to, or definitive of, the action itself. Collins advocated the use by sociologists and historians of science of 'depth interviews ... with the object of a more full participation in mind', but he recognized that the 'method of full participation can rarely be attained in practice'.[17] Nevertheless, he upheld the goal of understanding and participation as an ideal that distinguished ERP from the scientific model of causal explanation and detached observation advocated by SSK.

Collins's programme for understanding the perspectives of practising scientists and his sense of the richness and complexity of these perspectives encouraged historians of science to develop more detailed accounts of the many factors that go into the production of scientific knowledge. It also led them to stress the unending process of controversy and consensus involved in the formation of scientific knowledge and the communities that produce it. In a similar vein, the anthropologist Clifford Geertz developed the technique of 'thick description' to interpret the activities of individuals living in agonistic cultures of great diversity and multiplicity. Downplaying the modernist notion of cultural unity and coherence, Geertz called upon anthropologists and, by implication, historians of science to traverse the diverse conceptual structures and systems of meaning that 'create the event or behavior as a discrete entity to begin with'.[18] As will be seen in Chapter 6, Golinski used this interpretive framework to triangulate the terrain of eighteenth-century chemistry and the Chemical Revolution by exploring the dynamic networks of meaning and cultural codes that shaped its evolving identity as an experimental science.

Ethomethodology and Laboratory Studies

Ethnographic studies of scientific work and life, published in the late 1970s and 80s, accelerated the replacement of SSK's 'macrosociological' accounts of the generation of scientific knowledge with more narrowly focused 'microsociological' descriptions of the internal construction and practices of the sciences. In keeping with ethnomethodology's interest in the reflexivity of actors as they construct and maintain 'meaning' in their everyday lives, ethnographers of science conceptualized the production and distribution of scientific knowledge not as a 'function of classically theorized social variables', such as class structure or social status, but in terms of 'the day-to-day shoptalk and methodic practices at particular laboratories'. The need for detailed descriptions of the individuality and integrity of these complex practices impelled some ethnomethodologists to challenge any 'disciplinary hegemony', even that of sociology itself, in the understanding of scientific knowledge and practices. Attuned to postmodernist nominalism, ethnographic studies of scientific life and work revealed, behind the rationally reconstructed veneer of conceptual coherence and disciplinary rigour on display in scientific textbooks and research journals, 'the situated and improvisational performance of actual practices in "messy" practical and interactional circumstances'.[19] These studies replaced accounts of the formal or material unity of science, expressed in terms of such abstract entities as 'facts', 'statements', 'experiments' and 'rules', with nominalist descriptions of how reason is 'spatialized' in 'the idiosyncratic, local, heterogeneous, contextual and multi-faceted character of ... practices' around the 'lab bench'. But ethnographers of science did not abandon entirely the modernist demarcation between science and nonscience, locating it not in the privileged epistemological content or form of scientific knowledge, but in the elaborate, tightly controlled discursive and material resources concentrated in scientific laboratories. Nevertheless, they did replace the modernist problematic of representation with the postmodernist problematic of intervention, construing the sociology of scientific knowledge as 'an investigation of how scientific objects are produced in the laboratory rather than a study of how facts are preserved in scientific statements about nature'.[20]

The ethnomethodological concept of 'indexicality' located the meaning and significance of linguistic expressions and practical activities not in any underlying general rules, principles, beliefs or assumptions, but in the observable and accountable behaviour of individuals embedded in specific organized activities, spatial settings and material situations. Accordingly, Lynch and his colleagues generated empirical descriptions of 'competence systems', or sequences of observable, local and informal practices which, while not 'thematized in formal accounts of scientific method', are incarnated in the self-composed activities that make up 'the details of a discipline's daily work', whether it involved performing an experiment, following a mathematical proof, constructing a theoretical explanation, or reading and publishing a textbook or research report. Seeking to reveal 'the temporal "building" and "building up"' of the final, finished and formal products of science – its objects, experiments, instruments and texts – out of the 'commonplace modes of inquiry and understanding that operate in a day-to-day research situation', ethnomethodologists linked the particularity of these practices to the incommensurability of its products. In accord with the nominalist principles of postmodernism, however, they substituted for Kuhn's idea of 'an incommensurability between *paradigms*' the notion of 'an incommensurability between *individuals* and the *things* they manufacture'. They sought to transpose the problematic of constructivism and the incommensurability of scientific theories from the arena of academic philosophical arguments about the nature of science in general to the domain of empirical descriptions of the 'technical and researchable features of actual researchers'.[21]

Denying any privileged status to the cognitive content and method of science, ethnographic studies of 'laboratory life' refused to pay attention to what laboratory members believed or thought about what they were doing. In contrast to Collins's view of the interpretive and participatory work of the sociologist of scientific knowledge, Bruno Latour and Steve Woolgar, in their influential study *Laboratory Life*, presented the ethnographer of science as an informed 'stranger' who, like anthropologists studying alien cultures, observe what is going on without being taken in by, or internalizing, members' ways of seeing and interpreting what they are doing. Used to facilitate greater objectivity in the observation of the overt practices and settings of specific laboratories, the behaviouristic device of ignoring the participants' self-understanding inevitably led the ethnographer to describe just what the outside observer finds intelligible in a laboratory: 'traces, texts, conversational exchanges, ritualistic activities, and strange equipment'.[22] This behaviouristic stance also harmonized with the Derridean claim that material practices of writing constitute the meanings and representations they ostensibly convey and the texts they construct and deconstruct. Thus *Laboratory Life* treated the laboratory as an 'extended text', or field of 'literary inscriptions' and the 'inscription devices' that produce them. Latour and Woolgar gave a textualist construal to constructivism:

> It is not simply that phenomena *depend on* certain material instrumentation; rather the phenomena *are thoroughly constituted by* the material setting of the laboratory. The artificial reality, which participants describe in terms of an objective entity, has in fact been constructed by the use of inscription devices.[23]

According to Latour and Woolgar, laboratory researchers study not independently existing things but 'literary inscriptions' produced by laboratory technicians working with 'inscription devices'. These devices, which include machines such as chart recorders, oscilloscopes, scales, etc, transform material substances into symbols, or 'inscriptions', such as spots on screens, points on scales and peaks on graphs, which researchers use to generate 'the occasional conviction of others that something is a fact'. A fact is constructed when a community of researchers is persuaded to accept a 'demodalized statement', such as 'TRF is Pyro-Glu-His-Pro-NH$_2$', without qualifications, such as 'the data indicates that ... ' or 'I believe the experiment shows that ... ' which characterized its indexical origins. The ethnographer of science describes the dense social and rhetorical activities involved in stripping modalities from previously qualified and contentious statements to produce the apparent stability and independence of facts. Striving to establish rhetorical hegemony over more and more people, scientists and laboratories do not '*in*form' us about the world, so much as '*per*form ... the world we live in'. Once widespread performative agreement is reached, the earlier conditions of contention and persuasion are 'forgotten' by the scientific community, which views 'the object', originally regarded as 'the virtual image of the statement', as 'the reason why the statement was formulated in the first place'. Constructed by events within the laboratory, a fact, or 'reality "out there"', is the result of a 'rhetorical *persuasion*' in which 'participants are convinced that they have not been convinced'.[24] This rhetorically induced amnesia enables the fact to be used as a 'black box', or unquestioned instrument in future research. Equating the content of the natural sciences with the same kind of literary and interpretive activities to be found in the humanities and the social sciences, ethnomethodologists of science traced the privileged status accorded to the natural sciences 'to the expensive equipment and institutional maneuvers that transform naturalistic statements into practically unassailable texts'.[25]

The idea of science as a rhetorical enterprise, conducted in an 'agonistic field', gained considerable traction among sociologically minded historians of science, who studied the 'materialities of communication' and related scientific texts not to their 'conceptual priorities', but to their 'sociotextual locations'. Textualism encouraged historians of science to replace traditional concerns with matters of meaning and interpretation with analyses of the materiality of signs, the technologies of inscription and the practices of writing. Golinski registered this shift in historiographical priorities when he proclaimed: 'It seems to me more promising to abandon the attempt to explicate formal conceptual structures, metaphysical or psychological, and to concentrate instead on the concrete practices involved in writing texts and producing discourse'.[26] Besides emphasizing the rhetorical function of methodological principles within and without laboratories, these scholars described such things as the use of narrative devices to describe experimental discoveries and create the 'found object' illusion of scientific discoveries; the deployment of styles of writing to convey the authority of authoritative texts; and the role of visual imagery in giving substance to abstract concepts of nature. Emphasizing the historical situatedness of scientific representations, they showed how key scientific terms pass through various discursive spaces, creating virtual witnesses of experiments and generating support for proposed explanations.[27]

But the textualist idea of 'the written sign' as 'an institution, backed up by other texts and embedded in a network of enforceable codes' blended in the minds of historians of science with Reception Theory's claim that the meaning of a text inheres in its interaction with its readers, and encouraged them to conjoin descriptions of textual strategies with accounts of the hermeneutical practices and objectives of readers and audiences. Golinski, once again, provided a useful summary of constructivist strategies when he argued for the necessity of blending rhetorical and hermeneutical analyses in order to place scientific discourse in its historical context. This context involves the *'conventions'* constraining the structure of discourse, the *'audience'* shaping its intent or signification, and the *'situation'* enabling links to be formed between conventions and the speakers and audiences that use and respond to them. Whether they focused on the symbolic or denotative function of the language of science, its role as a vehicle of meaning or its rhetorical function, sociologically minded historians of science viewed language not as an unproblematic, transparent means of representing reality, but as a discursive practice, or performative act, constitutive of the content as well as the form of scientific representations, if not the world they represent. These historians registered the nominalist sensibilities of postmodernism, calling into question, for example 'the static opposition between the figural and the literal' and drawing attention to Nietzsche's claim that knowledge 'is nothing but working with the favorite metaphor, an imitation which is no longer felt to be an imitation'.[28]

Discourse Analysis and Reflexivity

Proponents of Discourse Analysis and the Reflexive Turn rejected as scientistic Latour and Woolgar's search for an ethnographic 'metalanguage' in which to describe, 'without resorting to any of the terms of the tribe', their 'direct observations' of laboratory practices.[29] The claim that the ethnographer describes just what he or she observes, not what members believe or think, is going on in the laboratory implied a lingering commitment to the problematic of realism that Discourse Analysts found untenable. Since sociologists and historians have access to the reality of scientific practice only through scientists' 'accounts' of these practices, proponents of Discourse Analysis argued that the realist interest in what actually happened in the development of science, shared by SSK and ethnomethodologists alike, rested on the unwarranted assumption that competing accounts of what happened could be objectively evaluated. But it is difficult to determine the real 'interests' at causal play in a given situation because the imputation of interests depends upon the competitive and contentious 'interest work' of the participants involved in the action.[30] Arguing that accounts shape or determine interests, and not vice versa, Discourse Analysts enjoined sociologists and historians of scientific knowledge to abandon SSK's realist project of determining what are the real interests shaping scientific development in favour of describing how scientists construct their accounts of these interests and their influence on the development of science.

Discourse Analysis treated scientific discourse not as an indication, or representation, of something else (reality), but as an object of study in its own right. Instead of providing definitive explanations of scientists' beliefs and practices in terms of their motives, interests or goals, Discourse Analysis showed how scientists used interpretive repertoires and accounting devices to produce different accounts of their world in different interactional settings. Among the many proposals for shifting the focus of historiographical attention from reality to discourse, Augustine Brannigan's constructivist account of the nature of scientific discoveries stands out. Brannigan replaced the realist, 'found object' model – which identifies scientific discoveries with mental events in the minds of discoverers – with the 'attributional model', according to which a scientific community attributes the honorific 'discoverer', or 'discovery', to an individual or event on the basis of contingent judgements in specific scientific and social contexts. David Miller exploited Brannigan's account to good effect in his description of the 'attributional processes' at work in the nineteenth-century retrospective controversy about who was the real 'discoverer' of the composite nature of water, Cavendish or Watt.[31]

While ethnomethodologists and Discourse Analysts focused on the reflexivity of scientists as they constructed and maintained everyday scientific knowledge, they paid no attention to the reflexivity involved in their own accounts of these constructions. Although reflexivity was an important tenet of the Strong Programme, sociologists of scientific knowledge generally regarded it as uninteresting and paid little or no heed to the destructive implication that, as a species of socially constituted knowledge itself, the sociology of scientific knowledge has no privileged claim to credibility or acceptability. Most sociologists occupied this unstable position by adopting a form of 'social realism', which combined realism in the social sciences with constructivism in the natural sciences: 'In other words, social forces are real, but natural ones are socially produced representations'.[32] But a handful of scholars, like Woolgar and Michael Ashmore took the 'reflexive turn'. Refusing to adopt the pragmatic compromise of social realism, they viewed reflexivity not as a problem to be avoided in order to proceed with substantive sociological inquiries, but as a challenge to be embraced on the road to a deeper philosophical and sociological understanding of science. While social realists undermined the hegemony of the natural science by denying them privileged access to their subject matter, they seemed to reaffirm the scientific idiom in the representative nature and function they ascribed to their own theory. Insisting that both science and the social studies of science are discursive disciplines, reflexivity theorists fell in with the deconstructionist tendencies of postmodernism, employing 'Derridean disruptions of conventional social science writing practices' to question or undermine the validity of the statements produced by these practices. Thus, they used 'multi-voice texts ... to replace the unitary, anonymous, socially removed authorial voice of conventional sociology with an interpretative interplay within the text', which shifted the focus of attention away from 'them' to 'how "we" conduct our inquiries about "them"'.[33] Prima facie incompatible with the unreflective realist orientation of mainline sociological and historical inquiries, the reflexive turn, nevertheless, reinforced the burgeoning interest of historians of science in the rhetorical and discursive practices of science and sensitized them to the difficulty, if not impossibility, of choosing between competing accounts offered by participants and analysts of these practices.

Disciplines and Networks

Some constructivist scholars criticized laboratory studies because they ignored wider social phenomena of interest to classical sociology, and because they offered no solution to the problem of construction. Arguing that there was no reason why the wider social world beyond the walls of the laboratory could not be 'viewed as an arena in which knowledge is constructed', they showed how the 'proclaimed universality of science' is 'the result of large-scale extensions of local forms of life'.[34] Rouse and Golinski deployed the Foucauldian notion of 'disciplinarity' in this interpretive endeavour. They tied scientific expertise to the extension through society of pedagogical and organizational practices and institutions designed to produce communal unity and uniformity through the rigorous division and control of the time, labour and instruction of apprentice scientists. Now the interest of historians of science in the formation of disciplinary identities, divisions and coordination in the scientific enterprise did not start with the rise of the sociology of scientific knowledge. It can be traced back at least as far as Comte's paean to the role of the division of labour in the development of science, which linked the unity and progress of science to the disciplinary specialization and coordination of narrowly-focused individual investigators serving, without necessarily comprehending or even caring about, the epistemic or social interests and goals of science as a whole. However, besides ushering in 'a remarkable expansion of interests in scientific disciplines, their origins, fusions, fissions, and extinctions', sociologists of scientific knowledge gave a constructivist and contextualist twist to this abiding interest. Whereas the positivist model of 'professionalization' in science linked the emergence of scientific disciplines to the teleology of progress and increasing autonomy in science, sociologist of scientific knowledge like Shapin emphasized the importance of the *construction* of disciplines and their boundaries in the development of science.[35]

While Schaffer traced the rise of the disciplinary structure of science to the Second Scientific Revolution of the early nineteenth century, which separated 'the disciplined training of scientists from the heroic discovery [of natural philosophers] for which no training was possible', Christie and Golinski extended the reach of the historiography of disciplinarity back in time, portraying eighteenth-century chemistry as 'a discipline and practice with a continuously ascertainable sense of its own identity'.[36] More recently, sociologically minded historians of chemistry like Lissa Roberts called for the 'history of [eighteenth-century] chemistry's disciplinary journey from its self-defined status as art to its recognized status as science'.[37]

In *Science in Action* and *The Pasteurization of France,* Latour developed his earlier ethnographic study of laboratory life into the Actor-Network Model of science, which offered an influential solution to the problem of construction that linked laboratory practices and the formation of scientific disciplines to a wider field of discursive and technical negotiations and interactions. According to Latour, scientists use both discursive and nondiscursive devices to extend beyond the walls of the laboratory the rhetorical hegemony that transforms a mere statement into an established fact. If readers resist the rhetorical pressure exerted by the technical literature in the field, and continue to reject written claims emanating from the practitioners of laboratory science, they are confronted with the laboratory instruments that produced the original 'inscriptions'. The inscriptions and the machines that produce them are eventually 'black-boxed', or circulated as unproblematic means for exhibiting the phenomena of nature. Facts, texts and machines pass along 'networks' of people and things, gaining authority as they move away from their point of origin and obtaining durability 'only through the actions of many people'. Latour's nominalist solution to the problem of construction tied laboratory knowledge to worldwide 'technoscience' through networks of 'associations' of local actors, or 'actants'. Latour turned his back on a generation of progressive and critical sociologists when he dismissed, as 'far fetched', the objections of 'my (semiotic) reader', who indignantly asks,

> Where is capitalism, the proletarian classes, the battle of the sexes, the struggle for the emancipation of the races, Western culture, the strategies of wicked multinational corporations, the military establishment, the devious interests of professional lobbies, the race for prestige and rewards among scientists? All these elements are social and

this is what you did not *show* with all your texts, rhetorical tricks and technicalities![38]

Identifying science with the rhetorical activity of building friendly associations and dismantling unfriendly ones, Latour rejected the familiar distinction between the rarefied technical content of science and its more familiar social context, claiming that 'technical literature ... is hard to read and analyze not because it escapes from all normal social links, but because it is *more* social than the so-called normal social ties' On this interpretation, technical papers are harder to read and criticize than nontechnical manuscripts because they contain more 'black boxes' and the increased authority thereby associated with the actions of more people. Claiming that the Actor-Network Model 'gives back to the word "social" its original meaning of "association"', Latour treated all other social categories, such as 'power', 'structure', 'culture', 'institution' and 'paradigm', as different ways of describing and summarizing 'the set of elements that appear to be tied to a claim that is in dispute'.[39]

Latour's nominalist ontology constituted both a continuation and a significant break with previous work in the sociology of scientific knowledge. Latour incorporated into his model Collins's analytical focus on scientific controversies and their negotiated resolutions, as well as Barnes's insistence on the contingent and constructed nature of the distinction between the internal microsociological and external macrosociological factors of science; but he also faulted social realists for their failure to adhere rigorously to the symmetry principle of the Strong Programme. SSK's uncritical acceptance of the social constitution of scientific knowledge violated the symmetry principle which, according to Latour, required them to be as curious about how 'society' is constructed as they were about how 'nature' is constructed. Indeed, they needed to show how nature and society are 'co-produced' by technoscience. Thus Latour argued that not only are social interests and relations too weak and flimsy to explain the formation of technoscience, they are themselves the products of the activity they are put forward to explain. Indeed, Latour insisted that any attempt to explain science and its development in terms of either social or natural causes, or any confluence thereof, traded on the very distinction – between nature 'out there' and society 'in here' – that is itself the product of the agonistic field of negotiations that creates the objects of technoscience. The Actor-Network Model rebuffed the disciplinary hegemony of sociology not only in the analysis of laboratory life, but also in the study of the extension of that life into the wider social milieu. According to Latour, sociology was part of the problem, not the solution.

In advocating 'one more turn after the social turn', Latour called upon his colleagues to reject the modernist dualism of subject and object associated with the problematic of representation and explanation. He also enjoined them to overcome the 'Great Divide' between nature and society which, he claimed, conditioned the ways in which both sociologists and philosophers of science thought about the relation between science and society, with the former merely reversing the causal arrow posited by the latter.[40] Replacing the modernist dualism of nature and society with a postmodernist undifferentiated ontology of practice, Latour argued that nature and society, and the gulf between them, are continually constructed and deconstructed in the practices and networks of technoscience. Scientists recruit in their pursuit of knowledge and 'truth' not only people and their texts, in the manner of literary intellectuals, but also nonhumans, such as molecules, microbes and machines: they construct 'heterogeneous linkages of people and things', humans and nonhumans. Thus, according to Latour, Louis Pasteur discovered the anthrax vaccine only by learning how to coordinate the manipulation of bacterial cultures in his Paris laboratory with his persuasion of French provincial farmers worried about anthrax.[41] He had to control both microbes and men. In the pasteurization of France, as in the formation of other technoscience networks, nonhumans functioned not only as passive objects, waiting to be manipulated and represented by scientists, but as active agents contributing to the strength and resilience of the heterogeneous networks and associations that make up the world of technoscience. The 'extended symmetry' principle of the Actor-Network Model upheld the ontological equality of human and nonhuman agents: 'Neither is reduced to the other; each is constitutive of science'.[42]

Latour deployed the vocabulary of semiotics, and especially the term 'actant', in order to think symmetrically about human and nonhuman agents in the networks of technoscience. Disavowing social realism in favour of formal semiotics, Latour characterized these networks not in terms of social-historical categories, such as 'power'. 'class' or 'role', but in accord with the textualist idea that the material world has no significance without linguistic form and the semiotic claim that the analytical principles of structural linguistics are equally applicable to material signs. On this account, 'actants' are material agents, both human (e.g. 'Pasteur') and nonhuman (e.g., 'microbes'), given linguistic form. 'Actants' are roles in a narrative, and the job of the semiotic historian of science is to trace, through past texts and commentaries, the ways in which such signifiers are 'inserted into a developing story that wove together a coherent and yet heterogeneous network of entities and agencies'.[43] While the formalist and textualist sensibilities of Latour's semiotic analysis deterred historians interested in the real-world activities of human beings, his tendency to mix textual analyses of semiotic 'actants' with substantive narratives of scientists' historical actions made his analysis of technoscience more attractive to them. Thus they encountered in the narrative dimensions of Latour's analysis voluntarist models of agency based on Machiavellian and Hobbesian accounts of human nature and Nietzschean and Foucauldian concepts of power, which depicted major scientists as 'animated by a will to power and domination' and their readers and prospective allies losing or preserving their will to resist this power.[44]

Latour's notion of the agonistic, power-purveying networks and associations of technoscience augmented the picture historians of science derived from Foucauldian sources of the role of disciplinarity, institutionalization, and standardization – or 'regimes of construction' – in the making and diffusion of scientific knowledge. More specific to Latour's influence on the history of science was his stress on the importance of material agency and the role of 'quasi-objects', or 'mediating objects', in the construction of scientific knowledge. Latour acquired the concept of a 'quasi-object' from Michael Serres's work and used it to express his sense of the melding of nature and society brought about by the heterogeneous networks of technoscience. Viewed as 'quasi-objects', the instruments and objects of science and technology, such as air pumps and steam engines, electric telegraphs and chemical balances, DNA and oxygen, quarks and resonance hybrids, are neither purely natural nor wholly social. Thus, while they are more social, fabricated and collective than 'the hard parts of nature', they are also too 'real, nonhuman and objective' to be mere 'projections' of social relations.[45] According to Latour, historians of science need to rethink the role of objects and instruments in the development of science, viewing them as machines that mediate between nature and society in a way that 'welds a collective, culture, or community'.[46] In contrast to Bachelard's philosophy of 'pneumo-technics', which treated scientific instruments as 'theories materialized', Latour's quasi-objects registered the heterogeneity and flexibility of the 'specific local settings ... in which phenomena and hardware are produced and interpreted'. Whereas Bachelard upheld a modernist view of scientific instruments as stable vehicles of the global rationality of science, Latour adumbrated the postmodernist vision of contingent and fleeting materializations of local and specific knowledge claims.

Latour's sense of 'the importance of *metrology*, the enterprise that works to secure the compatibility of standards of measurement in different locations', informed historical studies of some of the 'regimes of construction' involved in the standardization of measurements, machines and physical constants in the sciences. But these studies also challenged Latour's view of 'metrological networks as ramifying from single sources' and establishing their hegemony over neutral terrains'.[47] They stressed instead the need for a more flexible interpretive model which allowed for greater variety in the constitution of networks; recognized the existence of multiple and contending networks with their own 'centres of calculation'; and allowed for the interaction between the networks of technoscience and more traditional social and political structures. More sensitive and nuanced applications of the Actor-Network Model to the history of science suggested that standardization in science is not a uniform process of technical and institutional consolidation, but a compromised, unintended and unforeseen product of conflicting systems of technical expertise operating in contested political landscapes.

A number of Anglo-American historians and sociologists of scientific knowledge registered more general, philosophical concerns about Latour's Actor-Network Model, finding the notion of nonhuman, or material, agency implausible, scientistic and whiggish, and stressing the inability of constructivist accounts in general to do justice to the role of the resistance or 'recalcitrance', of nature and society to the hegemonic practices of scientists and technologists. While critics of Latour's semiotic account of agency recognized the importance of parallels and a complex '*intertwining*' between human and material agency in the modern world, they rejected any notion of an 'exact symmetry', equivalence, or interchangeability between humans and machines, citing the traditional view that machines are completely lacking in the kind of '*intentionality*', or purposive, future-oriented behaviour, characteristic of human beings. Another line of criticism drew attention to a historiographical dilemma created by the attempt to treat material agents in their own terms, and not sociologically, as the products of human agents.[48] Since only scientists have the instruments and conceptual resources to identify authoritatively the properties of material agents, historians and sociologists who attempt to describe the intrinsic properties of material objects and their roles in the resolution of scientific controversies must favour one side of the debate, namely the winning side. Thus the so-called 'extended symmetry' thesis championed by Latour not only ends up violating the symmetry principle of the Strong Programme, which requires analysts to remain neutral when describing scientists' competing accounts of the material world, but also smuggles in retrospective and whiggish sensibilities anathema to the sociology of scientific knowledge. Finally, a slew of historians, philosophers and sociologists of scientific knowledge faulted the constructivist sensibilities associated with the Actor- Network Model for failing to provide '*a dialectic of resistance and accommodation*', wherein theoretical, phenomenal, material and institutional constraints frustrate the best laid plans of even the ablest scientists and technologists, requiring significant revisions in their original goals and interests, as well as in their material and institutional instantiations.[49]

Science as Practice

Andrew Pickering responded to the problems of agency and objectivity in constructivism with a model of 'science-as-practice', based on a postmodernist reformulation of Kuhn's 'conception of normal science as a process of open-ended, exemplar-based modeling'. Pickering's notion of a 'teleological principle' inherent in the practice of science rendered otiose Interest Theory's search for extrinsic interests to explain the development of science; and his notion of material agency as *'temporally emergent* in practice' removed the charge of retrospective realism and scientism levelled at the Actor-Network Model of material agency.[50] Linking the temporality of scientific practice to an open-ended process of 'delicate tuning', or reciprocal adjustments between material and human agency, Pickering argued that the future properties and contours of the material agents, or machines, of science are never known in advance. Given that sociologists are interested in 'the *real-time* understanding of scientific practice', and not its retrospective reconstruction, they are in no worse shape than scientists when it comes to identifying the properties and powers of material agents. Thus contrary to the claims of its critics, the constructivist account of material agency can escape the charge of retrospective scientism without 'returning to the SSK position that only human agency is involved' in scientific practice.

Rejecting SSK's exclusive focus on human agency, Pickering also incorporated the modernist notion of 'science-as-knowledge' into a *'performative* image of science'. He claimed that the sociology of scientific knowledge could be rescued from the 'methodological horrors' of 'the reflexive approach to science studies' only if it replaced the 'representational idiom', which views scientists as 'disembodied intellects making knowledge in a field of facts and observations', with the performative view of science as 'a field of powers, capacities, and performances, situated in machinic captures of material agency'. According to Pickering, science involves the 'continuation and extension' of the everyday business of coping with the active forces and agencies of nature, and *'machines'* are crucial to how it does this: 'Nature is continually *doing things',* and science is continually developing machines to cope with the consequences of what it does. Viewed from a performative perspective, the representational dimension of science is 'threaded through the mechanic field of science', and can be understood only if science studies moves away 'from a pure obsession with knowledge and towards a recognition of science's material powers'.[51]

Pickering's concept of the 'mangle of practice' identified 'scientific *culture'* with the 'field of knowledge ... [and] ... the 'made things' of science' and 'scientific *practice ... as the work of cultural extension'.* Pickering is interested in the everyday 'practices' of science not in themselves, but only insofar as they 'are among the resources for scientific practice and are transformed, or transformable', by the 'constant *telos'* of that practice, which is the achievement of *'coherence'.* Stressing the temporality of scientific practices, Pickering argued that a stable configuration of science's multiple practices – intellectual, instrumental, institutional and social – is achieved through 'the reciprocal tuning of machines and disciplined human performances'. This process takes the form of 'a *dialectic of resistance and accommodation',* where resistance denotes the frustration of 'an intended capture of agency in practice' and accommodation involves an active human response to resistance through the modification of original goals and interests and the machines that embodied them. The 'dance of agency', or the 'mangle of practice', involved in this dialectic produces an endless reversal of passive and active roles, in which scientists actively construct new machines, passively monitor their performances – while 'material agency actively manifests itself' – only to resume an active response in an appropriate revision of the original 'modelling vectors'. Although this dialectic is dynamic, endless and 'unpredictable', a scientist usually acquires enough confidence at moments of consolidation and coherence to 'provide an argument, a reason to believe in his findings'. Pickering insisted that 'to speak of coherence here is ... a way of explicating the 'logic of science' (not of denying it any role)'. In this manner, he subsumed modernist models of the 'statics of knowledge' (the 'taken-for-granted statics of theory and evidence') under a dynamic vision of 'science-as-practice'.[52]

But Pickering's 'logic of science' is decidedly context-laden; it betokens an 'ultra-relativistic image of science', in which goals, resistances and accommodations are seen as fleeting, 'radically situated' moments of coherence in the development of a scientific culture which, though marked by detectable patterns, is ultimately contingent and unpredictable. Pickering's relativistic sensibilities readily morphed into a postmodernist assault on the problematic of normativity when he expressed his preference to 'stop speaking of relativism' altogether and 'to speak instead of the full-blown '"historicity" of science', thereby emphasizing how all its parts 'are bound to the wheel of what happened'. In a similar postmodernist vein, Pickering rejected the modernist problematic of realism, which concerns the 'correspondence' of the theoretical 'terms' of science' to the 'furniture of the world', in favour of the performative philosophy of 'pragmatic realism', which asked '*how the connections between knowledge and the world are made*'. Pragmatic realists used the term 'realism' to 'point to a constitutive role for 'reality' – the material world – in the production of knowledge', no matter how mediated by the culture and practice of science that role might be.[53]

The same postmodernist and deconstructionist sensibilities informed Pickering's notion of agency. Pickering's account of agency emphasized the '*parallels*' and 'constitutive *intertwining*' that exist between human and material agents, calling attention to the repetitive quality and temporal emergence common to both of them and to the way in which 'they collaborate in performances'. Presenting this account as a posthumanist vision of agency, Pickering replaced the modernist 'black-and-white distinction of humanism/antihumanism', evident in the clash between the humanism of sociologists and the 'antihumanist idiom' adopted by 'scientists and engineers', with a '*posthumanist* space', in which '[t]he world makes us in one and the same process as we make the world'. The deconstructionist sensibilities of postmodernism also influenced Pickering's sense of the temporality of scientific practice. While the contrast he vividly drew between his own interest in the temporality of scientific practice and the emphasis given by science studies to 'atemporal cultural mappings' may seem to distance Pickering from the postmodernist problematic of the spatialization of reason, his 'basic sense of emergence, as a sense of brute chance happening in time', only reinforced the postmodernist tendency to randomize history.[54] While interesting and significant, Pickering's 'engagement with the issue of temporality' fails to provide historians with a balanced sense of necessity and contingency needed to offset postmodernist nominalism with a just recognition of the enduring dimensions, as well as the fleeting moments, of history. Pickering's notion of science-as-practice is, in many ways, the apotheosis of postmodernism in science studies, with all the costs (and benefits) that cultural movement bequeathed to the historiography of science.

Leviathan and the Air-Pump as Interpretive Paradigm

The works of Steve Shapin and Simon Schaffer provided an important bridge between the sociologists' abstract models of science and the concrete concerns and interests of historians of science. Their joint study of Robert Boyle's 'experimental life', *Leviathan and the Air-Pump*, quickly became a classic in the field, providing historians with an exemplar of the sociological paradigm. Taken as a whole, their work testifies to the transition in the sociology of scientific knowledge from the modernist view of science-as-knowledge to the postmodernist idea of science-as-practice. It provided historians of chemistry with general guidelines and specific suggestions about how to approach eighteenth-century chemistry and the Chemical Revolution.

Prior to the publication of *Leviathan and the Air-Pump*, Shapin functioned as the historical point man for SSK in general and the Edinburgh School in particular, drawing attention to the 'social use', or 'instrumental', model of the social constitution of science. However, Shapin's early studies of British science in the seventeenth, eighteenth and nineteenth centuries stayed within the narrow confines of the Mertonian framework. Predicated on the assumption that '*science as people think of it and as they use it* is every bit as historically important as science as scientists conceive of it', these studies said next to nothing about the cognitive content of science or the way in which the social use of science influenced the formation of scientific knowledge.[55] It was only in the late 1970s and early 1980s, when he examined nineteenth-century phrenology and the seventeenth-century Leibniz–Clarke Correspondence, that Shapin articulated SSK's 'ostensive' model of science and society, upholding the modernist notion of science-as-knowledge and its explanation in terms of pre-existing professional and social interests.[56] At the same time, however, Shaffer was gesturing, in his historiographical reflections, towards the postmodernist notion of science-as-practice and the constitutive role of scientific practice in the formation of society. Thus he characterized the 'specificity' of eighteenth-century natural philosophy and Priestley's chemistry not by reference to any 'system of concepts', but in terms of the experimental and discursive 'practices' in which Priestley and other practitioners of natural philosophy used rhetoric and machines, such as 'the air pump, condensing engine, and electrical machine', to gain and maintain the assent of audiences to 'self-styled matters of fact'.[57] The melding of Schaffer's constructivist sensibilities with Shapin's 'social use' model gave *Leviathan and the Air-Pump* its distinctive perspective on how Boyle's experimental philosophy did not reflect or produce society in Restoration England, so much as interact with, or 'cultivate', emerging social interests.

Leviathan and the Air-Pump explored the nature and historical origins of 'experimental practices' in science. It deconstructed whiggish 'canonical' accounts of 'Robert Boyle's researches in pneumatics', dismissing them as 'member's accounts' produced by historians who share the culture Boyle founded and treat his results as 'self-evident'. Unfortunately, 'the self-evident method' closes off important questions about the historical origins and epistemological status of 'a culture's routine practices', which can be asked and adequately answered only by 'playing the stranger'. Unlike the 'genuine stranger', who is 'simply ignorant' of a culture's beliefs and practices, the historian or anthropologist who plays the stranger is in the advantageous position, compared with members immersed in a culture's beliefs and practices, of knowing 'that there are alternatives to those beliefs and practices'. The historian acquires this superior knowledge through 'the identification and examination of episodes of *controversy* in the past' which involve disagreements over beliefs, practices and values subsequently taken as canonical. Focusing on the controversy between Boyle and Thomas Hobbes, who rejected not only Boyle's experimental results, but also the very idea of experimental knowledge and the social uses Boyle made of it, Shapin and Schaffer did not 'rewrite the clear judgment of history' – Hobbes lost the debate in the English scientific community – but showed that 'there was nothing self-evident or inevitable' about it. The judgement that favoured Boyle was the product of a negotiated settlement, which the historian can understand only by stressing 'the fundamental roles of convention, of practical agreement, and of labor in the creation of the positive evaluation of experimental knowledge'.[58]

Leviathan and the Air-Pump is a rich blend of SSK and 'the work of [the] British and French microsociologists of science', as well as that of the pioneering Fleck. Following the path of social realism, part of the narrative line of *Leviathan and the Air-Pump* moves from the external social context of science to its internal cognitive content, showing how the new culture of experimental natural philosophy resulted from transferring to the laboratory the conventions, codes and values of gentlemanly behaviour designed to ensure civil peace in Restoration England. The other, microsociological line of inquiry showed how the wider dissemination of disputes involved in the production of 'matters of fact' in Boyle's laboratory shaped society and its demarcation from science. Shapin and Schaffer also deployed 'Wittgenstein's notions of "language games" and "forms of life"', as well as Foucault's account of the power-knowledge nexus, in their concept of 'scientific method as integrated patterns of *practical activity*' in science *and* society. Thus they argued that 'solutions to the problem of knowledge are embedded within practical solutions to the problem of social order, and that different practical solutions to the problem of social order encapsulate practical solutions to the problem of knowledge'.[59] Shapin and Schaffer identified three kinds of 'technologies' Boyle used to facilitate elsewhere the replication of phenomena embedded in the technical and social practices of his own laboratory. Boyle's simultaneous production and communication of knowledge involved a *'material technology'* embedded in the construction and circulation of the airpump through the acquisition and transmission of the requisite material skills and devices; a *'literary technology'*, which provided an effective and persuasive way of making the phenomena produced in the air pump known to those who were not direct witnesses of them; and a *'social technology'*, which deployed prevailing assumptions about gentlemanly 'trust' to define how natural philosophers should deal with each other and evaluate knowledge claims.[60] Boyle's production of 'matters of fact' involved the proliferation of trustworthy 'direct witnesses' (social technology) to the correct performance and replication of the air pump (material technology) and the creation of 'virtual witnesses' through the deployment of a literary technology designed to create in the minds of distant readers the impression of 'being there'.

Boyle used these technologies not only to win acceptance for particular factual claims, but also to erect an important boundary between 'fact' and 'hypothesis', between genuine experimental knowledge which carried with it 'moral', if not apodictic, certainty, and the realm of mere interpretive guesswork and speculative opinion which exploited uncertainty to breed dissensus. While Boyle offered the procedures of experimental philosophy, which emphasized the achievement of consensus among suitably qualified individuals, as a source of social cohesion in the chaotic aftermath of the English Civil War, Hobbes regarded as divisive, rather than cohesive, Boyle's model of civil society as a balance of independent powers governed by gentlemanly ties of trust and toleration. Rejecting Boyle's model of fallible knowledge and consensual order, Hobbes argued that deductive certainly in science and genuine freedom and security in society required the deployment of infallible means of persuasion and demonstration grounded in incorrigible, not consensual, sources of truth and authority. Thus, while Boyle invented the laboratory as the rightful source of experimental knowledge, Hobbes invented Leviathan, or the State, as the only legitimate source of social order and political rights. According to Latour's reading of *Leviathan and the Air-Pump*, the conflict between Boyle and Hobbes generated 'the modern constitution of truth', which demarcated science from society, or 'the *representation* of things through the medium of the laboratory' (Boyle's work) from 'the *representation* of citizens through the medium of the social contract' (Hobbes's accomplishment).[61]

The historiographical tensions in *Leviathan and the Air-Pump* between macrosociological realism and microsociological constructivism carried over into the subsequent work of Shapin and Schaffer, and its reception in the scholarly community. In *A Social History of Truth*, Shapin glossed *Leviathan and the Air-Pump* in terms of the social realism of SSK, emphasizing the dimension of human agency and the interpretive move from the wider social context to the internal scientific content. Blending Collins's 'core-set' interest in microsystems of practitioners' trust and the Edinburgh School's concern with the effect of broader social factors on the internal content and dynamics of science, Shapin argued that the fortunes of seventeenth-century experimental philosophy hinged upon successfully 'exploiting common assumptions about gentlemanly decorum and credibility' to build 'relations of trust and credit' among natural philosophers.[62] Shapin's stance here contrasted with the more dominant constructivist theme of *Leviathan and the Air-Pump,* which focused on the interpretive move from the 'internal' world of science to the 'external' world of society, insisting that 'the Mertonian internal/external distinction' was constructed, not pregiven.[63] It was produced, or constructed, in the process of transforming the inherently private knowledge of the secluded laboratory into a form of public knowledge with universal validity. An adequate solution to the problem of construction required historians of science to examine the role of material and human agency in these liminal states of knowledge production and their extension throughout society.

Shapin himself developed some of these constructivist themes in his publications in the 1990s. Stressing the constructedness of the 'external/internal distinction', he insisted that the job of sociologists and historians of science was neither to abandon the distinction in the name of disciplinary purity nor simply to find ways of bridging a pregiven gap between science and society. Rather their job was to describe the rhetorical, technical and political tools scientific actors used to construct this and other distinctions and boundaries regarded as essential to 'what is to count as knowledge and the proper means of securing it'.[64] Shapin also translated the traditional problematic of 'knowledge from nowhere' into the language of constructivism, drawing parallels between the traditional epistemological idea of a pure mental state of knowledge and the sociological paradigm of the laboratory as 'a building set apart' from the public arena for the production of knowledge with a claim to universal validity. Noting that knowledge, like power, is inscribed in space and not in time, Shapin enjoined historians of science to describe the physical and social boundaries deployed to maintain laboratories and other sites of knowledge production as private and exclusive sources of genuinely public knowledge. Spatializing finitism's view of knowledge as local and situated, Shapin looked to the development of a 'vigorous project in the geography of knowledge' to show how epistemological, disciplinary and social distinctions map onto the physical space of the sites of knowledge production, as well as onto the physical and social spaces through which it spreads so efficiently.[65]

But Shapin was prepared to go only so far in his dalliance with constructivism. Besides defending, against the criticisms of Latour and Discourse Analysts, the legitimacy of SSK's deployment of 'sociological descriptive, interpretive, and explanatory categories', he upheld 'allegedly "modernist" or "humanist" dualisms' associated with the problematic of realism and representation.[66] Schaffer, in contrast, was more readily disposed to deploy the antirealist language of constructivism, referring to a *matter of fact* as an 'artifact' that 'posited certain forms of social organization' and dealt with 'a set of specific political problems'. On this view, Boyle's mechanical philosophy was not about nature, but 'part of a set of claims about the proper audience for natural philosophy and the proper behavior of experimenters'.[67] But the main difference between Shapin and Schaffer concerned the question of agency, with Shapin thinking exclusively in terms of human agency and Schaffer gravitating towards Latour's view of material agency. Latour maintained that in *Leviathan and the Air-Pump,* Shapin and Schaffer practised 'the anthropology of the object', making issues of knowledge, authority, reality and power in Restoration England 'turn *around the object,* around *this* specific leaking and transparent air pump'. In this vein, Schaffer went on to treat Priestley's opposition to Lavoisier's chemistry 'as a consequence of the eudiometer' and linked Lavoisier's successful response to his ability 'to learn and undermine this technology'.[68] Similarly, in his papers on Georgian Mechanics and Enlightened Automata, he noted how important it was that 'the demonstrator become habituated to the machine' and explored 'the technico-politics of the Age of Reason' in the relation between 'machinery viewed as human and humans managed as machines'.[69] By blending innovative themes based on the anthropology of the object with traditional interests in the anthropology of the subject, Shapin and Schaffer generated an impressive and dynamic body of scholarship with interesting, important, though not entirely unambiguous, implications for the discipline of the history of science. They bequeathed to historians of eighteenth-century chemistry and the Chemical Revolution a tensile interpretive framework, forged by the dynamic unfolding of the discipline of the sociology of scientific knowledge and the underlying transition from the modernist problematic of science-as-knowledge to the postmodernist notion of science-as-practice.

Summary and Conclusion

Sociologists of scientific knowledge generated a broad spectrum of models of the social constitution of science. These models instantiated forms of philosophical nominalism, which emphasized the specificity of scientific knowledge, the particularity of its practices and the locality of its places of production. This spectrum of interpretive strategies represented a movement of the discipline of the sociology of scientific knowledge from the modernist problematic of science-as-knowledge to the postmodernist paradigm of science-as-practice. Within this movement of thought, Foucault's interactionist account of the relation between science and society occupied an intermediate position between the modernist 'ostensive' model of the causal dependence of science on society, promulgated by SSK and the Edinburgh School, and the postmodernist 'performative' view, favoured by constructivism, of the coproduction of science and society by the practices of technoscience. The Interest Model was a classic instantiation of the modernist problematic of science-as-knowledge, with its early users identifying the causal interests of scientists with 'external', macrosociological interests, and its later practitioners also recognizing 'internal', microsociological skills as explanatory interests within the scientific community. Focusing exclusively on 'core sets' of internal interests within the scientific community, the Bath Relativists developed the Empirical Relativist Programme which replaced the problematic of representation and the unity of science adopted by the Edinburgh School with particularistic accounts of the practices, instruments and institutions involved in the replication of scientific experiments.

Ethnographers of science adopted ERP's focus on the microsociological and controversial dimensions of science, but sought to describe rather than interpret or explain the indexical activities involved in the mundane constructivist practices of science, Blending ethnomethodology with Derrida's material semantics of writing, Laboratory Studies, spearheaded by Latour and Woolgar, encouraged the idea that science is a rhetorical enterprise conducted in an agonistic space. But the idea of science as a rhetorical enterprise, shaped not by matters of meaning but by the materialities of signs, the techniques of inscription and the practice of writing, blended in the minds of historians of science with the hermeneutics of Reception Theory to produce an interest not only in the construction of texts, but also in their reception and assimilation by readers and audiences.

While Discourse Analysts and Reflexivity Theorists reinforced the tendency of Laboratory Studies to identify science with its rhetorical and discursive practices, other constructivists stretched the arena of construction beyond the narrow confines of the laboratory. They responded to 'the problem of construction' with ideas of the extension or circulation of 'local forms of life'. Shaped by Foucauldian notions of disciplinarity and Latour's Actor-Network Model, these responses eschewed traditional sociological categories, such as class, structure and power, in favour of a monistic ontology of agonistic associations or networks of individuals, both human and nonhuman, held together by the discursive and nondiscursive practices of technoscience. This perspective sensitized historians of science to the role of 'regimes of construction' – of disciplinarity, institutionalization and standardization – and 'material agents' in the making and diffusion of scientific knowledge. Pickering incorporated the concept of material agency into a performative model of science-as-practice, according to which scientists situate themselves at the intersections of a dynamic 'mangle' of distinct local practices, governed by a dialectic of resistance and accommodation between human and material agents and powers. This 'dance of agency' is independent of extrinsic interests and, though endless and unpredictable, produces moments of consolidation and coherence labelled 'knowledge'. Pickering's performative philosophy of pragmatic realism incorporated the relativist and antirealist sensibilities of constructivism into a posthumanist vision of agency and a postmodernist sense of the randomization of history.

Shapin and Schaffer mingled themes associated with the modernist problematic of science-as-knowledge and the postmodernist paradigm of science-as-practice in a tensile body of scholarship that provided historians of science with a powerful and influential sociological paradigm. This paradigm focused on scientific controversies and linked solutions to the problems of knowledge to practical solutions to the problem of social order. Shapin and Schaffer enjoined historians of science to identify the material, social and literary mechanisms, or 'technologies', scientists used to resolve controversies, define their discipline and delineate the boundaries and relations between science and society. They bequeathed to subsequent scholars a complex interpretive legacy shaped by the slow, tortuous transition from the humanism of modernism, in which man reigned supreme, to the posthumanist space of postmodernism, based on the emergence of man-the-machine.

6 POSTMODERNIST AND SOCIOLOGICAL INTERPRETATIONS OF THE CHEMICAL REVOLUTION

The most immediate impact of postmodernist nominalism and sociological finitism on the historiography of eighteenth-century science and natural philosophy took the form of a pervasive interest in the 'specificity' of historical practices and a devaluation of the role of global traditions in the development of science. The globalist view of the cognitive priority of research traditions construed eighteenth-century science and philosophy as a group of coherent bodies of theory and practice, composed of a set of discourses formed and unified by their allegiances to one or other of the cognitive traditions associated with the names of Leibniz, Descartes, Locke, Wolff, Kant and, above all, Newton. But sociologically inclined historians of science dismissed the 'tradition-seeking method' as 'profoundly unhistorical' and portrayed 'a great deal of diversity and a low degree of consensus' in the cognitive features of eighteenth-century science.[1] While a number of scholars drew attention to the variety in Newtonian matter theory, the importance of anti-Newtonian beliefs, and the eclectic nature of eighteenth-century scientific thought in general, Schaffer identified 'natural philosophy' as a mode of discursive and experimental practice distinct from science and philosophy. In the same vein, Bensaude-Vincent viewed the Chemical Revolution not as a specific instantiation of the platonic form of a scientific revolution, but as a local event, peculiar to late eighteenth-century France and 'inappropriately universalized' by its participants, as well as by subsequent historians and philosophers of science.[2] The historiography of specificity placed at the core of its analysis of eighteenth-century chemistry in general and the Chemical Revolution in particular the nominalist idea that universality does not flow directly from the nature of science, but is the outcome of the standardization, or circulation through society, of inherently local linguistic and material technologies deployed in particular sites of knowledge production.

Nominalist Interpretations of the Chemical Revolution

The Specificity of Chemical Theory and Practice

The need for scholars to attend to the 'specificity' of chemical theory and practice in the eighteenth century was highlighted by difficulties attending attempts to identify and delineate the research tradition(s) of phlogistic chemistry.[3] Other than the shared aim of explaining the property of inflammability in substantive terms, the various phlogiston theories promulgated during this time lacked the continuity in metaphysical and methodological presuppositions necessary to the formation of a coherent research tradition. Indeed, it was the presuppositional diversity of phlogistic chemistry, rather than an internal inconsistency or explanatory vacuity of a unified body of theory, that warranted Lavoisier's notorious observation that phlogiston 'is a veritable Proteus which changes its form every minute'.[4] Emphasizing the multiplicity and diversity of the paradigms Lavoisier confronted, Holmes challenged the longstanding orthodoxy that the 'central confrontation' in the Chemical Revolution was between Lavoisier's oxygen theory and 'the reigning paradigm' of phlogiston, descended from Stahl and championed by Priestley.[5] During the 1780s, Priestley was not a conservative defender of the Stahlian tradition, but 'the leader of a revolutionary movement in physics and chemistry', and the Chemical Revolution involved Lavoisier's challenge to Priestley's position of leadership within the 'new doctrine of airs'. Holmes's reconfiguration of the Chemical Revolution also called for a recognition of the 'specificity' of eighteenth-century 'pneumatic chemistry', which must be viewed not in its modern guise as a subdivision of general chemistry, but in its eighteenth-century grandeur as an interdisciplinary activity that encompassed physics, chemistry and medicine. Pneumatic chemistry, and not chemistry *per se*, determined the ontological domain of the Chemical Revolution.

Following Guerlac's lead, Holmes viewed the Chemical Revolution as a synthesis of continental analytical chemistry and British pneumatic chemistry. But he argued that whereas the continental tradition cleaved to a distinct disciplinary identity, pneumatic chemistry was pursued by 'people who were not identified primarily as chemists' and whose results were 'not necessarily seen by contemporaries as more particularly belonging to chemistry rather than physics, or medicine'. This was particularly true of Priestley. Indiscriminately interested in the physical, chemical and medical properties of the many airs he prepared and isolated, Priestley's phlogistic account of the 'provisions' in nature, such as vegetation and the agitation of seas and lakes, to offset the 'vitiation' of the atmosphere caused by respiration, combustion and putrefaction expressed medical and social concerns, as well as a broader, theistic view of a harmonious and benevolent nature. Priestley's phlogistic account of the complementary roles of respiration and vegetation in the balanced economy of nature loosened phlogiston from its traditional Stahlian identification with the principle of inflammability. According to Holmes, it was the resulting 'doctrine of airs', and not the traditional chemical doctrine of Stahl, that Lavoisier had in mind, and with which he aligned himself, when he referred to 'a revolution in physics and chemistry'. It was the loss of his position of 'leadership and adulation' in this revolutionary movement, rather than 'the demise of an old chemistry that he had not practised', that Priestley could never accept'. Noting that in Paris and Germany, Lavoisier's 'new chemistry' did indeed encounter the 'old chemistry' of Stahl, Holmes called for a more decentred view of 'the Chemical Revolution as a set of different kinds of events within different contexts'.[6]

Bernadette Bensaude-Vincent and Ferdinando Abbri captured the diversity and specificity of eighteenth-century chemistry in an 'interactive model' of the Chemical Revolution. Considering the fortunes of Lavoisier's chemistry outside of France and England, they argued that the Chemical Revolution was not an 'Anglo-French affair', centred on the chemistry of gases and the phlogiston-oxygen debate, but a 'European space of multiform critical discourses', in which Lavoisier contended with a variety of competing theories, besides the phlogiston theory, across domains of inquiry that included pneumatics, mineralogy, metallurgy, geology and physiology, and in a range of different scientific, cultural, national and economic contexts. Abbri thus modified the centre-periphery diffusion model of the Chemical Revolution with a 'geographical' metaphor designed to highlight the 'the significant scientific developments' that took place in the peripheral sites and 'their influence on the science of the great political centers'.[7]

But Abbri's interactionist version of the centre-periphery model remained focused on Lavoisier, and viewed his 'relationships' with the European periphery as confirmation of 'the extraordinary relevance and complexity of the French chemist's enterprise, its roots and its heritage'. John Perkins and Larry Holmes offered a more decentred articulation of the centre-periphery model however, suggesting ways of studying the responses of eighteenth-century chemists to the Chemical Revolution other than in terms of 'national differences'. Focusing on the provincial cities of Nancy and Metz, Perkins showed how a complex array of local 'social, cultural, economic, and political forces' constructed a 'public image of chemistry' in the French provinces 'as enlightened, rational, ameliorative, and utilitarian'.[8] Holmes used Collins's notion of informal 'core sets' of individual practitioners to further decentre and particularize the centre-periphery model of the Chemical Revolution when he canvassed

> the important insights we might attain if we define the 'different institutions of chemistry' and the 'individual or collective attitudes' that differentiated the new nomenclature of Lavoisier's chemistry, not along geographical boundaries alone, but also along the parameters that defined the relationships of individuals and groups to

the core discipline of chemistry that was forced to adapt to the challenges posed by Lavoisier and his followers.[9]

In a similar vein, Patrice Brett's conceptualization of Lavoisier's 'scientific sociability' focused less on the enduring identity, or structure, of his 'solitary' mind than on the ways in which he constructed 'his social and scientific image' among the different audiences and communities with which he was involved.[10]

Deconstructing the Chemical Revolution

Deconstructionist tendencies implicit in the historiography of specificity surfaced in its application to the interpretation of Priestley's science. Questioning the biographer's assumption of 'a continuity of themes through the course of an individual's life', Crosland, Schaffer, Christie and Golinski criticized postpositivist globalism for replacing the diachronic movement of history with synchronic attempts 'to encapsulate Priestley's work within a synoptic conceptual or metaphysical structure'. They replaced all talk of Priestley's enduring identity, scientific or philosophical, with descriptions of the different ways in which he constructed his scientific and social image in the communities and audiences with which he interacted. Claiming that 'there are many Priestleys', these interpretive strategies excluded any notion, static or otherwise, of Priestley as a coherent historical actor.[11] If postpositivist globalism submerged the dynamic mutability of Priestley's science beneath an overly rigid conceptual structure, postmodernist historians lost sight of its direction or developmental unity in a bewildering array of partial and fragmented practices and interventions. Either way, the historical Priestley was short-changed.

The historiography of specificity also 'deconstructed' the traditional concept of 'the Enlightenment' and postpositivist scenarios of its influence on the Chemical Revolution. In their anthology *The Sciences in Enlightened Europe*, William Clark, Golinski and Schaffer criticized traditional idealist accounts of the Enlightenment because they posited a 'single mentality', or 'spirit of the age', spreading like 'the glow of illumination' from the metropolitan 'centre' to the 'colonial' peripheries of society and culture. Calling this 'intellectual uniformitirianism' into question and stressing the need to study the Enlightenment as 'a social, rather than a metaphysical, phenomenon', Clark, Golinski and Schaffer canvassed analyses that stressed the specificity of local contexts and the inability of global concepts such as 'the Enlightenment' to deal with their geographical diversity and temporal variability. They sought not just to show that Enlightenment science was different in different places, but also to reveal how 'the construction of cultural unity in eighteenth-century Europe' involved a struggle between the 'concrete forces of unification' and 'those of local differentiation'. Adequate accounts of this struggle called for 'a series of [historiographical] reductions from the transcendental to the mundane; from the mind to the body; from humanity to society; and from the universal to the local'.[12]

Alfred Nordmann offered an analysis of the relation between the Enlightenment and the Chemical Revolution that instantiated the historiographical 'reduction' from the universal to the local. Arguing that all knowledge is local and all 'Enlightenments in the plural' are parochial, he challenged the postpositivist view that the differences between Lavoisier's chemistry and Priestley's natural philosophy can be understood as diverse manifestations of a unitary system of Enlightenment thought. Nordmann used the failure of George Lichtenberg – a representative figure of the German Enlightenment – to recognize Lavoisier – a paragon of the French Enlightenment – as a great scientist to argue that it is time to abandon a unified conception of the Enlightenment. Whereas the Germans favoured a 'procedural notion of Enlightenment', according to which 'the truth emerges by itself once it is submitted to public scrutiny', the French emphasized the need to inculcate a 'passion for truth' through techniques of representation and persuasion. Thus Lavoisier used the production of spectacular effects in public demonstration to *stage* the Chemical Revolution, 'rather than let the public enlighten itself'. In Lichtenberg's eyes, Lavoisier sullied disinterested reason with a distasteful triumphalism and hypothesis-strewn mandatory nomenclature. Where postpositivists saw a unified science and Enlightenment, Nordmann found 'incommensurable chemistries ... accompanied by incommensurable conceptions of "Enlightenment"'.[13]

Nordmann's account of the Chemical Revolution is more nominalist than Abbri's challenge to the centre-periphery model of this event. Whereas Abbri aimed at a more sensitive account of the 'multiform interactions which took place' between the centre and peripheries of European science, Nordmann decentred the Enlightenment and the Chemical Revolution entirely. After exploring the 'incommensurability' between Lavoisier's and Lichtenberg's Enlightenments, Nordmann suggested that 'a consideration of Priestley may well give us three Enlightenments and thus an apparently repulsive plural of a term that – like "truth" – permits of no plural'.[14] The problem with this suggestion is that Priestley *shared* Lichtenberg's 'procedural notion of Enlightenment' and disinterested reason, while he eschewed all forms of idealism, Kantian or otherwise, opting for a form of Lockean empiricism similar to that Condillac bequeathed to Lavoisier. In stressing the differences and discontinuities between local sites of knowledge production, Nordmann's nominalist account of the Enlightenment and the Chemical Revolution lost sight, in a dispersion of incommensurable theories and cultures, of the enduring traits of scientific change.

The historiography of 'specificity' also involved postmodernism's 'spatialization of reason', which substituted synchronic categories of spatial dispersion for the diachronic vocabulary of temporal change and development. Instead of seeing the Enlightenment as a dynamic, intellectual and cultural, break with the past, Roger Chartier found it 'in the multiple practices ... aimed at the management of space and populations'.[15] Similarly, Bensaude-Vincent portrayed the Chemical Revolution in general and the spread of the 'new nomenclature' in particular not as 'a process of change resulting from a paradigmatic shift', but as 'the expression of a "collective" will' dispersed across 'a diverse and contrasted landscape, shaped around various poles'. This vision of 'the particularities of place rather than the universality of space' replaced the old model of the Chemical Revolution as a unified and coherent process of temporal development, starting at the Parisian centre and moving out towards the cultural peripheries, with a 'polymorphous and multipolar network model', which emphasized the multiplicity and autonomy of 'local' cultures of chemistry, actively serving local interests and objectives. But instead of treating 'local particularisms' as 'symptoms of [developmental] immaturity', Bensaude-Vincent's 'historical geography' treated them as 'focal points for the concentration of resources', as 'temporary stabilizations of 'the identity of the discipline'.[16] Incorporated into this 'geographical perspective', the historiography of 'specificity' not only reduced the 'Chemical Revolution' to a mere label, or discursive device, to further the agonistic strategies of local practitioners; it also replaced the diachronic language of temporal development, essential to the idea of change of any sort, with talk of 'a European cartography for chemistry', to be used for 'mapping' local sites and the 'network channels' of communication between them. Reflecting postmodernism's loss of any sense of enduring time connecting the past to the present, the historiography of 'specificity', in some of its more radical formulations, treated the Chemical Revolution not as a dynamic process in time, but as a static network, or pastiche, of geographically distributed and differentiated events, sites or 'localities' of knowledge production.

Texts, Instruments, Practices and Audiences in Eighteenth-Century Chemistry

Rhetoric and Textual Mechanisms in the Chemical Revolution

Postmodernist discussions of texts, instruments, practices and audiences in the production and distribution of scientific knowledge informed the accounts historians of chemistry gave of the role of rhetoric and writing in eighteenth-century science and the Chemical Revolution. But not all of these accounts adhered strictly to the constructivist line, being content to hang more conventional narratives and Mertonian theses on the mantle of constructivism. The works of Crosland, Charles Bazerman and Wilda Anderson fall into this transitional framework. Reacting against postpositivist theoreticism, Crosland offered 'a practical perspective on Joseph Priestley as a pneumatic chemist', in which he claimed that Priestley's identity as 'a compulsive writer ... [brought] together his science and his theology', eventually driving 'him to experiments' and forming 'the basis of his subsequent reputation as a chemist'.[17] Crosland made no attempt to relate this bald claim to the scientific content or rhetorical function of Priestley's texts, merely using the rhetoric of rhetorical analysis in a straightforward description of certain aspects of 'Priestley's career'. Bazerman, on the other hand, provided a close-grained analysis of the 'textual mechanisms' deployed by Priestley in his *History and Present State of Electricity* to 'integrate past, present, and future work in this [embryonic] field'. But besides paying little attention to the specific historical contexts and scientific controversies surrounding the texts he analysed, Bazerman subsumed the daily controversies and dissensus emphasized by constructivist scholars under a Mertonian view of the long-term teleological emergence of cooperation and consensus in the development of science.[18]

Wilda Anderson took a step closer to a constructivist narrative in her 'literary analysis of a group of related texts, mostly about chemical subjects, from the last half of the eighteenth century in France'. Focusing mainly on Macquer's *Dictionnaire de Chymie* of 1776 and the *Méthode de Nomenclature Chymique*, published by Lavoisier *et al* in 1787, she described the 'literary procedures' – including 'arguments ... semantic fields ... literary devices ... and conventions' – used by different authors 'to transmit reliable knowledge'.[19] Although she referenced the views of Foucault on epistemes, discourse and disciplines, Bachelard's idea about disciplinary boundaries, and Derrida's *De la Grammatologie,* her analysis did not take the form of a 'deconstruction' of the texts and disciplinary distinctions of eighteenth-century science. Accepting the traditional distinction between scientists, philosophers and 'the *literateur*', her analysis zeroed in on some of the 'conventional targets of intellectual history' and the history of chemistry, namely the relation between Lavoisier's reform of the chemical nomenclature and the preoccupation of Enlightenment philosophy with the relationship between knowledge and language.

In this manner, Anderson explored the transformation of Macquer's 'philosophical chemistry', based on 'an individual operation performed on a language', into Lavoisier's 'scientific chemistry', in which 'language sets the parameters for writers'. According to Anderson, it was Lavoisier's rhetoric, not his experimental method or results, which brought chemistry from the library – where, in keeping with the Cartesian spirit of systems, it imposed order and intelligibility on empirical data – into the laboratory where the chemist, as the observer of nature, was the servant rather than the creator of science. Part of a more general transformation, in which 'the universal philosophy of the Enlightenment splits into a myriad of self-sufficient disciplines', Lavoisier's 'institutionalization of science as its own speaking voice' represented the emergence of a true discipline, in which authors take on the mantle of 'the anonymous practitioners of science

... who speak with a collective voice in the name of a neutral, objective, natural science'.[20]

Anderson's narrow concept of discourse, which was insufficient to the broader purposes and objectives of constructivism, was the source of her blinkered perspective. While Anderson glossed her analysis as a way of revealing 'the nature of the generation of knowledge as a result of discursive procedures', she identified these procedures with the 'tools of poetic innovation, of rhetoric in the largest sense, as well as the standard tools of logical exposition'.[21] Compatible with the representational model of science, this narrow, literary construal of discourse was a far cry from the postmodernist, constructivist idea of discourse as a form of performative intervention in the world. Thus, while Anderson focused on the internal literary structure and devices of some of the core texts of the Chemical Revolution, constructivism called for a consideration of the form, function and content of these texts and devices in relation to the construction of facts and the constitution of the discursive, technical, institutional and social practices which, taken together, made up the Chemical Revolution.

The Instruments and Communities of British Pneumatic Chemistry
Eschewing all ideas of the synoptic unity of Priestley's mind, Simon Schaffer's constructivist account of Priestley's role in the Chemical Revolution focused 'on issues of interpretation of texts, replication of experiments, and the winning of assent to self-styled matters of fact'.[22] Schaffer described the various roles Priestley performed – as a writer, instructor, preacher, instrument maker and merchant, natural philosopher, polemicist and political activist – in relation to the diverse 'communities of practitioners' in which he lived and worked. These 'communities' included the didactic context of the Dissenting Academies, experimental natural philosophers attached to the Royal Society and the Club of Honest Whigs, scientific-instrument makers and merchants, industrialists and natural philosophers attached to the Lunar Society, the all-important 'community of pneumatic physicians ... concerned with the relation of life, health, and atmosphere', as well as the wider political coalitions and religious congregations in which Priestley took on the mantle of 'Rational Dissent'. This nexus of practitioners and communities provided the backcloth for Priestley's 'interventions in natural philosophy', which involved the disruption of a set of discursive and experimental practices tied to a form of 'public' display, or 'theatre', with persuasive moral and political intent. This theatre endorsed the prevailing social order with displays of natural powers – especially electricity, which was 'dramatic, effective, and ... "miraculous"' – as the separate and immediate interventions of the power of God. Priestley's scientific and philosophical texts of the late 1760s and 70s broke the connection natural philosophers made between God's direct power, the performer and the established political order. These texts evoked two

'readings' among Priestley's contemporaries associated with different 'practices' and 'contexts'. In the 'practice of *performance*', radical public lecturers used the 'display of active powers to show the corrupting effects of all governments on their citizens, not to show the divine origin of the power of governments'. The other reading of Priestley's texts, based on his desire to reveal not the power, but the rationality, of God's creation, emphasized 'the doctrine of *system*, which licensed a practice of association between enlightened intellectuals, who were capable of comprehending the true system of natural and civic philosophy'. According to Schaffer, 'the struggle between radical performers and theorists of the system', between theatrical demonstrators and expert members of elite clubs and societies, shaped 'the politics of knowledge from the 1770s'.[23]

Schaffer analysed Priestley's role in the Chemical Revolution by shifting the interpretive focus from Priestley's 'concept' of a 'divine system of benevolence' to 'the social relations and practical techniques of work' adopted by the 'community of pneumatic physicians' who centred their concern with 'the relation of life, health and atmosphere' on 'Priestley's pneumatic model of the system' of nature. Priestley persuaded this community to employ his 'eudiometer', which used the diminution in volume of a mixture of common air and nitrous air (NO), to measure the purity, or 'goodness', of the atmosphere, as they struggled to formulate new environmental controls to curb the diseases ('fevers') of modern institutional populations, in hospital ships, jails, military camps and towns. Schaffer tied Priestley's identity and burgeoning reputation to the spread of '[t]he techniques of eudiometry, and the transmission of standard forms of the equipment'. During the 1770s, Priestley's eudiometric technology was developed, modified, and given new cultural and political meaning in Austria and Italy, where it was integrated into Enlightenment programmes of anticlerical reform crafted by experts and supported by the power and prestige of enlightened despots like Leopold II and Joseph II. But the failure of subsequent attempts to obtain uniform results with a standardized version of the eudiometer, and the eventual realization that the nitrous air test did not reveal all the bad qualities of an air, but only its degree of phlogistication, or lack of oxygen, meant that medical eudiometry was almost completely abandoned in the late 1780s. Still the atmosphere remained a key site of medical management and social control and, according to Schaffer, this wider social, medical and policy context placed the eudiometer at the core of pneumatic chemistry and made mastery of its technique the *sine qua non* of a good chemist.[24]

Consistent with the stress on 'material agency' evident in his discussion of the role of air pumps, condensing engines and electrical machines in natural philosophical 'displays', Schaffer claimed that the 'fight' over the 'right way of making eudiometric trials ... touched on all the social and intellectual troubles of pneumatics'. Treating '[t]ransmission of skill with the test as a vector for the transmission of Priestley's theoretical language and phlogistic cosmology', Schaffer placed Lavoisier in the more defensive role of defending himself against Priestley's 'war against French chemists ... launched as a consequence of the technology which adequately embodied his phlogistic cosmology'.[25] Focusing on the 'tragic-comedy of oxygen discovery in 1774/75', Schaffer noted that, following Priestley, Lavoisier used the nitrous air test to distinguish between respirable and nonrespirable airs, which he interpreted along the same phlogistic lines as Priestley. Treating respiration as the release of phlogiston from the blood into the lungs – a process of phlogistication – Priestley's phlogistic cosmology identified the respirability, or 'virtue', of an air with its degree of dephlogistication, as measured by the diminution in volume with nitrous air. But Priestley's eudiometric techniques and practices also embodied his theistic assumption that common air is the best possible air for human respiration, which led both he and Lavoisier to assume that common air diminished more than any other kind of air in the nitrous air test. Initial eudiometric trials with the air obtained by heating *mercury calcinatus per se* supported this conclusion; and it was only after subsequent trials showed that the new air diminished more than any other air in the nitrous air test that Priestley and Lavoisier recognized it as a new kind of air, which Priestley dubbed 'dephlogisticated air' and Lavoisier called 'oxygen'. According to Schaffer, it was only by disengaging 'from the technology which he had painfully learnt to use' that Lavoisier, 'in a range of experiments on the synthesis of atmospheric air of spring 1776', was able to show that respiration involved the fixation of oxygen in the blood, not the release of phlogiston from it. Apparently, Lavoisier needed to 'learn and undermine' the 'technology of the nitrous air test for phlogistication' in order 'to build a new theory of air, respiration, and life'.[26] Schaffer provided little or no textual evidence to support his prioritization of eudiometry in the development of Lavoisier's research programme. Indeed his valorization of 'material agency' resulted in a narrow, almost trivializing concept of the Chemical Revolution, focused on a specific instrument and the skills of local communities of practitioners in using it. His account left no place for the transformative unity of a larger historical event, treated by previous historians as epochal and revolutionary in its range and significance.

The Public (Enlightenment) Discourse of Chemistry

Jan Golinski incorporated Schaffer's analysis of the material technology of the Chemical Revolution into a scenario that gave priority to human agency over material agency, stressing not so much the practical interaction between humans and machines as their cultural encoding in systems of meaning and communities of practitioners. Following Geertz, who called upon historians and anthropologists to explore the diverse conceptual structures and communities of meaning that create human events and behaviour, Golinski explored eighteenth-century chemical discourse in relation to the different 'interests' and 'identities' of the groups and individuals who used it.[27] Analysing a series of texts from Boyle to Dalton, Golinski sought to specify 'the social relations between chemist and audience' and 'the symbolic discourse and cultural codes' used to cement this relation. Golinski treated the language and instruments of chemistry not as unproblematic means of representing reality, but as sets of specific rhetorical, aesthetic and dialectical devices used by individual scientists to construct and control scientific communities in accord with their aims and interests.

In *Science as Public Culture: Chemistry and Enlightenment in Britain, 1760–1820*, Golinski ran together two distinct senses of the term 'science as public culture', one derived from the constructivist 'problem of construction' and concerned with the scientific community's validation of the claims of individual practitioners, and the other based on Habermas's view of the rise of the 'public sphere' and concerned with popular science and its consumption by the lay public. Considering the problem of construction, Golinski cited *Leviathan and the Air-Pump*, according to which 'the formation of the experimental way of life ... involved the constitution of relatively private spaces for experimentation and ... declaredly public spaces for communicating its findings'. He thus argued that the private knowledge and phenomena of eighteenth-century chemists became the public property of the scientific community to the extent that they succeeded in constructing networks of individuals, or 'audiences', held together by the circulation of scientific texts, laboratory instruments and skills. Stressing Latour's 'performative' point that since the seventeenth century 'experimental natural philosophy contributed to the remodeling of public life as a whole', Golinski's solution to the problem of construction merged with the second sense of 'science as public culture', which concerned not science *as* culture, but science *in* culture, and which denoted the way in which wider social relations could be reorganized through the making of facts, the replication of experiments and the circulation of instruments and texts.[28] Golinski's account of the formation of eighteenth-century chemistry as a public culture ran together considerations of the formation of specialist chemical communities and the construction of wider public audiences for chemistry.

Golinski viewed the 'Enlightenment' as an appropriate unifying mechanism in the transmission of knowledge from one locale to another locale 'provided it is understood as a concrete historical process and not as the diffusion of disembodied ideas'. Eschewing the extreme nominalist idea that the Enlightenment lacked 'an overall identity', Golinski focused on 'the particularity of the local circumstances of the *reception* of Enlightenment culture, rather than on the general feature of that culture which gave it widespread appeal'. He focused on 'the experience [rather than the idea] of enlightenment'. This focus was shaped in part by the earlier studies Roy Porter, Larry Stewart and Simon Schaffer made of 'the relation between natural philosophy and Enlightenment public life' in Britain. But whereas these scholars dealt with the relatively amorphous field of experimental natural philosophy, Golinski focused on the emerging discipline of chemistry. Claiming that eighteenth-century chemistry is best understood in terms of its 'intrinsic' features, rather than in terms of 'extrinsic' factors, such as matter theory and natural philosophy, which 'influenced' but did not 'constitute' it, Golinski (and Christie) eschewed Schaffer's focus on the relation between Priestley's chemistry and eighteenth-century natural philosophy; they called upon historians to delineate eighteenth-century chemistry as 'a disciplinary practice with its own identity'.[29] Recognizing the mutability of chemistry's disciplinary identity, Golinski explored the different ways in which its identity and boundaries were defined and redefined – in texts, lectures and the proliferation of 'phenomeno-technics' – in diverse 'public realms' constituted by different visions of the Enlightenment.

Golinski argued that while the Scottish Enlightenment provided a narrow, conservative context for the public definition of chemistry as a distinct discipline, Joseph Priestley and the English provinces forged a new egalitarian and entrepreneurial Enlightenment, encompassing a greater range and diversity of audiences for the new science. Embodying the democratic values of the English Enlightenment, Priestley's rhetoric centred on plain 'historical' descriptions – devoid of a specialized language and nomenclature – of simple, easily replicable experiments. It was designed to stimulate the participation of audiences in the production of useful and pious knowledge of God's creation, without imposing any system of authority, institutional or theoretical, upon them. Enlightened pneumatic chemistry and medicine reached its culmination in Britain in the Pneumatic Institution, founded by Thomas Beddoes in Bristol in 1797 in order to explore the effects of different kinds of air on a variety of common ailments. Unfortunately, the hilarious goings on with nitrous air (laughing gas) played into the hands of conservative forces, including Edmund Burke and the editors of the *Anti-Jacobin Review*, who viewed the Enlightenment as a whole as a period of follies, illusions and vain enthusiasms which needed to be brought to a speedy end. This conservative backlash provided the context in which Humphry Davy developed a new model of chemistry based on Lavoisier's elitist programme for reforming the theoretical language, experimental practice and institutional organization of chemistry as a public science. A conflict thus arose at the end of the eighteenth century between an older model of chemistry, associated with Priestley and based on readily accessible doctrines, simple instruments, replicable experiments and a participatory audience and public, and the new French model tied to the deployment of powerful instruments, as exemplified in Davy's voltaic pile, complicated experiments, such as Lavoisier's experiments on the composition and decomposition of water, a specialized theoretical language, and a passive audience and public receptive to the disciplinary dictates of experts and specialists. Capturing in the language of the sociology of scientific knowledge distinctions drawn by postpositivist scholars between Lavoisier's authoritarian, demonstrative science and Priestley's egalitarian, empirical chemistry, Golinski placed this conflict at the core of the Chemical Revolution.[30]

The Water Controversy: Revolution versus Tradition

The constructivist valorization of the study of controversies in science distinguished Golinski's interpretive focus on the conflict of the 1780s from that of his postpositivist predecessors who, interested in the formation of Lavoisier's revolutionary ideas in the 1770s, regarded the process of public persuasion Lavoisier and his allies undertook ten years later 'as an aftermath to the main events'. In contrast, Golinski argued that 'Lavoisier's system as a whole was articulated in this debate', which hinged upon his 'radical break' with 'the "*longue durée*" of the eighteenth-century chemical laboratory'.[31] Armed with the oxygen theory of combustion, respiration and acidity, Lavoisier took the fight to phlogiston. He won over his French colleagues, Berthollet, Baumé, Guyton and Fourcroy, with his 1785-public experiment on the analysis of water into its constituent gases (oxygen and hydrogen) and their synthetic recombination to form water again. Fortified by the support of his French colleagues, Lavoisier launched a broader assault on the phlogiston theory in *Refléxions sur le Phlogistique*, published in 1786, and an even more effective propaganda campaign, based on a new nomenclature and system of chemistry and organized around the publication of the Méthode de Nomenclature Chimique in 1787 and the *Traité Elementaire de Chimie* in 1789.

Focusing on the reception of the new chemistry in Britain, Golinski drew attention to postpositivist accounts of the phlogistic renaissance in Scotland and England in the 1770s and 80s, where Adair Crawford, P. D. Leslie and James Hutton used phlogiston in their explanations of numerous physiological and geological phenomena, and Cavendish, Kirwan and Priestley treated it as a material substance subject to the same laws of chemical affinity as any other substance.[32] As will be seen more fully in the next chapter, British phlogistic chemists were not only confident in their own empirical and theoretical accomplishments; they also had no problem in dismissing the oxygen theory as an immature, ill-conceived and easily refuted alternative to phlogistic explanations of combustion and calcination. But the 'discovery' of the composition of water in 1783–4 removed the more glaring empirical anomalies of the oxygen theory and produced a dramatic reversal of scientific fortunes. In his large-scale public experiments on the synthesis of water from hydrogen and oxygen and its subsequent decomposition in red-hot iron tubes, Lavoisier sought to 'demonstrate' the compound nature of water and the validity of the oxygen theory. Since the doctrine of the 'decomposition of water' brought the oxygen theory from the brink of cognitive extinction to the high ground of a general system of chemistry that threatened 'the reality of phlogiston', the British phlogistic chemists had no alternative but to launch a major assault on Lavoisier's experiments on the analysis and synthesis of water. The empirical dimensions of the ensuing dialectic will be considered in the next chapter; the following account will focus on the nonempirical features of the Chemical Revolution emphasized by sociologists of scientific knowledge and their postpositivist predecessors.

Whereas postpositivist historians linked the debates between Lavoisier and his British interlocutors about the 'facts' and 'interpretations' of his experiment on the analysis and synthesis of water to a wider framework of metaphysical, epistemological, methodological and political ideas, Golinski sought to show how they were really 'debates on how scientific practice is to be carried on'.[33] According to Golinski, when British chemists questioned the facts and interpretations Lavoisier put forward in these experiments, they were also asking more fundamental questions about the instruments Lavoisier used to generate them, the experimental set-up in which he embedded them, the language he used to describe and communicate them and the audience he evoked to witness and replicate them. Seeking to model chemistry on a form of proof derived from geometry and algebra, Lavoisier's famous 1785-experiment took the form of a 'demonstrative experiment', in which precise and accurate measurements revealed the invariance of reactants and products integral to the idea of a (chemical) equation. This method of proof required the use of expensive instruments and complex apparatus, carefully calibrated and designed to eliminate extraneous impurity-effects. The result was a mode of experimentation dependent upon the refined skills of disciplined practitioners, witnessed directly and indirectly by an audience of passive recipients of expert knowledge, and described in the language of a clearly articulated theoretical system.

British chemists like Priestley, Kirwan, James Keir and William Nicholson accused Lavoisier of laying claim to an implausible degree of exactness and precision with his admittedly superior balance. They questioned the relevance of such claims to the debate between the Phlogistians and Antiphlogistians by citing the presence of imponderable substances – phlogiston and caloric respectively – on both sides of the growing divide. They also questioned the plausibility and reliability of an experiment that, as Priestley put it, required 'too much correction, allowance, and computation' in deriving empirical results that supported the oxygen theory. Priestley regarded Lavoisier's expensive and complex demonstrative experiments as an assertion of private rights and usurped authority, in the manner of the Reign of Terror, over 'what should be the public apparatus of experimental philosophy'.[34] What Lavoisier perceived as an improvement in the demonstrative power of chemistry, Priestley and the English chemists viewed as an attack on the autonomy of reason and the independent judgement of individual practitioners performing their own experiments and sharing their results in an active community of epistemological equals. They preferred an inductive chemistry based on numerous simple experiments performed by a variety of practitioners to a few, difficult, demonstrative experiments performed by a theoretical elite. As Larry Stewart argued, the English preference for 'simple and affordable instruments ... suggested a shattering of elitism' which 'attracted the attention of reformist elements in numerous taverns, coffee houses, clubs and societies' in late eighteenth-century London, caught in the cross-fire between the 'French Republic' and 'American revolt' on the one hand, and on the other, 'repeated assertions of monarchy, oligarchy, and political stability'.[35]

Similar concerns surfaced in the controversy that followed the publication of the new nomenclature in 1787, which British chemists rejected because it served the hegemonic interests of a particular theory rather than the shared concerns of a community of individual experimentalists devoted to a purely factual discourse. Priestley's 'radical vision of language as the common property of a democratic and egalitarian scientific community' was also a variant on a century-long concern of eighteenth-century English linguists with the notion of 'common usage', or 'custom', which they used to oppose the importation of French words, linguistic principles and institutional practices into the English language. Golinski concluded that although the French authoritarian model of chemistry prevailed in the reactionary political climate of the 1790s, Priestley's more egalitarian sensibilities lingered on in the preference of English chemists for a provisional, factual form of knowledge; it also persisted in Martin van Marum's important replication of Lavoisier's results in relatively simple experiments on the analysis and synthesis of water, in which he balanced the desire for precision and accuracy with considerations of the ease and cost of replication. Golinski thus linked adequate interpretations of the Chemical Revolution to 'an understanding of how science is constructed as "public knowledge"' based on a combination of macrosociological considerations of the 'institution, popularization, and other external aspects of science' and microsociological analyses of the 'techniques, instruments, and discourses' deployed by the relevant 'community of practitioners'.[36]

Instruments and Audiences in French Chemistry

Lissa Roberts and Bernadette Bensaude-Vincent developed constructivist analyses of the Chemical Revolution which related Lavoisier's use of the ice-calorimeter and the balance to the rhetorical, experimental and theoretical practices that accompanied them and the audiences for which they were intended. The proliferation of imponderable substances – in the form of electrical and magnetic fluids, phlogiston and the matter of heat, or caloric– in physics and chemistry created a special problem for the quantifying spirit of the eighteenth century, especially in the burgeoning science of heat. Since, by definition, the imponderable matter of heat, or caloric, was not detectable by the balance, Black and Lavoisier had recourse to other physical instruments to measure its 'chemical' effects. While they used the thermometer to measure the degree, or intensity, of heat of a body, which they identified with the density of the matter of heat contained in it, Lavoisier and Laplace invented the 'ice-calorimeter' in order to measure the amount of heat involved in a chemical reaction. This device determined the heat released during a chemical reaction by measuring the water produced in a surrounding chamber packed with ice, which was in turn surrounded by an insulating chamber also packed with ice. Roberts showed how the identity and function of the new instrument changed dramatically with changes in the 'integrated network' of theoretical, experimental and rhetorical 'practices' in which it was used and embedded.[37]

In accord with Heidegger's practical hermeneutics, Roberts blended the views of Galison, Rouse, Pickering, Collins and Holmes into an eclectic interpretive framework which emphasized the constitutive role of contextually shaped and variable laboratory instruments and language in the formation and development of scientific knowledge and inquiry, understood as both cognitive and operative. Highlighting 'the changing contexts that gave shape to the instrument's use and meaning', Roberts noted that Lavoisier and Laplace introduced the calorimeter in 1783 as a 'machine', six years before naming it. Introducing their machine 'to an audience wedded to the public rhetoric of polite theoretical agnosticism', and wishing to avoid theoretical controversy with Adair Crawford and William Irvine about the nature of heat and its role in chemical reactions, they emphasized the experimental usefulness of the machine, presenting it as a superior source of reliable phenomena, 'indifferent to causal interpretations'.[38] But when Lavoisier described the same device in his great work of theoretical syntheses and propaganda, the *Traité*, he named it the 'calorimeter' and used it as instrumental confirmation of the caloric theory of gases, and as a way of weaving 'the materiality of caloric into the fundamental structure of both his theory and his experimental practice'. Attached by Lavoisier to 'the new chemistry's assertions of systematic authority', the calorimeter evoked not only concerns about its effectiveness as an experimental tool, but also the 'larger issue of how the very practice of chemistry should be disciplined'.

Once again, Lavoisier's British interlocutors accused him of abandoning the inductive, egalitarian mode of experimental inquiry in favour of a 'coercive discourse', which 'wielded quantitative data as a battle-axe of authority'. As the Chemical Revolution eclipsed the period of theoretical agnosticism in the early 1780s, the calorimeter changed 'from a tool of empirical investigation to a demonstration device', invoking a field of knowledge and practice rather than a set of reliable phenomena. By 1793, however, Lavoisier recognized, in a context of growing political pressure and a sense of foreboding doom, the problematic, time-consuming nature of calorimeter experiments and 'freed his systematic views from their dependence on the calorimeter's technical merits', letting it stand as an 'emblem' rather than an indispensable experimental or demonstrative tool of the new chemistry. By recalling these 'local contexts' and the 'actively mediative role of instruments and language' in them, Roberts supported the constructivist thesis that 'the fundamental legacy of the Chemical Revolution' was not a theoretical or methodological innovation, but 'a new way of conceiving and articulating our relation to the world of scientific investigation'.[39]

Lavoisier and the Construction of the Chemical Revolution

Bernadette Bensaude-Vincent offered in *Lavoiser: Mémoires d'une Révolution* a constructivist account of Lavoisier's role in the Chemical Revolution that went further than Roberts in questioning the revolutionary nature of Lavoisier's achievements and legacy. Bensaude-Vincent claimed that the Chemical Revolution had no essence; it instantiated no philosophical ideal of a 'scientific revolution'. It was an entirely local event peculiar to late eighteenth-century France, which Lavoisier and his supporters overlaid with the essentialist rhetoric of a 'scientific revolution'. Like A. Levin, who argued that the Lavoisians alternated between rhetorics of 'revolution' and 'reform' depending upon the changing fortunes of these terms in the volatile political culture of France in the late eighteenth century, Bensaude-Vincent claimed that Lavoisier self-consciously crafted a rhetoric of 'order and precision' to fashion his amorphous, opportunistic research efforts to meet the requirements of the Paris Academy which, in contrast to the English audience of gentlemanly investigators, consisted in a small peer group of trained specialists.[40] Bensaude-Vincent did not deny the reality of the Chemical Revolution, but insisted that its form and degree of rupture with the past were entirely the products of local conditions that historians and philosophers have inappropriately universalized. Once historians recognize the specificity of the Chemical Revolution, Bensaude-Vincent maintained, they can recognize their disagreements about its meaning and implications for what they are: 'historical artifacts'.[41]

Within the constructivist historiography of localism, Bensaude-Vincent focused on the role of instruments, texts and language in the construction of Lavoisier's chemistry. She denied that experimental physics was the source of Lavoisier's work with the balance, which she linked instead to activities well beyond its use in the laboratory: the question of the balance goes 'beyond the epistemological level to implicate the social dimension as well'. Appealing to the constructivist views of Latour, Michel Serres and Norman Wise, Bensaude-Vincent treated the balance as a 'quasi-object', or force of mediation, which through its circulation 'creates the consensus that welds a collective, culture, or community'. The balance was not just 'a costly and sophisticated precision instrument', but the material embodiment of a principle of balance, or equilibrium, which inspired and integrated the multiplicity of Lavoisier's reforming views and practices in chemistry, physiology, economics and politics. Functioning as a local 'mediating object', the balance did not contaminate science with ideology, as the defenders of the arationality assumption would claim, but enabled 'the social context' to play a 'positive role' in the constitution of Lavoisier's chemistry.[42]

Bensaude-Vincent's discussion of the reception of Lavoisier's new chemistry in France and the rest of Europe stressed, against the retrospective judgements of hindsight, that the outcome of this process was not a foregone conclusion and did not result in the immediate formation of a coherent research programme, or 'research school'. Rejecting the standard view that the *Traité* was a 'capstone' of the Chemical Revolution', which brought it to a 'triumphal conclusion', Bensaude-Vincent emphasized its limitations and inadequacies as a textbook, drawing particular attention to its compromised logic and content, which involved an unstable mixture of Condillac's 'simple-to-complex' procedural logic and the encyclopedic logic of the traditional natural-history approach to chemistry.[43] While Bensaude-Vincent shared Donovan's view of Condillac's limited impact on the methodological and cognitive features of Lavoisier's chemistry, she recognized the important role his philosophy played in shaping Lavoisier's pedagogical practices. But, accepting Holmes's view that Lavoisier did not revolutionize the whole of chemistry, she also stressed the fragmentary, local tenor of the *Traité*, viewing it as the temporary product of passing circumstances, with which Lavoisier soon grew dissatisfied and began to move beyond. Presenting 'a more complex view of the Chemical Revolution', in which controversy and dissensus underlay apparent doctrinal conformity, Bensaude-Vincent showed how, writing in the fractious atmosphere of the French Revolution, Lavoisier, Antoine Fourcroy and Jean-Antoine Chaptal produced rival textbooks which, while supporting the oxygen theory, promulgated decidedly different views of the logic of chemistry, the structure of its pedagogy and the standardization of its practices, as well as its social and technological import, institutional organization and the identity of its founder. Framed by his perceived martyrdom at the hands of the Terror, Lavoisier's claim to the mantle of 'father of modern chemistry' soon prevailed over Fourcroy's claim to collective paternity; the Founder Myth enabled Laovoisier's disciples to legitimate their pedagogical schemes and functioned as a symbol of French national pride.[44]

Working in the wake of Bensaude-Vincent's seminal analysis of the historical origins of the Founder Myth, and lending partial support to the hermeneutical identification of an event with its subsequent effects and interpretations, Christoph Meinel and Mi Gyung Kim showed how the image of Lavoisier as 'the father of modern chemistry' was forged in the 'rhetoric and strategy of self-fashioning' deployed by nineteenth-century chemists and academicians concerned with the disciplinary identity of chemistry as defined along methodological, professional, moral and national lines.[45] Thus Kim showed how, in their different 'performative contexts', Jean-Baptiste Dumas, Adolphe Wurtz and Marcelin Berthelot furthered their personal and professional interests by developing variations on the theme of Lavoisier as the methodical, thoughtful experimentalist, who avoided the shortcomings and pitfalls of the aimless empiricism and excessive speculation associated respectively with the natural-history approach to chemistry and the rival physicalist tradition in the Academy. These chemists also used the image of Lavoisier as the 'tragic hero' to rehabilitate his political image and to highlight the moral qualities of the future masters of chemistry.

These essays are laden with constructivist overtones, which emphasize local agents and 'performative' contexts rather than global structures and traditions. Thus Kim's claim that 'Dumas crafted the rhetoric of "method" to avoid the pitfalls of both fact-mongering and excessive speculation' prevalent in local practices ignores his indebtedness to the positivist *tradition*, which avoided the extremes of empiricism and rationalism in a methodological definition of science. Operating within these parameters, Kim offers a persuasive account of how, despite the gap between Lavoisier's vision and the subsequent development of chemistry, he came to be regarded as 'the father of modern chemistry'. Whereas modernist, and especially positivist, historians of science based normative hagiographies on realist readings of claims to privileged authorship and origin, Kim, in keeping with her constructivist and postmodernist sensibilities, deconstructed narratives of the patrimonial legitimation of knowledge by showing how they were constructed to serve the specific social interests and practices of local practitioners and their audiences.

Disciplinary Practices in Eighteenth-Century Chemistry and the Chemical Revolution

The rise of the sociology of scientific knowledge produced constructivist versions of the longstanding interest of historians of science in the formation, development and demise of scientific disciplines. Sociologically inclined historians of science replaced earlier scenarios of the teleological emergence of scientific disciplines with accounts of their construction, maintenance and transcendence by 'interested' agents and their specific practices. Resurrecting the traditional notion of the Chemical Revolution as the moment when chemistry achieved disciplinary autonomy by breaking with 'related arts and sciences', these historians conceptualized this autonomy not in terms of the methodological or theoretical uniformity and independence envisaged by their positivist and postpositivist predecessors, but in terms of Foucault's notion of 'disciplinarity', according to which scientific disciplines are both branches of knowledge and modes of social action. They traced the formation of communal unity and uniformity, inherent in the construction of disciplines and their boundaries, to the 'swarming' throughout the chemical community of locally formed disciplinary practices. This approach to eighteenth-century chemistry and the Chemical Revolution canvassed the idea of the multiplicity and complexity of the dynamic network of theoretical, experimental, discursive, institutional, cultural and social practices that constituted the wide and varied field of eighteenth-century chemistry.

Didactic Discipline or Investigative Practice

Christie and Golinski claimed that postpositivist historians of science, who focused on the influence on chemistry 'of activities such as speculative natural philosophy, matter theory, epistemology, methodology and theology', not only shifted 'the focus of analysis away from chemical practice to non-chemical fields of discourse'; they also presupposed 'the relationship between such [extrinsic] factors and the practice of chemistry' that the historian of chemistry 'problematized'.[46] In place of the postpositivist distinction between the 'internal', cognitive dimensions of science and its 'external', social causes, Christie and Golinski developed a historiography based on the distinction between the 'intrinsic' and 'extrinsic' features of science. They claimed that the nature and identity of 'chemical practice' depended upon and varied with its intrinsic, cognitive *and* social, features, which they distinguished from the extrinsic, cognitive *and* social, forces that influenced its development. Highlighting 'the human activities of practicing and talking about chemistry', Christie and Golinski outlined a two-stage research strategy for historians of eighteenth-century chemistry: first, identify the factors 'maintaining a discipline and practice with a continuously discernible sense of its own identity'; then, delineate the 'extrinsic forces from other discursive spheres' which conditioned and influenced, but did not constitute, the development of that practice.

Christie and Golinski adopted Owen Hannaway's thesis that chemistry originated as 'a didactic tradition at the beginning of the seventeenth century', and they claimed that eighteenth-century chemistry was best understood by 'stressing the intrinsic features of a developing tradition seen to inhere in a community of texts devoted to didactic discourse'. Hannaway emphasized the 'self-generating power' of the 'didactic form of chemistry', which entered the Enlightenment as 'a methodized discourse, conscious of its curricular identity but still in search of its own theoretical principles and problematic'; and he treated eighteenth-century chemistry, including the phlogiston and oxygen theories of combustion and the affinity theory of chemical reactions, as 'largely the search' for these organizing principles.[47] Giving a constructivist twist to Hannaway's interpretive framework, Christie and Golinski called upon historians of eighteenth-century chemistry to explore the different ways in which 'the discipline of chemistry was conceived and defined' in terms of interactions between its 'didactically methodized text[s]' and the various theoretical, philosophical, institutional and social forces 'applied in the furtherance of these didactic and definitional endeavors'.[48]

Christie and Golinski used this interpretive framework in their suggestive account of a number of significant issues and episodes in eighteenth-century British chemistry. Shunning postpositivist accounts of the reduction of chemistry to Newtonian matter theory, they emphasized the 'didactic context' of affinity theory and the work of the Scottish chemists, which were 'antireductionist and distinctly *chemical*'. In a similar vein, they faulted postpositivist scholars for failing 'to place Priestley in the context of any chemical tradition beyond the most cursory outline of British pneumatica'. Finally, while stressing the plurality and historical variability of responses to Lavoisier's work, Christie and Golinski endorsed Hannaway's view that 'Lavoisier's politically most effective move in the consummation of his revolution [was] his decision to embody the new chemistry in an elementary textbook'. This move brought together in the new nomenclature Condillac's philosophy of language and the tradition of didactic texts in chemistry. More generally, according to Christie and Golinski, the Chemical Revolution did not involve 'the definitive constitution of a science' – and certainly not '*tout court*, by extrinsics' – but 'particular moments of dialectical tension' between a pre-existing science, with its intrinsic didactic characteristics, and the extrinsic constraints of its different intellectual and cultural contexts.[49]

In contrast to Christie and Golinski, Larry Holmes argued that intrinsic to eighteenth-century chemistry was its identity as an 'investigative enterprise', not a 'didactic discipline'. This interpretive disagreement registered Holmes's historiographical resistance to the shift of attention, engineered by science studies, 'from the encounter between investigators and the piece of nature they hope to come to understand' to the interaction between investigators and their audiences.[50] Wishing to avoid the extremes of philosophical idealism, which focused on the disembodied ideas of science, and sociological constructivism, which assimilated science to society, Holmes insisted that the ideas and institutions of science functioned to facilitate the 'investigative enterprise' pursued by scientists at the workbench. According to Holmes, the job of the historian of science is to follow closely the dense, meandering investigative pathways along which scientists spend most of their working days, without getting lost in the rarefied air of conceptual abstractions or the surrounding thicket of social antagonisms.

In the 1980s, Holmes articulated his new historiographical vision against the backdrop of postpositivism, using the terminology of 'research programmes' and 'guiding assumptions' to describe Lavoisier's long and tortuous 'conceptual passage', while insisting that the ideas Lavoisier deployed were neither entirely autonomous nor socially determined, but 'emerged form' his sustained exploration of nature. In the 1990s, however, Holmes replaced the terminology of 'concepts' and 'traditions' with that of 'practices' and 'day-to-day investigative lives', stressing, for example, that 'Lavoisier had to practice a science based on such an implicit "law" [of the conservation of mass] before the latter became self-evident to him'.[51] Holmes now presented Lavoisier as a 'practical reasoner', identifying enduring experiments, not cognitive moments, as the meaningful units of the investigative enterprise. Holmes reconstructed Lavoisier's 'extended series of experiments', or 'investigative pathway', not in terms of the exfoliation of an underlying research programme, but by reference to an 'experimental system' that determined the 'trajectory' of his creative discursive and experimental practices. Adopting Pickering's notion of the 'mangle of practice', which bracketed the wider social context in a 'dialectic of resistance and accommodation', Holmes developed an account of the Chemical Revolution which replaced the Kuhnian extremes of revolutionary fissures and the continuous accretions of normal science with the image of a myriad, or 'mangle', of experimental and discursive practices subject to the contingent, open-ended patterns of 'coherence' generated by the dynamism of an investigative community.

From Manipulative Art to Instrumentalist System: Continuity and Discontinuity

Lissa Roberts shed light on the Chemical Revolution by linking 'eighteenth-century chemistry's history of linguistic organization and reform' to the 'broader, praxical' transformation engendered by 'disciplinary attempts ... to establish chemistry as a science'. Roberts incorporated Shapin's agency-based notion of literary technology, understood as an author's expository means of mobilizing assent for matters of fact, into an antihumanist, Heideggerian framework, which treated human actions as enmeshed in and structured by networks of anonymous practices. She analysed a 'series of chemical tables' in terms of 'the underlying structures' which, 'together with the changing structure of chemistry's other technologies', produced 'a particularly structured space of possibilities wherein knowledge of nature was generated, organized, and articulated'. Roberts approached Lavoisier not as an individual agent of change, but as 'both a central figure and symbol of an entire network of change' involved in the structural transformation of 'the discourse and disciplinary practice of eighteenth-century chemistry'.[52]

Roberts identified three developmental stages in a century-long movement from the depiction and practice of chemistry as a manipulative art, with no claim to a systematic understanding of nature, to its emergence as a science which identified 'nature with both the process and the product of chemical manipulation'.

Between the 1720s and 40s, chemistry was viewed and practised as a practical art serving other disciplines and oriented towards the systematization of useful knowledge. It could lay bare isolated phenomena in the laboratory, but it was not designed to encompass nature's causal structure. But the turn in disciplinary practice from art to science began when some chemists of the period linked greater control over the mysterious workings of nature to the accumulation and expansion of the chain of isolated phenomena. This was the purpose of the 'synoptic' affinity tables envisioned and constructed by Peter Shaw and Étienne Geoffroy. Not designed to encompass nature theoretically, these tables provided 'ready-to-hand', analogically organized surveys of the field which would 'free successive generations from the empirical task of reinventing the wheel'. Still, by making the science of chemistry 'first and foremost a set of statements about nature', these tables buried '[c]hemsitry's artisanal roots'. The next stage in chemistry's disciplinary development stretched from the 1750s to the 'appearance of 'revolutionary' chemistry in the late 1780s'. This stage was dominated by 'three overt investigative methods and structures of presentation', including the cataloguing of solvent action and the improvement of affinity tables by French and Swedish chemists and the construction of narrative histories of experiment by pneumatic chemists like Priestley. These 'distinct organizational foci' shared an 'underlying structural view of nature and chemistry's relation to it', according to which the gap between laboratory practice and scientific knowledge could be closed only if practitioners cooperated in the accumulation of individual facts in the hope that nature herself would eventually provide the criteria for transforming these facts into knowledge.[53] But cooperation soon gave way to conflict, not only over individual facts but more fundamentally over their significance and incorporation into the discipline of chemistry. A wholly different approach to the discipline was needed.

Lavoisier's revolutionary chemistry supplied this new approach by fusing 'laboratory activity and knowledge of nature into a unified goal of chemistry'. Lavoisier presented his new nomenclature and theory of chemistry as 'a structured method of discourse', which 'actively and unitarily shaped the field of chemical research from the start rather than retrospectively naming its individual components'. Just 'how the "new" chemistry collapsed the power of nature into that of laboratory production' is evident, according to Roberts, in the *Traité*, which displaced knowledge of the qualitative characteristics and material nature of its central elements, oxygen, caloric and hydrogen, with the enunciation of their causal roles, as the generators of acidity, heat and water, in 'the field of consequently structured production'. While the practical arts linked nature to a potentially infinite range of interpretive possibilities, science circumscribed the domain of artful practice that structured nature. Lavoisier's revolutionary chemistry 'referred all chemical activity back to "natural cause"', which in turn pointed 'towards a consequently structured world of effective production'.[54]

Roberts viewed the Chemical Revolution in terms of 'instrumentalism', a perspective that denotes not only the increased use of 'sophisticated instruments, measurement, and empirical testing in science', but also the phenomenological idea that as agents of human practice, the instruments of science are constitutive of the objects of science and the knowledge acquired in studying them. Condillac's theory of language was, of course, an exemplary representative of instrumentalism, and Roberts used the familiar account of how Condillac's influence shaped Lavoisier's nomenclature into a method or instrument of discovery – which not only signified the known, but also how to get from the known to the unknown – 'to expose the instrumentalist nature of the "new" chemistry in all its facets'. What distinguished the Chemical Revolution from prior instances of the control and manipulation of nature was not only the construction of 'an instrumentally conceived language', but the integration of this language into a 'network of scientific practices', or an 'instrumentalist system'.[55]

The rise of instrumentalism involved the 'Death of the Sensuous Chemist'. The new chemistry required its practitioners to 'subordinate and discipline' their bodies and senses 'in the service of machines'. It replaced the artisan-chemists' reliance on the particular evidence of their refined senses with 'determinative evidence that was transposable across qualitative and spatial borders'. The sensible examination of the qualitative characteristics of bodies gave way to the quantitative measurements and precise determinations of laboratory instruments. As unmediated sensory evidence played less of a public role in science, instrumental and manual dexterity replaced sensory refinement. Not only did Lavoisier construct a systematic instrumental space to monitor and measure insensible gases and imponderable caloric, he moved away from 'direct sensory determination' in all levels of his 'recorded practice'. He initiated in chemistry a process of standardization which, by harmonizing with other attempts in French society – such as the introduction of the metric system – to combat the arbitrariness of local traditions by 'institutionalizing the process of decontextualization through standardized measures', buttressed the new chemistry's claim to authority. While Fourcroy stressed the link between Lavoisier's experimental procedures and 'chemistry's new doctrines', Roberts preferred a constructivist to a whiggish construal to this claim. Roberts concluded that since 'we never "face" the world except through the mediating presence of instruments', the introduction of new instruments in the history of science does not involve 'our increased familiarity with nature', but only a change in 'the structure and constitution of what we take to be science'.[56]

Theoretical Chemistry and the Workshop Tradition
Ursula Klein's account of the origins of modern chemistry contrasts sharply with Roberts's image of a great divide between Lavoisier's scientific chemistry and the

workshop traditions that preceded it. Blending constructivist interests in the relation between investigators and their audiences with Holmes's focus on the interaction between investigators and the natural world, Klein drew attention to the transformation of 'the material culture' of 'the chemical-pharmaceutical and metallurgical workshop tradition' into a 'new social and epistemic space', in which chemical operations, once performed for practical, commercial and pedagogical reasons, were now put to 'epistemic' use. Championing the scientific status of pre-Lavoisian chemistry, Klein linked the emergence of modern chemistry to the formulation of 'the concept of the chemical compound' by chemists of the Paris Academy in the opening decades of the eighteenth century. But Klein rejected Duhem's familiar claim that the emergence of this important idea depended upon the incorporation into chemistry of the atomistic sensibilities associated with the rise of the corpuscular philosophy. Questioning this interpretation because it presupposed 'the classical dichotomy between the work of the mind and that of the hand', Klein treated the formulation of the modern concept of the chemical compound as an 'intellectual moment' in 'the development of an operational chemistry' closely tied to the 'workshop and engineering tradition' in seventeenth- and eighteenth-century chemistry.[57]

Like Holmes, Klein stressed the disciplinary identity and autonomy of the experimental practice of chemistry in relation to contemporaneous theoretical and philosophical developments; unlike Holmes, she linked the development of this 'pragmatic chemistry' to the wider 'interplay of material operations, intellectual work, and new social conditions' in the constitution of 'the science of chemistry'. According to Klein, 'the foundation of teaching institutions and of the Paris Academy' provided academicians like Geoffroy and William Homberg with an independent platform from which to launch a 'radical intellectual change' that shaped a large part of the experimental practice of chemistry in the eighteenth and nineteenth centuries. This radical change, evident most clearly in Geoffroy's affinity table, involved a moment of continuity and discontinuity, in which the material practices, operations and results of the chemical workshops were put to new, epistemic ends. Shifting the focus of attention away from the 'phenomenology of operations' to 'the nonobservable activities of the substances underlying the operations', Geoffroy's table provided not only a new conceptual framework in which to order and classify the results and recipes of apothecaries and metallurgical workshops, but also 'a new epistemic object – namely the abstract chemical transformation – and the embedding of it in a new conceptual system'. Central to this conceptual system was the idea of 'chemical combination, compound and reaction'.[58]

Klein argued that the authors of seventeenth-century 'chemical-pharmaceutical books of recipes' had no coherent conception of the 'chemical transformations underlying their operations'. Influenced by Paracelsus's philosophy of 'natural mixta', according to which 'natural bodies', though generated from a mixture of principles or sources, are homogeneous throughout, they thought and operated in terms of the 'enhancement' of qualities, or the extraction of 'essences', rather than the analysis of substances. This model shaped the dominant claim of chemists at the Paris Academy that the 'fabrication of salts' involved 'the extraction and enhancement of the essence of metals and alkalis' used in their production. Towards the end of the seventeenth century, however, an alternative technology of salt fabrication, which utilized 'acid dissolutions of metals and alkalis', came to the fore, along with a new interpretive model that stressed the penetrating power of acid particles, which cuts 'the dissolved bodies mechanically into pieces and holds them by love or force'.[59] Though animistic and asymmetrical in its focus on the dissolving power of the acid particles, this transitional model in the development of the concept of the chemical compound allowed some degree of struggle, or reaction, between weak acids and alkalis; above all, it involved the idea of 'the conservation of the basic substances' in a chemical transformation.

Jettisoning entirely animistic explanations, Geoffroy's table envisioned the symmetrical elective interaction between enduring constituent substances, the combination and separation of which underlies chemical transformations. According to Klein, the formation of this crucial concept involved the recognition by practical chemists of the 'reversibility of metallurgical operations and of the pharmaceutical production of salts'. These operations were reversible because they were done with substances that could be transformed and recovered in successive operations, and this was possible because the substances used in these operations were pure and homogenous, and therefore unalterable in their physical properties and chemical behaviour. Geoffroy delimited the field of pure and homogeneous substances from the larger field of natural substances not conceptually, but practically in terms of criteria established in the commercial, metallurgical and chemico-pharmaceutical practices of his day. According to Klein, attempts by eighteenth-century chemists to expand this conceptual framework beyond the mineral realm proved frustrating because 'the distillation and extraction of plant and animal materials could not be reversed'. The result was the juxtaposition of independent investigative pathways in chemistry until Lavoisier, who deployed modes of classification based on compositional rather than observable parameters, succeeded in extending the new conceptual framework to the realm of organic chemistry.[60]

Chemistry's Disciplinary Break with Pharmacy

Sharing Klein's interest in chemistry's pharmaceutical roots, Jonathan Simon also reinforced Roberts's sense of the revolutionary nature of Lavoisier's chemistry by stressing its disciplinary break with pharmacy. Whereas Klein saw an important continuity between the science of eighteenth-century chemistry and the pharmaceutical and metallurgical workshop traditions that preceded it, Simon charted 'the disciplinary split between chemistry and pharmacy'. Thus he showed how the period between the publication of Nicolas Lemery's *Course de Chymie* in 1675 and Venel's article on chemistry in the *Encyclopédia,* in 1753, saw 'the growth of philosophical chemistry', or 'an independent theory-oriented discipline' in which 'preparations for practical and medicinal purposes' gave way to 'preparations for their own sake, or for the sake of chemistry alone'. Lavoisier took up where Venel left off, replacing hitherto vague notions of chemistry with a 'clear vision of a systematic scientific discipline' and its disciplinary break with pharmacy. It was left to Fourcroy 'to reestablish affiliations, but this time with chemistry as the dominant science and pharmacy as a subservient art'.[61] To this end, he launched a new journal entitled *La Médicine éclairé par la Science Physique,* in 1791, and eliminated pharmacy from the dictionaries of the arts and sciences in the *Encyclopédia Méthodique* on the grounds of giving more space to chemistry, 'the necessity of which the [present] state of the science has made obvious'.[62]

Simon approached the problem of the disciplinary identity of chemistry obliquely; he asked not how chemists conceived their own discipline, but what were the disciplines they 'cast out of their science'? Focusing on the textual dimensions of Lavoisier's chemistry, Simon noted that while the *Traité* entertained an alliance between its philosophical chemistry and metallurgical practices, it contained no reference to pharmaceutical texts. Lavoisier's concept of a multitude of simple substances as the elementary constituents of chemicals also highlighted the mineral, as opposed to the animal or vegetable, realm of nature because inorganic chemicals are relatively simple 'and yet represent the greatest diversity of elements and their combinations'. But the prospect of applying Lavoisier's method of compositional analysis to organic compounds did not appeal to pharmacists, whose 'extracts' or 'simples' served the purpose of preparing commonly used medicaments, not identifying the principles that 'made up all bodies'. According to Simon, Lavoisier put aside commercial and pharmaceutical interests in the properties of bodies and the transformation of matter in favour of a 'chemico-philosophical system' which shifted from the idea of elements as the bearer of properties to the notion of the balance, or equivalence, of simple substances in the reactants and products of a chemical reaction.[63]

Simon used the disciplinary approach to the interpretation of eighteenth-century chemistry to present 'the [C]hemical [R]evolution as a multifaceted transformative process rather than as a single intellectual event'. Adopting a Foucauldian notion of disciplinarity, 'as a set of formalized constraints to be internalized by the chemist', Simon 'took up Crosland's invitation to examine the authority that underwrote' the new nomenclature, 'taking into account both the implicit and explicit sources of the authority for the publication of the *Méthode de*

Nomenclature Chimique'. Besides appealing to the 'institutional authority' of the Academy, which suffused their combined text and individual careers, the authors of the *Méthode* used the 'philosophical authority' of Condillac to displace 'the authority for the nomenclature into nature'.[64] The disciplinary identity of the new chemistry, as an independent science vis-à-vis the dependent art of pharmacy, also depended upon Fourcroy's institutional reforms, which 'transferred the ultimate authority over pharmacy from the guild' to 'a state-administered regime centered on a few newly-reformed schools of pharmacy'. Noting that these changes 'could only be fully realized in the reforming ferment of the Revolution', Simon stressed 'the vital role played by the French Revolution in establishing the form of modern chemistry'.[65] While Simon's disciplinary analysis of the Chemical Revolution remained mainly within a framework of microsociological parameters, his conclusion pointed in the direction of a more macrosociological dimension to this important event in the history of chemistry.

A Genealogy of the Chemical Revolution

Mi Gyung Kim's analysis of the Chemical Revolution also stressed the development of 'an authentic discipline' of operational chemistry in France in the first half of eighteenth-century. The immediate goal of her study *Affinity, That Elusive Dream: A Genealogy of the Chemical Revolution* is to combat 'our historical amnesia' about the important role of chemical affinity in the development of this disciplinary tradition. This amnesia resulted from the legitimacy that Lavoisier's perceived status as 'the father of modern chemistry' gave to his omission of the topic from the *Traité*. Dismissing Lavoisier as a 'philosophical chemist', who used chemistry to develop 'a true representation of nature' rather than to produce 'new medicaments', Kim insisted that the more likely candidates for the leadership of eighteenth-century chemistry were Geoffroy and Macquer, who produced a 'theoretical discourse of affinity' based on a unique combination of practical training in pharmacy and the elite discourse of the Academy.[66]

An appreciation of the interests and achievements of eighteenth-century practical chemists, denigrated by philosophical chemists as undisciplined mystics or sooty empirics, required a new approach to the Chemical Revolution based not on the disciplinary relationship between chemistry and physics, but on the 'changing associations' between chemistry and medicine. While medical schools and teaching physicians provided a social identity for chemistry as a public science in France, 'pharmacy maintained the disciplinary identity of chemistry as a material culture', and provided it with a 'plethora of analytic techniques'. According to Kim, the disciplinary identity of chemistry in the French Enlightenment involved a 'series of debates' triangulated by the elite discourse of philosophical chemists, the publications of physicians, and 'the vast and sophisticated material culture of the apothecaries'. Kim's 'genealogy of the Chemical

Revolution' described how the 'culmination of these debates ... transformed an apothecaries' trade into a public science'.[67]

Kim bestowed several layers of meaning on the term 'genealogy'. As a 'commonsense term', she used it to liken the 'active collective memory' of elite institutions, such as the Jardin du Roi and the Académie des Sciences, to 'family traditions', which provided 'relatively stable contexts of performance and reflection' for theoretical articulation and innovation. Influenced by Foucault, Holmes and Pickering, she also used the term to describe the use of analytical techniques in the contingent stabilization or disruption of the diverse material, social, literary and theoretical resources that constitute a scientific tradition. She gave historical substance to Foucault's sense of the infusion into the techniques and discourse of knowledge of relations of power and authority in her discussion of the 'dynastic mechanisms' within the Academy, along with 'techniques of persuasion and organization' designed to elicit the support of the broader public, which shaped the 'fabric of chemical theories'.[68]

According to Kim, seventeenth-century chemists bequeathed to the eighteenth century a stable material culture of laboratory practice, a strong didactic tradition and a 'philosophical chemistry' – developed by Robert Boyle and subsequent natural philosophers – designed to edify Enlightenment audiences and win for chemistry 'a respectable place among the other sciences'. But this philosophical chemistry 'did not guide the practice of apothecaries and metallurgists whose labor molded chemistry as a prototype for modern laboratory sciences'. Interested in tracing the evolution of this practical chemistry, Kim called attention to 'the discourse of a *theoretical* chemistry' aimed not at 'explaining phenomena as part of the general order of the universe', but at meeting 'the demands of the laboratory'. This theoretical chemistry generalized the 'experiences' of apothecaries and metallurgists 'with concrete experimental systems' into two 'broad theory domains ... of composition and affinity', reflecting the apothecaries' need to produce 'medically salient' compounds and the metallurgists' technique of sorting metals according to 'their differential dissolution in acids and fire'. Overall, the French chemical tradition involved a dialectic between philosophy's concern with ultimate 'principles' and 'attraction' in nature, theory's focus on the 'composition' and 'affinity' of chemicals, and the deployment of 'substances' and 'operations' in practice.[69]

Kim's account of the investigative pathways followed by eighteenth-century French chemists balanced considerations of continuity and discontinuity, while stressing the specificity of the Chemical Revolution. She embedded the revolutionary moment of the Chemical Revolution in the diachronic movement of a larger theoretical tradition, with nodal moments constituted by extra-theoretical factors. The first nodal moment occurred at the beginning of the eighteenth century in the context of a reorganized Paris Academy. It involved the stabilization of the conflicting demands of the corpuscular hypothesis, Aristotelian and Paracelsian theories of elements and principles, new developments in techniques of chemical analysis and the provision of a social niche for academic chemists. It resulted in establishing in the eighteenth century 'two contrasting approaches to the problem of composition: principles and affinity'. The second nodal moment, which occurred after an interlude of some fifty years of theoretical, experimental and linguistic developments, 'coincided roughly with the period known as the Chemical Revolution', in which Guyton de Morveau sought to quantify and rationalize the affinity tradition and Lavoisier 'introduced algebraic and quantitative precision' into the principles tradition. But, according to Kim, the subsequent collaboration of Lavoisier, Guyton de Morveau, and the Arsenal Group produced an 'elite' chemistry, concerned with the dynamic interplay between heat and affinity in the 'constitution', rather than the 'composition', of bodies. Although these elite views reached a wider audience through nomenclature reform, they did not have a significant impact on the way in which nineteenth-century European chemists practised chemistry, which was based on Berzeliuz's stochiometry rather than the French concerns with constitution. Kim thus endorsed Bensaude-Vincent's claim that the Chemical Revolution was 'a "French" (even a Parisian) affair, writ large by the participants' rhetoric and by our historiographical tradition'.[70]

Kim claimed that when historians start to listen to the voices of 'apothecaries, physicians, mineralogists, and industrial chemists', the 'very notion of the Chemical Revolution will be altered beyond recognition'. Historians needed not only to listen to new voices, but also to recognize the complexity of the situation, which involved 'cognitive, methodological, epistemological, and institutional transitions'. Kim thus rejected the exclusive focus – of Hannaway, Holmes and Golinski, respectively – on the teaching, laboratory, or public culture of eighteenth-century chemistry, and offered instead 'a composite, historicized identity for the chemist', who operated in a multiplicity of 'locales', around which the historian should 'move ... and "play" the stranger'. The situation was further complicated by the 'social hierarchy' within which eighteenth-century chemists flourished and is best understood, according to Kim, by replacing the abiding concern with the emergence of chemistry as a scientific discipline with the image of a gradually 'evolving material culture that acquired', in numerous transformative moments, 'various philosophical languages, theoretical structures, and social niches over the course of time'. Historians need to scrutinize the multifaceted 'disciplinary metaphor' of 'chemical affinity' through the lens of 'labour', not the 'perpetual mirage' of philosophy. Texts on chemical philosophy and matter theory are only a miniscule portion of chemical discourse; they express what 'chemists wished to be and what we want them to be', but not what, in fact, 'they were'. As chemical practitioners endured, against the background of 'their direct contact with the upper and learned classes of society', the pains, sorrows, frustrations and 'chronic ailments' of laboratory labour, their 'errant dreams' and 'straying fancies' created 'another image for themselves, the counter-identity of a laborer'. It is, Kim concluded, the duty of historians of chemistry to reverse this process by gaining, through 'a deep immersion in the body culture of chemistry', an appreciation of 'the intensity of labor involved in chemical practice'.[71]

Chemical Apparatus in the Eighteenth-Century

The sense of chemistry as 'labour' informed recent studies of the physical and operational characteristics and development of the instruments and apparatus of eighteenth-century chemistry. Levere and Holmes treated the relative neglect by previous historians of the instruments and apparatus used in the eighteenth-century laboratory as the combined effect of 'the dearth of material evidence' – due to the fragility, disposability and modularity of chemical apparatus – and a historiographical focus on the theories, careers and institutions of chemists and chemistry. The assumptions, shared by positivists and postpositivists alike, that instruments are unproblematic means for testing or illustrating claims arrived at by reason also robbed them of epistemological significance and favoured the ahistorical conclusion that they were employed in the past in the same way they are in the present.[72] Constructivism challenged this assumption, however, arguing not only for the constitutive role of instruments in scientific theory and practice, but also for the variety of uses and types of instruments. But constructivism's focus on the use of instruments to confer authority, create audiences and mediate between science and the wider culture overlooked instrumental developments 'not necessarily in phase with the vicissitudes of experiment and theory'.[73] Consistent with the postmodernist view of science as a multiplicity of diverse practices, some historians bracketed the wider context of use and focused instead on the physical design of instruments and the operational manipulation of apparatus.

While Crosland emphasized the 'simplicity' of the apparatus used by British pneumatic chemists and Holmes provided a detailed account of the evolution of Lavoisier's chemical apparatus, Levere, Bensaude-Vincent and Golinski explored the conflicting themes of stability and rapid change in the instrumental development of gasometers, eudiometers, hydrometers and thermometers in eighteenth-century chemistry.[74] More specifically, Johann Prinz provided a retrospective description of the apparatus Lavoisier used to found modern spyroergometry, 'a medical diagnostic procedure to continuously register respiratory gas metabolism during ergometer exercise', while Peter Heering used the 'so-called replication method' to repeat the work of Lavoisier and Laplace with the ice calorimeter.[75] Echoing the earlier work of Maurice Daumas, scholars like Levere, Bensaude-Vincent and Golinski called attention to the indispensable role of highly skilled instrument makers in the production of Lavoisier's new and important precision instruments. But Daumas treated Lavoisier's chemistry 'not so much as a chemistry of precision as a chemistry of method', in which new 'instrumental possibilities' derived epistemic significance from a new precision-oriented 'state of mind'.[76] In contrast, constructivist analyses located the meaning and significance of Lavoisier's general methodological pronouncements in the particularities of his precision instruments and practices.

The historiography of science-as-practice requires historians to be 'acquainted not only with the literary but also with the material sources of science'. The suggestion is that by reaching beyond the usual written and iconographic historical sources and examining whatever remains of past chemical apparatus, historians can get closer to the material reality of science. In order to overcome the problem of access associated with the incompleteness and dispersion of the material objects of science, Beretta and Scotti designed the 'project *Panopticon* Lavoisier' to reunite in virtual reality the (posthumously) dispersed objects – manuscripts (in archives), instruments (in museums), and books (in libraries) – of Lavoisier's scientific collection.[77] Reminiscent of Sarton's pedagogical vision of the 'lecture-"laboratory"', which sought to unite 'the academic and the museum sides of science history', *Panopticon Lavoisier* assigns witnessing a fundamental role in the history of science, enabling audiences to visualize and verify for themselves the interpretations of historians.[78] *Panopticon Lavoisier* has a greater chance of realizing its research and pedagogical goals than Sarton; but if electronic media make manuscripts and instruments more accessible, they also, through the ease of electronic manipulation and the ephemerality of its objects, make them less reliable as touchstones of objectivity. They further distance historians and readers from the past, leaving them to dwell in the clean, convenient, but ersatz realm of electronic simulation rather than the messy, recalcitrant world of primary sources and the reality they purportedly represent.

Heering's 'replication method' avoids this problem; it requires historians not only to examine, describe and reconstruct past apparatus, but also to use it to repeat the experiments that were done with the apparatus.[79] The aim of the replication method is not, in the manner of the scientist, to evaluate, through replication, the results of the experiment, but to examine reflexively the experimental skills required to perform the experiment as it was originally described. But, although the experience gained from the 'so-called replication method' can augment information gained from literary sources and opens up the possibility of new sources and lines of inquiry, it cannot, anymore than the information gained from literary sources, be applied to the historical situation without interpretation. Indeed only an adequate interpretation of science and its relation to society, which recognizes how instruments get their meanings and are rendered intelligible to the historian through their use, can prevent the study of past instrumentation from lapsing into a sterile antiquarianism. This is the sense to give to Lakatos's notorious claim that 'history of science without philosophy of science is blind'. While sociologically minded historians of science appealed to sociology, rather than philosophy, for interpretive enlightenment, the next chapter will use history to open our eyes.

Summary and Conclusion

Influenced by postmodernist nominalism and sociological finitism, sociologically minded historians of eighteenth-century chemistry reduced the cognitive 'traditions' delineated by their postpositivist predecessors to the circulation through society of the inherently local and situated practices involved in the particular sites of knowledge production. The historiography of 'specificity' highlighted the particularity, plurality and historical uniqueness of phlogistic chemistry and the 'doctrine of airs'. It also engendered decentred models of the Chemical Revolution, which interpreted it not as an 'Anglo-French affair', centred on the chemistry of gases and the oxygen–phlogiston debate, but as an interaction or conflict between relatively autonomous local cultures of chemistry, each with its own principles, practices and procedures. Deconstructionist tendencies associated with the historiography of specificity surfaced in a number of interpretive guises, including strategies designed to replace any reference to Priestley as a coherent historical personage with talk of many 'local' Priestleys; attempts to identify a multitude of local incommensurable Enlightenments accompanied by incommensurable chemistries; and the suggestion that the new chemistry was an entirely French affair inappropriately universalized by the rhetoric of its participants and subsequent commentators. In keeping with postmodernism's spatialization of reason, the historiography of specificity viewed the Chemical Revolution not as a process of change involving a paradigm shift, but as a static network of geographically distributed and culturally differentiated sites of local knowledge production.

A number of scholars characterized these sites of knowledge production in terms of the texts, instruments, practitioners and audiences involved in their construction and circulation. While a few scholars focused exclusively on the literary structure and devices of the core texts of the Chemical Revolution, constructivists related the form, function and content of these texts to procedures and practices for the construction of facts, the replication of experiments and the formation of chemistry as a public science. While Schaffer, Roberts and Bensaude-Vincent explored the role of material instruments, such as the eudiometer, ice-calorimeter and balance, as mediating objects used by chemists to persuade audiences and create communities, Golinski prioritized human over material agency in an account of the discursive practices Enlightenment chemists used to establish science as public culture. Emphasizing the plurality and particularity of these practitioners, audiences and communities, Bensaude-Vincent portrayed Lavoisier as an opportunist operating in the hierarchical culture of the French Academy and, along with Meinel and Kim, reduced his claim to be the father of modern chemistry to a rhetorical effect produced by the enthusiasm, ambition and patriotic ardour of followers, disciples and subsequent commentators. Contrasts between the coercive, hierarchical culture of Parisian chemistry and the egalitarian, gentlemanly world of British chemistry lent credence to the constructivist claim that the fundamental legacy of the Chemical Revolution was neither theoretical nor methodological, but sociological. It changed not chemists' relation to nature so much as their relations to one another.

This sociological orientation redirected traditional interest in the disciplinary identity of eighteenth-century chemistry along Foucauldian lines, which tied the formation of cognitive and social unity and uniformity to the circulation through society of locally formed disciplinary practices. Constructivist historians delineated a range of perspectives on the disciplinary identity and boundaries of eighteenth-century chemistry, stressing either its didactic or investigative characteristics, or drawing attention to the boundaries that separated it, as a dynamic set of multiple practices, not from physics and natural philosophy, but from medicine and pharmacy. These analyses mitigated or undermined the traditional view of Lavoisier's revolutionary break with chemistry's prescientific past. Talking about many 'chemical revolutions', rather than *the* Chemical Revolution, they described particular moments of dialectic tension between chemistry's intrinsic disciplinary identity and the extrinsic constraints of its different cultural contexts. The historiography of science-as-practice, deployed by Christie and Golinski, Holmes, Kim, Klein, Roberts and Simon, identified the discipline of chemistry with its material, or practical, rather than its theoretical, culture. Bracketting the macrosociological dimensions of eighteenth-century chemistry, it focused on the microsociological features of its material culture, stressing its identity as a form of work, or labour, embodying specific skills, utilizing particular instruments and occurring in situated institutional organizations. This historiography encouraged historians of chemistry to acquaint themselves not only with the literary, but also with the material sources of science. While the idea of a relatively independent domain of instrumentation in science encouraged some historians to focus on the physical design and manipulation of the instruments and apparatus of chemistry, in the absence of a hermeneutics of practice, it threatened the history of science with a sterile antiquarianism.

7 THE CHEMICAL REVOLUTION AS HISTORY

Steven Shapin summed up the contrasting implications of postpositivism and postmodernism for the historiography of eighteenth-century science in the following words:

> Against an older view that the 'new science' (and especially the 'Newtonianism') of the early and mid-eighteenth century was the underpinning of 'the Enlightenment', we now have a developing perspective which points out the existence of a number of species of natural knowledge and a number of opposed 'Enlightenments'.[1]

While Shapin and his constructivist allies championed the 'developing perspective' at the expense of the 'older view', some of his colleagues worried that the emphasis that sociological contextualism gave to the 'contextuality and contingency involved in scientific research' problematized the obvious 'intersubjective character' of that research, denied the conditions of generalizability necessary to historical explanation, encouraged an unproductive relativism inimical to a normative sociology of scientific knowledge, and pointed towards 'an extreme, obviously absurd form of positivism and solipsism'.[2] Considering these problems through the lens of eighteenth-century studies, Golinski claimed that the 'essential problem appears to be that of specifying the unity of the Enlightenment in a way which complements the analyses of the separate contexts in which it occurred, without obliterating their individuality'.[3]

An adequate solution to this 'essential problem' has important consequences for our understanding of the Chemical Revolution as an integral part of the Enlightenment. The solution advanced in this chapter recapitulates an earlier, dialectical sense of the inextricability of the moments of continuity and discontinuity, identity and difference, in the phenomenon of change, scientific, social or historical; it incorporates modernist concerns with the unity of science and postmodernist preoccupations with its disunity, or diversity, into an account of the dynamic unity-in-diversity that constituted the Chemical Revolution. Such an account is beyond the reach of sociological contextualism, requiring a more robust sense of the constitution of science by society and a more balanced view of the role of traditions and individuals, structure and agency, in the development of science and society. Above all, this robust form of contextualism prioritizes history over ontology or epistemology, treating the unity or specificity of scientific practices as a historical contingency, not a philosophical or sociological principle. It substitutes a fundamental sense of the dynamic mutability of that contingency for the historiographies of *stasis* associated equally with the formal identities of modernism and the spatialized dispersions of postmodernism.

Robust Contextualism

Golinski used the notion of the Enlightenment as a 'shared social experience' to articulate a middle ground between in the idealist view of 'the Enlightenment as a unified mind' or 'single cultural movement' and nominalist accounts of 'a multiplicity of specific contexts, each one constituted by numerous (local and temporary) social factors.'[4] But Golinski's developed position hinged on a false, or at least a contentious, dichotomy – between embodied practices and abstract thought – derived from the uncritical assumption that contextualism is a form of nominalism. However, this assumption was itself the historically contingent outcome of the form of realism, or universalism, promulgated by postpositivism and forcefully rejected by sociologists of scientific knowledge. Accepting the postpositivist identification of theories and concepts with the domain of transcultural rationality, sociologically-minded historians of science identified the social context of science with the specificity of its material circumstances, experiences and practices, thereby excluding any reference to the Enlightenment as a coherent movement of thought, culture, or society. These sociological scholars used epithets like 'practice', 'praxis' and 'practical' to lend credence and legitimacy to their nominalist materialism; but in the process they did violence to the core meaning of philosophies of practice, which posited not the ontological or epistemological priority of concrete matter over abstract thought, but the constitutive role of practice, whether collective or individual, in the formation of the social and material world and our comprehension of it. Consistent with this notion of practice, Marxian contextualism, or historical materialism, recognized that nominalism is not essential to contextualism, and that concepts and theories can be embodied in collective, or class, as well as individual, forms of practice. On the Marxist perspective adopted in this chapter, the Enlightenment is contextualized as a general and abstract movement of thought grounded in a socioeconomic structure of the appropriate generality and abstractness.

The robust form of contextualism adopted in this chapter faults constructivism for its 'internalist contextualism', which focuses more on 'the social context of scientists in the society of science' than in the world at large. While constructivism's attention to the microsociological, or internal, interests and skills of scientists served to illuminate the hitherto murky recesses of laboratory life, it failed to take adequate account of the larger contexts in which scientific construction takes place, treating them as the aggregation or circulation of the interests and skills at work in the laboratory. The upshot of this narrow perspective was a failure to reference the structural or global sources of institutional power, ideology and funding in science and a tendency to neglect its place in the class formation and hegemonic culture of the broader society. Those constructivists, like Shapin, Schaffer and Golinski, who made a more concerted effort to relate science to its broader social context also held a rather restricted view of that context, focusing on the political and religious while ignoring the economic dimensions of the situation. In contrast, the following account of the Chemical Revolution will give historical substance and historiographical structure to earlier, pre-constructivist claims, made by Thackray and Donovan among others, that the Chemical Revolution emerged out of 'a diversity of intellectual traditions' and 'a complex interplay of cultural, social, intellectual, and economic factors'.[5]

Emphasizing the need to make a clean break with whiggish sensibilities in all their manifestations, robust contextualism replaces the lingering retrospective tendency to plot the unfolding identity of a fixed entity in the past with a strong sense of the relational unity and complexity of particular historical situations. For example, while robust contextualism endorses Holmes's attempt to rescue eighteenth-century 'pneumatic chemistry' from its retrospective identification as a branch of general chemistry, it criticizes his continued analysis of it as an interdisciplinary configuration of physics, chemistry, and medicine, which carry their retrospective baggage. Robust contextualism also faults the intrinsic–extrinsic historiography developed by Christie and Golinski for its impoverished concept of the disciplinary identity of eighteenth-century chemistry and the link between this concept and their retrospective interest in the seventeenth-century origins of modern chemistry. By identifying the intrinsic, or defining, features of the discipline of eighteenth-century chemistry with its didactic form and function, they relegated to the extrinsic, or incidental, level what were in fact the defining features of Priestley's chemistry. More generally, the importance placed by the historiography of disciplinarity on identifying, against a background of shifting cultural contexts, the intrinsic features of a continuous developing tradition is at odds with robust contextualism's sense of the relational unity of the so-called constitutive and contextual factors of science in a given historical situation. Robust contextualism treats the disciplines of eighteenth-century chemistry not as rigid structures dominating a unified and immobile scientific field, but as domains of development articulated within a complex and dynamic field of inquiry encompassing 'scientific' and 'nonscientific' factors. Identifying and distinguishing these factors contextually and not retrospectively, robust contextualism takes seriously the claim, floated by Christie and Golinski, that chemistry is 'historically relative'.

Robust contextualism offers a balanced and historicized account of the role of traditions and individuals, structure and agency, in the development of science. It eschews both nominalist reductions of public science to the aggregation of private practices and realist claims that scientific progress, or 'maturity', involved the imposition of the conceptual unity and uniformity of a shared tradition on the doctrinal diversity and disagreements of earlier individualistic modes of natural philosophy.[6] According to robust contextualism, it is a matter of historical contingency, not developmental necessity or philosophical stipulation, whether or not a given scientific practice is 'specific' or paradigm-based. In prioritizing history over ontology or epistemology, it stresses, for example, important differences between the scientific practices of Priestley and Lavoisier without, in any way, evaluating or downplaying them. Thus it draws attention to the contrast between Priestley's individualistic sense of speaking 'from my own observations' and Lavoisier's participation in a 'community of opinions in which it is often difficult for every one to know his own'; but it neither devalues Priestley's individualistic mode of discourse nor deconstructs Lavoisier's sense of community.[7] Rather, robust contextualism places at the core of its account of the Chemical Revolution the historically constituted conflict between the specificity and individuality of Priestley's response to the generality and unifying power of the research programme of the 'Antiphlogistians'.

A number of scholars shared a strong sense of the explanatory inadequacies of the concept of 'immutable' structures or traditions inherent in structuralist formalisms. They argued that the associated 'antimony of objectivist determinism [of static structures] and … subjective free-play [of spontaneous agents] … allows no possibility for historical agency on the part of individual or collective human subjects'.[8] According to this line of criticism, structuralism cannot grasp history as a process, in which historical agents both make and are made by history. History is neither the direct product of spontaneous human action, nor is it the mechanical output of static structures operating behind the back of human actors. Historical structures and agents are interdependent. While structures need the practical mastery of members to achieve their 'objective' effects, agents acquire strategies, means and objectives from their position in a structural field of inquiry and action. Structures are not external constraints on actors, but 'facilitating conditions, or conditions of the possibility of action', which are dynamically reshaped by that action.[9] Given that individuals share common structural conditions, their actions can be seen as collective, and the unintended consequences of the clash of individual wills can be seen as the outcome of the clash of collective wills, or of class conflict. Understood as protean and not pregiven, historical structures operate as dynamic sets of material and ideational constraints underlying and delimiting a diverse range of local practices. Robust contextualism treats the Enlightenment and the disciplines of eighteenth-century chemistry as protean structures, formed and maintained by collective wills. It offers a balanced sense of the dynamic interplay between the flexible structures and active agents that forged the different and diverse communities and practices of late-eighteenth-century chemistry into a dynamic totality known as the Chemical Revolution. Given the immense complexity and diversity of this event, however, the following account will focus on the dialectical interaction between Priestley and 'the Lavoisians', leaving to future studies the difficult and necessarily more extensive task of dealing with the Chemical Revolution as a whole.

Complexity is a crucial feature of this dynamic totality, characterizing both its empirical profile and ontological status. Pursuing this line of criticism, Larry Holmes drew attention to the failure of existing accounts of the Chemical Revolution to appreciate the 'complexity of the event'. Instead of focusing on 'one or another of a group of subproblems' – such as the discovery of oxygen, the theory of caloric, the reform of the chemical nomenclature, the deployment of the balance, or the rhetorical structure of the *Traité* – as the defining thread' of the Chemical Revolution, Holmes deployed the notion of an 'interpretive enterprise' in order 'to show how the various thematic strands that historians have isolated as critical factors were interwoven' in the dynamic unfolding of Lavoisier's career. Insisting that the Chemical Revolution was a 'complex multidimensional episode', an integrated network of theoretical, experimental, discursive, organizational and cultural factors, he called for the integration of 'scholarly essays that highlight specific topics' into a 'story that must someday be told on a grander scale'.[10] Robust contextualism takes up where Holmes left off, extending his microsociological image of the complexity of science to the macrosociological level.

The stress this 'grander' narrative places on the complexity of historical situations or contexts is integral to robust contextualism's sense of the priority and autonomy of history. As Jonathan Simon noted, if the Chemical Revolution is viewed as 'a multifaceted transformative process rather than a single intellectual event', it can be closely connected, but not reduced, to 'the social and political changes that took place in eighteenth-century France'.[11] The idea of complexity is an integral part of a nonreductive form of historical materialism, which prioritizes history over metaphysics by replacing 'vulgar', or 'orthodox', Marxist views of the sociocultural superstructure as a direct and unmediated consequence of a teleologically unfolding economic base with the image of a field of mutually but also unevenly determining forces. In line with Engels's and Althusser's critiques of the base-superstructure model of historical materialism, robust contextualism recognizes the interaction of all social elements and stresses that while 'the economic movement finally asserts itself as necessary', in many historical developments and struggles, superstructural elements 'preponderate in determining their *form*'. An appreciation of the complexity involved in the ontological priority and autonomy of history also avoids the annihilation of 'temporality' associated with both the linearity and homogeneity of modernist time and the 'overwhelming spatiality' of postmodernism's 'mediascape'.[12] The following account of the Chemical Revolution ties the historicity of the event to its ineluctable complexity and temporality.

Beyond Modernism And Postmodernism: History Rules

The Priority and Autonomy of History

As Bruno Latour noted, 'history did not really count' for the moderns and postmoderns.[13] Instead of grasping the Chemical Revolution as a product of history, a specific mode of temporality, modernist and postmodernist historians of chemistry viewed it as a scientific discovery, a moment of rationality or a matrix of practices and interests that happened to have occurred in the past. They subsumed history under the disciplinary interests and categories of science, philosophy or sociology, and they failed to develop an adequate account of the moments of continuity and discontinuity in scientific change. They produced 'unhistorical' histories to the extent that they failed to capture the ineluctable complexity and temporality of historical events.[14] While positivist and postpositivist historians encompassed scientific change within an overarching identity, or series of identities, postmodernist historians lost sight of the patterns of historical change in a bewildering array of local actors and specific situations. But the dispersing thrust of postmodernism did not move beyond the modernist framework of progressive development, merely using anachronistic references to confuse and shock the '"modernist" avant-gardes'. Modernist temporality, however, with its image of history as 'an ordered front of entities sharing the same contemporary time', could not do justice to the dynamic autonomy and irreducibility of historical events, 'which pertain to all sorts of times and possess all sorts of ontological statuses'.[15] Robust contextualism blends the unity and uniformity of modernist temporality with the diversity and multiplicity of postmodernist spatiality in a dialectical sense of a historical event as the unitary, but complex, organization of distinct, but interrelated, temporalities.

An adequate account of scientific and historical change requires not only a just recognition of the complexity and temporality of historical events, but also a keen sense of the priority and autonomy of history. Reflecting on the modernist problematic, Latour argued that the autonomy of history and the reality of time were occluded in a world in which 'everything had to be contained between the poles of [pre-existing] Nature and Society'.[16] Like Marx and Engels and the pragmatists, Latour insisted that nature, no less than society, is subject to the change wrought by human practice, or labour. History, or historicity, is everything; it encompasses both the history of nature, or natural history, and the history of men. In this sense, Marx and Engels 'kn[e]w of only a single science, the science of history'.[17] A strong sense of humanity's primordial temporality was not the sole prerogative of Marxism however. It also provided a platform for Heidegger's analytic of finitude and Nietzsche's genealogy of morals. These accounts of temporality melded in the mind of Foucault to produce an interesting account of the priority and autonomy of history, which was also shaped by Althusser's struggle to balance the demands of Marxian temporality with the structuralist principles of synchrony and stasis. While considering Foucault's model of history for purposes of context and contrast, the following account of the priority and autonomy of history will focus mainly on Althusser's model of historical materialism and the Marxist tradition that informed it.

Marxist philosophy of history, whether in its historical origins, philosophical principles or political objectives, involved a robust sense of the historicity and temporality of human existence. The concept of the priority and autonomy of history distinguished the Marxian from the Hegelian dialectic, which otherwise nourished it. Whereas the Hegelian dialectic made history a mere instance of the 'closed ontology' of Reason, Marx emphasized the 'independence and openness of historical development'. Thus he derived the capitalist relations of production not 'from the self developing Idea', but from its antithesis to the prevailing relations of production. An acute sense of temporality was integral to the critical and revolutionary dimension and force of Marx's philosophy and sociology, which replaced Reason's necessary and unalterable determinations with History's contingent and transformable events. In contrast to the 'undifferentiated immediacy' of postmodernist 'pastiche', as exemplified, for example, in Latour's exclusive focus on the directly discernible 'associations' between individual actors, Marx used the historical dialectic to puncture illusions produced by immediacy, revealing the reality of class exploitation underlying the appearance of individual freedom and equality in capitalism. Questioning, in accord with the rationalist tradition from Galileo to Popper, the immediate evidence of his senses, Marx argued that the illusion of contractual equality between a 'single capitalist and a single laborer', begotten by the money-form of wages, vanishes completely when it is realized that 'it is the labor of last week, or of last year', appropriated by the capitalist, that pays the worker's 'labor-power this week or this year'. For Marx, the reality of capitalist exploitation, concealed at the level of immediate associations between individual workers and individual capitalists, is revealed by the theoretically-illuminated dynamics of temporality, 'which encompasses the class of capitalists and the class of laborers as a whole'.[18]

This is a dialectical temporality, which involves not only a complex interaction between 'otherwise unchanging objects', but, ultimately, a change in the very form and substance of objectivity.[19] As Critical Theory emphasized, Marxian temporality takes the form of an 'immanent criticism', which avoids the unpalatable extremes of contextual relativism and universal objectivism by treating critical ideals and standards as somehow implicit in 'the ideological opinions and practices that constitute social reality'. Subject to the transforming dynamic of temporality, the disciplines of philosophy, sociology and psychology are limited in their scope and validity by chronological and cultural boundaries identified and delineated by the science of history, which is itself governed by the same limitations of temporality. In Foucault's words:

> ... psychology, sociology, and philosophy ... are never directed at any thing other than the synchronological patternings within a historicity that constitutes and transverses them; and ... the form successively taken by the human sciences, the choice of objects they make, and the methods they apply to them, are all provided by History, cease-lessly borne along by it, and modified at its pleasure.[20]

The Complexity and Autonomy of History

In the 1960s and 70s, Althusser and Foucault used the notion of complexity to characterize the priority and autonomy of history. Though opposed to the essentialist or historicist vision of the unity, linearity and homogeneity of a sin-gle, absolute historical time, Althusser and Foucault also rejected the pluralistic orientation of the Annales authors, who affirmed 'the existence of different temporal strata and rhythms – the political, the economic, the geographical – without attempting to establish any systematic links between them'.[21] Seeking to avoid the spurious homogeneity of Hegelian and humanist conceptions of his-tory without lapsing into the fragmented pluralism of the Annales perspective, Foucault drew on structuralism's antireductionist recognition of the 'specificity' of the different levels and domains of reality and inquiry. He expressed his view of the autonomy of history and the discipline that studies it by characterizing the domain of 'discursive practices' as irreducible to – and, indeed, presupposed by – the domain of objects, the thoughts of individuals, the interpretations of cultures, or the determinations of institutional and social structures. Like John Searle's 'speech acts', the enunciations, or 'statements', of discourse are 'things done with words', like promises, commands and gestures. But unlike Searle, Foucault was interested not in 'the competence of a speaking subject', but in the 'body of anonymous historical rules' that determine in a specific time and place 'the conditions of operation of the enunciative function'.[22]

Foucault distinguished the rules of discursive formations, which govern the 'statements' that are actually made, or uttered, in a given historical configuration, from the rules of grammar and logic, which govern the far larger domain of gram-matically and logically possible sentences and propositions. Wedged between the 'primary relations' of society and the 'secondary relations' of language and logic, the 'preconceptual' relations of discourse require a mode of historical analysis which is irreducible to rational reconstruction or causal explanation. Foucault coined the term 'historical a priori' to capture his nominalist interest in the specific characteristics of actual discursive and social formations, rather than the universal conditions of their possibility. Resisting the assimilation of discursive formations to global theories or worldviews, Foucault treated them instead as 'fluid system[s] of disparate yet interlocking practices'.[23] Discursive formations function as 'systems of dispersion', characterized not by a common ground or unique origin, but by rules for the formation of statements that encompass different, possibly conflicting and certainly complex, systems of objects, concepts, theories and enunciative modalities.

While Foucault's model of history shared some of the features of postmodernist nominalism, Althusser's conception of the economy as determinant in the last instance remained squarely in the modernist problematic of realism, or universalism. Althusser articulated his conception of the priority and autonomy of history in 'a theory of the different *specific levels of human* practice ... based on the specific articulations of the unity of human society'. Avoiding both the simple unity of the Hegelian totality and the fragmented autonomy of the pluralistic parts, Althusser viewed the social whole as essentially complex, with each instance, part or level not reducible to an underlying determining essence, but enjoying a 'relative autonomy' with respect to the other instances, parts or levels. Althusser coined the term 'decentred totality' to denote the idea of a social formation in which each instance or level – such as the economic, political, ideological and scientific – possesses its own autonomy and causal efficacy within a 'hierarchy of effectivity'.[24] This hierarchy determines not the specific content or temporality of its levels or parts, but their 'locus of effectivity', their relations of cause and effect, dominance and subordination, order and arrangement. Striking a realist tone, Althusser identified two factors, one 'structural' and the other 'conjunctural', responsible for the articulation of hierarchical totalities. He argued that the trajectory of each instance or level of a social formation is determined not directly by the economic base, as in orthodox Marxism, but is 'overdetermined' by the conjunctural interaction of the other instances or levels, which it in part reciprocally determines. However, this reciprocity is also structured by the economic base, or 'structure in dominance', which determines which level, economic or otherwise, in the social formation is the 'dominant' level and how it distributes effectivity between the other levels. The economy is 'determinant in the last instance' in the sense that, whatever the specific form or trajectory of each instance or level in the social formation, they work together and achieve a recognizable social unity according to parameters laid down by the underlying economic relations.

Althusser replaced the modernist and historicist notion of the continuity and homogeneity of historical time, associated with the simple unity of the Hegelian totality, with the idea of 'a complex and *differential* temporality'.[25] Whereas the parts or levels of a simple totality are equally expressive of the determining whole, sharing a generic origin and teleological objective, those of a complex totality are inherently unequal and uneven in their impact and development. In a complex totality, the development of the different parts or levels is not a mere epiphenomenon of a 'primary' part or level, as in Orthodox Marxism; it does not conform to a shared pattern in a unified time, but is essentially 'uneven', involving the 'relative autonomy' of the different parts and the time-scales according to which they develop. Like the postmoderns, the Althusserian historian recognizes that the historical movement of a social formation does not involve 'an ordered front of entities'. But whereas the postmoderns constructed collages of explicitly anachronistic and contemporary entities, the job of the Althusserean historian is to portray 'polytemporal' assemblies of diverse modes of temporality – the new, old, progressive, regressive, cyclical and stagnant – no one of which is more outdated or contemporary than any other one. These different temporalities should not be measured 'as backwardness or forwardness *in time*', according to a 'single ideological base time', because they are not independent of the wholes they constitute. This unevenness is not an accidental feature of the whole, resulting from a mere conjunction of the parts, but reflects a 'certain type of *dependence* with respect to the whole': 'Unevenness is not measured against a time within which the social formation exists, but in terms of the contrasts and contradictions, the various relations of subordination and dominance, *within* the structure of the social formation itself'.[26]

Althusser thus points in the direction of a relational ontology and methodology, which equates wholes, essences and causes with nothing more than the relations between their parts, appearances and effects. Just as reality is not something underlying appearances, but is the structured relations of appearances, so, according to Althusser's Structural Marxism, the causal role of the economy is not an entity separable from its superstructural effects, but is identical with their relational structure. As Jameson put it, the structuring role of the economy 'is an absent cause, since it is nowhere empirically present as an element, it is no part of the whole or any of the levels, but rather the entire system of relationships among the levels'. Focusing on 'the level of the articulation of the component structures in the whole', the Althusserian historian traces the determination of the economy in the last instance by delineating 'the mode in which the elements of the whole are articulated upon each other'.[27]

So motivated, the Althusserian historian of the 'grander scale' endorses Michael Friedman's recent claim that 'a given temporal slice or historical episode ... has the meaning it does only in the context of a number of temporally extended traditions that interact, as it were at precisely this focal point'.[28] The job at hand is to show how the Chemical Revolution was a complex, multidimensional system, constituted by patterns of interaction between numerous and diverse levels, parts, elements, traditions. As we have seen in earlier chapters, many of these elements, levels and traditions have already been identified and characterized in the scholarly literature; they include empirical objects and information, theoretical strategies and constructs, experimental techniques and practices, methodological and epistemological principles, political formations, linguistic conventions, pedagogical and professional organizations, and social, cultural and economic institutions, values and regularities. The historian of the 'grander scale' must first characterize the internal content of these elements, levels or traditions in a way that recognizes their relative autonomy and internal modes of temporality and development. The patterns of interaction between these elements and levels, each with its own specific history, can then be spelled out not by reducing the content, form or existence of any one or more of them to one or more of the remainder, but by placing them in a relational complex, structured according to either a realist (Althusserian) 'hierarchy of effectivity' or a nominalist (Foucauldian) 'system of dispersion'. Either way, the more familiar issues of the discovery of oxygen, the phlogiston–oxygen debate, the opposition between Priestley and Lavoisier, the deployment of the balance, the definition of a chemical element, and the reform of the chemical nomenclature will be treated not as nodal points, or crucial events, in a uniform process of temporalization, grounded in the emergence of truth and reason or the assertion of power and interest. Rather, they will be referred, in their specific modes of temporality and connectedness to the field of history, understood as a relational complex, or series of relational complexes, irreducible to 'the law of an alien development', whether scientific, philosophical or sociological. As a complex system, or system of systems, the Chemical Revolution was neither a dispersed collection of disparate events, as the sociologists claim, nor a linear sequence of defining moments, as on the philosophers' models, but a dynamic pattern of 'multiple existences' with a definite shape and duration. It is the task of the *historian* of the Chemical Revolution to identify this pattern and determine its duration.

The remainder of this study will provide a prolegomenon to this 'grander' history by integrating existing accounts of the Chemical Revolution into a more specifically articulated Althusserean model of eighteenth-century thought, society and culture. While linking the conflict between Priestley and Lavoisier to the long-term structural changes that marked the transition from feudalism to capitalism, this account is sketchy about the more specific and local features of the social formations and political contexts within which Priestley and Lavoisier lived and worked. These issues and problems can be adequately addressed only in a future book-length application of this interpretive model to the minutia of eighteenth-century science and society, and this is dependent on a greater interaction and cooperation between historians of science and general historians. As it stands, this chapter offers the general historian a schematized account of the complex nexus of eighteenth-century science and society from the perspective of a historian of science concerned with both the historiography and history of an important moment in the development of modern chemistry. It is designed to win the assistance and cooperation of general historians in the challenge it poses to the hegemony of scientific, philosophical and sociological models of the history of science in general and the Chemical Revolution in particular. Consistent with the overall focus of this study on secondary rather than primary historical sources, the primary evidence needed to support substantive historical claims will be found in the secondary sources cited below.

Feudalism, Capitalism, and the Enlightenment

The Dissolution of Feudalism

Althusser's model of the relative autonomy of the superstructure and the role of ideology in the maintenance and reproduction of a social formation provides a powerful tool for mapping the Chemical Revolution onto a broader sociocultural dynamic associated with the Enlightenment, the roots and development of which can be linked to 'the long-drawn-out transition from feudalism to capitalism'.[29] According to Althusser's Marxist model, the persistence of a social formation depends upon the reproduction of its conditions of economic production, which include the requisite quantity, quality and distribution of labour power and its relations of production. The conditions of reproduction are secured by extra-economic practices and institutions associated with the superstructure. These practices and institutions, referred to by Althusser as 'state apparatuses', operate by political force or ideological persuasion. The 'ideological state apparatuses' – which include schools, religion, the family and other cultural and political organizations and activities – 'function massively and predominantly by ideology'.[30] As Althusser noted, the 'ideological state apparatuses' underwent an important transformation with the rise of capitalism, when a 'relatively large number of ideological state apparatuses' replaced feudalism's '*one dominant Ideological State Apparatus, the Church*,' which appropriated to itself not only religious, but also educational, cultural, political and familial functions. The dissolution of the unitary world of feudalism, associated in Chapter 4 with the rise of modernity, engendered the formation of new ideological apparatuses, and especially that of 'the Enlightenment', which not only challenged the Church head on, but also served to stabilize and reproduce the conditions of production of eighteenth-century capitalism.

The dissolution of feudalism threw Europe into an extended period of crisis, which affected society, politics and culture from the late medieval period to the end of the seventeenth century. Ingredients in this crisis included the rise of the strong nation-state, the invention of printing and the discovery of the New World, as well as the Protestant Reformation and the subsequent wars of religion. While Althusser treated 'all ideological struggle, from the sixteenth to the eighteenth century', as *concentrated* in an anti-clerical and anti-religious struggle', Shapin saw this period of turmoil as 'the immediate occasion', through the generation of a pervasive *'skepticism* about current systems of knowledge', for 'changed views of knowledge and its role in ensuring or subverting order'.[31] Other scholars, however, widened the scope of the connection between the rise of modern science, or new forms of knowledge, and the dissolution of feudalism. While Hans Blumenberg viewed the Copernican (Scientific) Revolution as merely one aspect of the birth of modernity, or the emergence of 'human self-assertion' that accompanied feudal dissolution, Norman Diamond argued that the attempt made by sixteenth-century Italian Humanists to elude the consequences and control the implications of their decaying society formed the basis of the Neoplatonic worldview underlying Copernican astronomy. In a more generalized and schematic way, Alfred Sohn-Rethel suggested that the application of science within production that accompanied the rise of modern science was initiated by the commercial revolution of the late Middle Ages, which destroyed the feudal mode of production based on the personal unity of head and hand in the individual producer. Similarly, Carolyn Merchant, Brian Easlea and Susan Bordo related the rise of modern, mechanistic science in the sixteenth and seventeenth centuries to a cultural response of control and domination, evoked by a pervasive sense of disorder and decay in nature, society and the self consequent upon the dissolution of the organic, feminine worldview of the Middle Ages and the Renaissance. Modern science's valorization of detachment, autonomy and separation, was a moment in a historical 'flight from the feminine', from memory of the union with the material world that characterized Medieval and Renaissance science and the integrated world of feudalism.[32]

That a conscious association between the rise of modern science and the dissolution of traditional society extended into the eighteenth century is evident in conservative reaction to the Enlightenment and the Chemical Revolution. Thus the arch-reactionary John Robison linked Lavoisier's chemistry, the metric system and the Revolutionary calendar to the obliteration of the past inherent in 'the unrestrained advocacy of rational analysis'.[33] More perspicaciously, Edmund Burke excoriated Enlightenment chemistry in the following terms:

> The geometricans and the chemists bring, the one from the dry bones of their diagrams, and the other from the soot of their furnaces, dispositions that make them worse than indifferent about those feelings and habitudes which are the support of the moral world ... While the Morvaux and Priestleys are proceeding with these experiments ... the analytic legislators and constitution vendors are quite as busy in their trade.[34]

Burke here opposed the dissolution of realities – nature and the mind – and sensibilities – the moral, the sensitive and the rational – involved in the process of dissolution and disinheritance that post-Enlightenment thinkers, as well as many subsequent commentators, saw as integral to the rise of modern science, the emergence of the Enlightenment and the birth of modernity. The image of science as a threat to tradition and the established order was intensified in Burke's mind by Priestley's 'seditious' support for the American and French Revolutions and talk of science 'laying gunpowder' under the old edifice of 'error and superstition'.[35]

The dissolution of feudalism was the flipside of the rise of capitalism in the seventeenth and eighteenth centuries. Generally regarded as the century of trade and, towards the end, as an 'age of revolution', the eighteenth century was a time of transition and turmoil, during which the emergent bourgeoisie challenged the hegemony of the feudal nobility and sought to define and control the embryonic working class. This was a complex situation, but for current purposes, it can be treated as occurring within a general framework of possessive individualism, designed to further the construction of an exchange economy based on the achievements of independent entrepreneurs in a preindustrial form of agrarian, commercial and manufacturing capitalism.[36] Fighting on two fronts, the emergent bourgeoisie deployed natural-rights liberalism to free the individual from aristocratic restraints and utilitarian liberalism to justify the imposition of bourgeois discipline and authority on the embryonic working class. Science and the Enlightenment played an important role in forging the consciousness of the new elite, which opposed ancient rank and privilege with industry, education and progress, and the plebian mentality of the urban mob and the crude rustic with rational politeness and amusement.[37] The characteristic Enlightenment link between chemistry, utility and progress and the positive, or forward-looking, aspect of the bourgeois struggle for ascendancy has been noted by Donovan, Golinski and Hufbauer, among others. The following analysis will focus on the negative, or liberationist, aspect of the Enlightenment – captured in the metaphor of the 'dissolution of feudalism' – and its impact on the Chemical Revolution.

The Enlightenment and the Rise of Capitalism

Marx used the metaphor of the 'dissolution of feudalism' to dramatize the contrast between the seamless web of feudalism and the patchwork cloth of capitalism. In feudalism, the economic, social, political and personal realms were integrated into a hierarchical unity, in which the individual was constituted by a web of relations to nature, society and God. This ideological apparatus served to stabilize an economic system based on a clear and overt division between 'necessary' and 'surplus' labour, between the labour-time needed to meet the subsistence needs of the direct labourer and that which was appropriated by the landlord in the feudal mode of production. This division necessitated the use of extra-economic means of coercion in the extraction of surplus value. The economic realm blended with the political realm in feudalism, where the 'exercise of power' combined the enforcement of political relations of subordination and domination with 'the economic performance of various tasks'.[38] In the feudal estate, the personal and social identities of the individual were inseparable, and the economic activity of the production and distribution of wealth embodied the hierarchical social relations of the system, which were simultaneously personal and political. With the rise of capitalism, this system of mutually constitutive parts gave way to a society based on clear-cut distinctions between the economic, social, political and personal realms. This change was effected by the emergence of wage-labour, which masked the distinction between necessary and surplus labour by the appearance of an exchange of equivalents between worker and capitalist ('a fair day's work for a fair day's wage!'). In capitalism, the appropriation of surplus value occurs entirely within the economic realm, without the enforcement of extra-economic forces. Thus, the integrated world of feudalism was dissolved and its parts transformed and reorganized into formally contained realms of distinct activities. Just as capitalism abolished the political character of civil society and the personal nature of political reality, so the organic unity of nature, society and the self in feudalism gave way to the mechanistic conjunction of autonomous, extrinsically related realms, including science, religion, politics and economics, each one of which was constituted by internal principles of identity and regulation.

Integral to this complex process of simultaneous social dissolution and autonomization was the emergence of *'civil society'* – which embraces 'the whole commercial and industrial life' of a people – from the local, parochial entanglements of 'ancient and medieval communal society'. The emergence of this autonomous economic realm involved the 'long and tortuous achievement of political rights by the bourgeoisie', along with the formation of a 'public space' for the ideological critique of traditional values and practices.[39] Responding to the systemic logic of the market, which established networks of commodity flows and transactions that undermined separate and local identities and communities, the bourgeois public realm of 'coffeehouses, reading clubs, and print journalism' provided a place of liberation, wherein Enlightenment champions of liberty, progress and autonomy deployed critical reason to dismantle traditional prejudices and local privileges. Claiming that 'Enlightenment books and periodicals did not so much distribute a universal human reason as provide tools for multiple constructions of meaning by their readers', sociologically inclined historians of science avoided the idea of the Enlightenment as 'some mind or spirit' by treating it as a 'macroscopic phenomenon' emerging from 'the microscopic or little enlightenments of the cafes, salons, societies, and clubs'.[40] In contrast, the following account treats the Enlightenment as a dominating ideological, or more specifically philosophical, movement or structure, with a complex identity that shaped, but did not determine, the conjunctural unity of ontological, epistemological, linguistic, methodological, empirical, theoretical, instrumental and institutional factors or levels that constituted the Chemical Revolution.

The Enlightenment notion of the self-defining subject, or epistemic self, was the ideological counterpart to the process of simultaneous dissolution and autonomization that shaped the development of eighteenth-century society. This notion contrasted vividly with Medieval and early-Renaissance thought, which defined the self in relation to the cosmic order, and equated reason with the eternal verities held in common by the human and the divine mind. Reason, the natural light of the mind, linked the cosmic order and the human subject in a relationship of reflection and systematic correspondence. The real task of knowledge on this view was the construction of metaphysical systems based on intrinsic *a priori* links between the knowing subject and the cosmic order, in which the method of proof and rigorous inference was used to spread the light of certainty over derived being and knowledge. In contrast and reaction to this perspective, Enlightenment thinkers rejected the idea of an intrinsic link between the self and the cosmos: the knowing subject encounters the world as something other, as a set of de facto contingent correlations. Drawing back from the world, the self-defining subject concentrated on the nature and limits of its own activity. The eighteenth century was characteristically an age of epistemology, psychology and methodology.[41]

Enlightenment thinkers no longer identified reason with a sound body of knowledge; they viewed it instead as an activity characteristic of man, as a method of inquiry which guided the discovery of truths. They embedded reason not in a metaphysical system, or a vision of the cosmic order, but in the operation of the methods of the empirical sciences. This, for Enlightenment thinkers, was the essence of Locke's rejection of innate ideas and Newton's method of analysis, which moved from phenomena to principles, and not vice versa. They insisted that science could progress independently of metaphysics, which like all other disciplines, should be subject to the method and results of the empirical sciences. In replacing metaphysical reason, grounded in things, with scientific reason, anchored to method, Enlightenment thinkers replaced the Renaissance doctrine of *signatures*, which posited an intrinsic link, of similarity or resemblance, between words and things, ideas and objects, with the view of language as a representational system of signs, which are independent of the things they represent. Within this linguistic framework, Enlightenment thinkers linked the method and language of science to the epistemological procedure of representation.

The Chemical Revolution captured something of the characteristic philosophical unity and force of Enlightenment ideology and culture to the extent that, despite their considerable intellectual and social differences, Lavoisier and Priestley shared an abiding commitment to the liberation of science from metaphysics, a strong desire to deploy the method of analysis in the understanding of nature, and a deep sense of the need to reform the language of science and reconstruct the nomenclature of chemistry. More specific to the realm of chemistry, Lavoisier and Priestley participated in a century-long process of ontological dissolution, wherein the elimination of Earth, Air, Fire and Water from the list of chemical elements robbed the Mediaeval cosmos of the unity and integration afforded by the doctrine of four contraries. But the fact that Lavoisier and Priestley pursued very different scientific objectives and reached fundamentally opposed conclusions undermines any notion of the Enlightenment as an abstract essence instantiated uniformly in diverse particular circumstances. Rather, in accord with Althusser's concept of the relative autonomy and uneven development of the different levels and factors of a decentred totality, the Enlightenment produced a unitary set of regulative principles and cultural constraints, articulated, according to a complex and differential temporality, upon a diverse range of disciplinary practices and substantive doctrines about the world. The international unity of the Enlightenment movement encompassed considerable diversity at the level of national and individual thought and practice. The dynamic unity-in-diversity that constituted the Enlightenment shaped the Chemical Revolution as a complex and dynamic conjunction of distinct disciplines and practices with a definite shape and duration. The components and dynamics of this conjunction will be more fully articulated in the next section.

The Enlightenment and the Chemical Revolution

The French and English contexts of the Enlightenment differed significantly. 'In France, the armies of Orthodoxy and Enlightenment faced each other' in a 'clarity of confrontation', in which a radical intelligentsia sought to replace one hegemonic system with another one. The need for systemic critique and organized opposition inherent in the assault on a strong and intrusive centre of intellectual and political authority accounted for the inclination of French philosophers toward theorizing, systematizing and an intellectual hierarchy. Given their relative political powerlessness, they were also more concerned with the purity of principle than the efficacy of practice. In England, however, 'triumphant Protestantism' had already smashed the centralizing power of Catholicism, and 'the eighteenth-century Church can scarcely be said to have an articulated ideology'. There was, therefore, no need for a radical intelligentsia to develop a systematic critique and organized assault on the bastions of clericalism and obscurantism, which had lost most of their power to 'blockade intellectual life'. Rather, there occurred in eighteenth-century England a multiplication and dispersal of rational activities and initiatives, oriented towards the interpenetration of theory and practice. Private enterprise and laissez faire took possession of the cultural economy before it took over the market economy.[42] These contrasting contexts nurtured correspondingly different philosophical sensibilities. The systematizing, critical and organizational orientation of French cultural activism shaped Lavoisier's stress on the theoretical activity of an active, hierarchical, corporate knowing subject; while Priestley's concept of a passive, individualistic, egalitarian, experientially oriented epistemic self was more suited to the fragmented compromise inherent in the English system of laissez faire.

So viewed, the Enlightenment functioned as the 'dominant' element in the formation of the Chemical Revolution as a relational structure, centred not on the conflict between the oxygen and phlogiston theories of combustion, or any other nodal point, but consisting in an ordered array of distinct cognitive and disciplinary elements and levels, each with a history of its own. More specifically, the dialectic between Priestley and Lavoisier involved a complex pattern of overlapping agreements and disagreements pertaining to theoretical issues between 'Phlogistians' and 'Antiphlogistians'; methodological debates between inductivists and hypothetico-deductivists; divergent epistemological evaluations of the knowing (sensory) mind; linguistic disputes about the relation between fact, theory and language; arguments over the role of qualitative and quantitative parameters in the development of chemistry; ontological shifts away from traditional doctrines of 'generic principles' and 'substantial forms' and towards the notion of 'simple substances'. As seen in the previous chapter, recent sociologically inclined historians of science have extended this dialectic to include the transition from a chemistry based on Priestley's valorization of accessible doctrines, simple instruments, replicable experiments and a participatory audience and public to the French model which tied the successful pursuit of chemistry to the deployment of powerful instruments, complicated experiments, a specialized theoretical language and a passive audience attentive to the voice of the specialist and expert. A full incorporation of these instrumental, practical and social dimensions into the model of the Chemical Revolution sketched in this chapter must await a future study. For the moment, they will be considered as subsidiary to the ontological, epistemological, methodological, theoretical, linguistic and institutional considerations outlined below.

The Ontology of Chemistry

The unfolding logic of the self-defining subject undermined essentialist views of the metaphysical unity of man and nature in a way that had significant consequences for the ontology of eighteenth-century chemistry and its emergence as an autonomous discipline. Sharing the epistemological and psychological assumptions of the *philosophes*, Lavoisier and Priestley equally devalued speculative notions and metaphysical systems as a hindrance to the development of science, which progresses through the rigorous analysis of sensory experience. Lavoisier dismissed established modes of chemical theorizing, in which the generic properties of bodies were explained by reference to a small number of ultimate 'principles' present in all substances, but beyond the reach of the senses or the analytical techniques of his time, as 'discussions of an entirely metaphysical nature'.[43] Consistent with the emerging logic of the self-defining subject, Lavoisier's ontological and epistemological reorientation coincided with a line of development endemic to eighteenth-century chemistry itself, in which the unisolable principle of Earth, Air, Fire and Water give way in the minds and texts of chemists to the class of specific earths, a multiplicity of chemically distinct airs, ordinary water and the momentary identification of phlogiston with inflammable air. These developments culminated in Lavoisier's pragmatic definition of an element as the end product of analysis and coincided with Priestley's characterization of material substances as 'things that are the objects of our senses, being *visible, tangible,* and having *weight,* etc'.[44]

The dynamic emergence of the Enlightenment notion of the self-defining subject from the womb of essentialist and substantive philosophies of the unity of man and nature resulted in a tensile ontological framework, which encapsulated the chemistry of both Lavoisier and Priestley. Lavoisier's notorious failure to make a clean break with the Stahlian tradition of 'real essences' he sought to overthrow is evident in his inclusion of caloric, the acidifying principle and the alkaline principle in his list of simple substances. Nevertheless, the main thrust of Lavoisier's ontological thought was innovative, and his followers purified his revolutionary views and aspirations. Priestley, on the other hand, was wedded more firmly to the essentialist logic of traditional chemical discourse. He soon abandoned his attempt to identify phlogiston with inflammable air; and in the early 1790s, he defended 'the hypothesis of [simple] water being the base of every kind of air, the difference between them depending upon the addition of some principles we are not able to ascertain by weight'.[45] Even then, however, Priestley remained true to the epistemological spirit of the self-defining subject, stressing the speculative and provisional nature of current attempts to determine 'the composition and elementary parts of all substances'. Thus, he laid 'no great stress' on these 'speculations', emphasizing the 'facts' relating to the observable properties and interactions of the substances of gross chemical experience.

The role of Enlightenment ideology in facilitating the transformation of the ontology of eighteenth-century chemistry encompassed a variety of intellectual influences and conceptual levels. On the level of discernible 'influences', Lavoisier acted in conformity with Locke's criticism of traditional metaphysics when he replaced the ontology of 'real essences', underlying and determining the observable properties of things, with the idea of elements as 'nominal essence', understood as 'simple substances', or 'the last point which analysis is capable of reaching'. The transition from a chemical ontology that grounded phenomena in underlying substances to one that focused on invariant relations of analysis and synthesis between observable things also suggests that the development of science in the Enlightenment involved the emergence of a new logic of concept formation in which the 'replacement of [metaphysical] substance by [empirical] law is a central issue'. Structural features of the Enlightenment ideology of the self-defining subject are also discernible in a striking similarity between the logical structure and ontological status of Lavoisier's chemistry and Newton's theory of gravitation. Newton and the Newtonians insisted against their Cartesian critics that although the concept of action-at-a-distance was unintelligible, or irreducible to divinely guaranteed innate ideas linking the knowing mind to the cosmic order, gravitation, as a universal phenomenon established by the method of empirical analysis, should be accepted as an explanatory principle. Though not an ultimate cause, gravitational attraction was to be accepted as a temporary 'last' element of nature. Similarly, besides eliminating from chemistry any reference to the ultimate 'constituent and elementary parts of matter', Lavoisier canvassed the provisional nature of his list of operational elements, which 'though they act with regard to us as simple substances', may be shown at some future date to be 'composed of two, or even of a greater number of principles'.[46] In detaching explanation from intelligibility, Lavoisier and the Newtonians upheld the self-defining nature of the knowing subject, which shares no common essence with the world, but which encounters it experientially as something other. Lavoisier's conception of a chemical element as the end product of laboratory analysis linked the development of chemistry as an autonomous discipline to the emergence of the Enlightenment notion of the self-defining subject.

Epistemology and Methodology

The longstanding positivist and Kuhnian view of an incommensurable break between Priestley's qualitative and Lavoisier's quantitative methodology in chemistry is no longer tenable. Not only did Priestley deploy the principle of the conservation of weight in his chemical theory and practice, he decried the reference to 'imponderables' in both the oxygen theory and the phlogiston theory. The difference between Priestley and Lavoisier lay not in the mere use of gravimetric experiments (methodology), but in the epistemological purpose of their use: whereas Lavoisier used gravimetric experiments to further his theoretical aims, Priestley put them to the empiricist end of criticizing theories. The epistemological differences between Priestley and Lavoisier hinged upon different construals of the Enlightenment ideology of 'analysis' in different philosophical and cultural contexts.[47]

Within a shared Lockean framework, which reduced the content and activity of the mind to sensations and their transformations, Priestley and Lavoisier upheld diverse conceptions of the nature and function of the knowing mind. Priestley adopted Hartley's view of the passivity of the epistemic subject; he viewed thought as the natural product of the mechanical law of the association of ideas, and he reduced discoveries in natural philosophy to the equal ability of all minds to accumulate and inductively order 'facts'. Lavoisier endorsed Priestley's devaluation of hypotheses that transcend the realm of observable facts; but he did not accept his inductivism, believing that 'by heaping up too great a number of experiments without order one might obscure the science rather than clarify it'.[48] Lavoisier adopted Condillac's view of the active knowing mind and the role of conjecture and refutation in the emergence of a new 'general theory'.

The contrasting epistemological allegiances of Lavoisier and Priestley were intertwined with disparate interpretations of the 'method of analysis', which was taken by the Enlightenment to be the proper scientific method, wherein truth is obtained by decomposing and recomposing ideas from the simplest components of experience. The 'analytic and historical method' deployed by Priestley was shaped by an empiricist methodology of enumerative induction, which contrasted with the rationalist procedure of 'mathematical analysis' underlying Lavoisier's experimental inquiries and theoretical conclusions. Condillac's claim that reasoning, as a species of analysis, is algebraic in character shaped Lavoisier's doctrine of the composite nature of water, which he formulated in the form of a 'true equation', or 'proof', approaching the logical rigour of mathematical 'demonstration'. Given the principle of the conservation of mass and the oxygen theory of combustion, Lavoisier found it impossible 'to doubt' the 'decomposition and recomposition of water' when 'we see that in burning together fifteen grams of inflammable air and eighty-five grams of pure air, we get exactly one hundred grams of water; and when we can, by decomposition, find again the same two principles in the same proportion'.[49] To doubt this conclusion, Lavoisier claimed, was to doubt the method of analysis, the only route to certainty in science.

But Priestley accused Lavoisier of achieving mathematical certainty in chemistry only by abandoning the method of analysis in favour of a 'synthetic style' of inquiry and presentation in which oxygen and hydrogen combined together to form pure water. Besides dismissing the acidic solution that Priestley always obtained in this experiment as an impurity-effect due to the presence of nitrogen in the reactants, Lavoisier developed sophisticated experimental procedure and elaborate laboratory apparatus in order to eliminate all impurity-effects from his results and to produce the idealized data necessary for the formulation of 'true equations in chemistry'. Priestley criticized these idealized experiments for using too much 'computation' and 'allowance' in advancing from experimental data to theoretical conclusions. Whereas Lavoisier used a preconceived theory of chemical composition to distinguish genuine products from impurity-effects in the reaction between oxygen and hydrogen, Priestley structured his phlogistic discourse to suit the claim that the passive reception of experimental data by the mind showed that nitric acid was as much a product of the reaction between dephlogisticated and inflammable air as was water. According to Priestley, the method of analysis required natural philosophers to provide the public with a faithful record of all their observations in the order and manner in which they occurred, and not, as in the 'synthetic' mode of discourse, to report the facts 'as if everything had been done to verify a true preconceived theory'.[50]

Within a shared epistemological framework, which opposed the method of analysis to speculation and system-building in science, Priestley and Lavoisier developed significantly different views of the relation between theory and experiment, the function of hypotheses in natural philosophy, and the relative priorities to be given to reason and experience in the development of chemistry. These differences informed the contrast between Lavoisier's epistemic preference for a few 'demonstrative experiments' – rather than the enumeration of 'probabilities' – and Priestley's methodological emphasis on the patient and impartial accumulation of observations. Incorporated into the contrasting British and French models of the practice and organization of chemistry, these epistemological and methodological differences involved different articulations of the Enlightenment ideology of 'analysis', which was designed to liberate the knowing subject from the web of similitudes that constrained pre-Enlightenment mind and society.

The Language of Chemistry

The contrasting strains of rationalism and empiricism in this moment of liberation also surfaced in Lavoisier's and Priestley's competing programmes for reform of the nomenclature of chemistry. These reforms involved different manifestations of a more fundamental transformation in the philosophy of language consequent upon the dissociation and autonomization of language, thought, experience and things in the critical solvent of Enlightenment analysis. This linguistic transformation is usefully represented in Foucault's discussion of the emergence of the Classical *episteme* from its Renaissance background. For Renaissance thinkers, language resided in the world as a system of signs enmeshed in the resemblances and similitude of things; and the doctrine of signatures maintained that the grammar of natural signs could be deciphered to reveal the syntax of being. The Classical *episteme*, which provided the underpinnings of the Enlightenment mind, severed this intimate link between words and things, replacing 'signatures', which inhere in and resemble things, with 'signs', which are distinct from the things they represent. The interpretation of the language of nature gave way to the representation of nature in the language of men, which was indentified with the analytical procedure of creating a systematic science by reducing the data of sensory experience to its representational components.[51]

Condillac summed up this line of thought when he declared that science is nothing but a well-made language. Lavoisier, accordingly, tied the reform of the chemical nomenclature to the logic and development of the oxygen theory; and, as Priestley complained, the former could not be utilized without understanding the latter. Priestley rejected Lavoisier's new nomenclature because it was based on 'principles ... not ... sufficiently ascertained'; he designed his own 'use of terms' to guarantee the permanence of a scientific language anchored to the epistemic bedrock of 'facts', rather than the shifting sands of 'hypotheses' and 'conjectures'. Thus he developed a chemical language that described the perceptible characteristics, circumstances, and transformations of perceptible substances, without any reference to underlying and imperceptible reaction mechanisms and chemical compositions.

Priestley's conception of a permanent scientific factual language was shaped by the more general Lockean view that language is a symbolic expression, or description, of independently existing mental patterns and processes.[52] But Priestley's rejection of Lavoisier's (Classical) identification of thought and language with the construction of a conventional (theoretical) system of signs did not lead him back to the doctrine of signatures, but to a third, intermediate position. He did not view words as inhering in and resembling things; nor did he view them as constituted by the system of knowledge in which they occurred. For Priestley, signs were anchored to facts, and derived their significance from experience. If the Enlightenment ideology of analysis dissolved the unitary vision of language, thought, experience and the world, empiricist variants preserved a link between language and experience, while their rationalist counterparts privileged the connection between language and thought (theory).

As a variation on the century-long British linguistic theme of 'common usage', Priestley's opposition to Lavoisier's new nomenclature was tied in with broader political issues and forces associated with the rise of the Enlightenment. Writers of various political stripes in eighteenth-century Britain used the notion of 'common usage', or 'custom', rather than an appeal to abstract rules and established regulations, to protect the linguistic freedom and integrity of the English people from a variety of threats and dangers perceived to emanate from both internal and external sources. While mid-century conservative grammarians identified this usage with that of the 'polite', as opposed to the 'vulgar', Burke equated it with 'traditional' usage and Priestley identified it with that of a community of free and rational men. In opposing the French nomenclature as a threat to the liberty and equality of his envisaged community of free and equal investigators of nature, Priestley gave expression to the rational and liberal principles of Dissenters, who also sought to remove the 'impositions' on their freedom of the established political order of eighteenth-century England. To the extent that Priestley's opposition to the new nomenclature involved a liberal and rational variation on the established theme of 'common usage', it constituted a radical and democratic transformation of the century-long conservative concern with the peculiar liberty of the English people and language and the threat posed to that liberty by the importation, or 'imposition', of French words, a French Academy and, finally, a French nomenclature for chemistry.

The contending linguistic views of Priestley and Lavoisier also represent two of the major responses in political philosophy to the problem of social order engendered by the dissolution of feudalism and the rise of capitalism. Echoing the seventeenth-century view that the unity of society required the imposition of a central authority to control and harmonize the pursuit of individual interests and passions, and endorsing the economic ideas of Turgot and those administrators who sought to reform the *ancien regime*, Lavoisier favoured a centralized political and linguistic order based on the sole authority of reason and upheld and administered by a holy alliance between king and savant. In contrast, Priestley's conceptualization of the linguistic usage of a community of free and rational individuals was shaped by the eighteenth-century policies of *laissez faire*, which developed a variety of arguments to link social cohesion to the politically unhampered pursuit of individual interests and passions.[53] It is interesting to note that at this crucial stage in the emergence of the bourgeois world, the three major philosophical responses to the problem of social order in that world were represented in the debate over the language of chemistry: While Burke upheld tradition, Priestley championed the individual and Lavoisier looked to the state. Clearly, the language of eighteenth-century science was intimately related to the language of eighteenth-century politics, both of which were shaped by the evolving socioeconomic order.

Empirical and Theoretical Dimensions of the Chemical Revolution

The unfolding logic of the self-defining subject and the Enlightenment ideology of 'analysis' underpinned a century-long process of ontological dissolution and autonomization, which robbed the medieval and Renaissance cosmos of the unity and coherence afforded by the four contraries, Earth, Air, Water and Fire. By the end of the eighteenth century, chemists had replaced three of these generative elements with, respectively, the class of specific earthy substances, a multiplicity of gases and ordinary water; while Fire lingered on in the attenuated form of the caloric theory of heat. Doubts about the traditional ontological status of Water engendered conflicts over its chemical identity, with Priestley linking his opposition to Lavoisier's view of the composite nature of water to his defence of the reality of phlogiston. This section will show how the empirical and theoretical dimensions of the controversy gave substance to the slow, sometimes hesitant, dissolution of feudal ontological unity in the analytical solution of Enlightenment nominalism.

During the 1770s, the oxygen theory focused on the problems of combustion, calcination, respiration and acidification. By 1778, Lavoisier had satisfied himself that 'the purest part of air' (Priestley's 'dephlogisticated air') was consumed by burning bodies, by metals during calcination and by animals during respiration; from its property of conferring acidity on nonmetals, he gave it the name '*gaz oxigène*'.[54] Priestley's response to these developments in the early 1780s centred not on an alternative theoretical system, but on the contrasting analogical features of a few 'principal facts', or crucial experiments, in the chemistry of gasses. Treating Lavoisier's experiments on the calcination and revivification of mercury by heat alone as a 'principal fact' in favour of the oxygen theory and against the phlogiston theory, Priestley responded in 1783 with a 'principal fact' of his own. Thus he argued that the total diminution of a volume of inflammable air confined over water during the revivification of lead by heating a sample of 'minium' (lead oxide) in the air was direct proof of the existence of phlogiston and its crucial role in the calcination and revivification of metals. Priestley also stressed the inability of the oxygen theory to explain either the fate of inflammable air in its explosion with vital air (which, contrary to the predictions of the oxygen theory yielded no acid) or the well-known phenomenon of the production of inflammable air from dilute metal-acid solutions. While Priestley traced this phenomenon to the phlogiston in the metal, Lavoisier's early view of the binary nature of acids and the simplicity of water and the metals obviated any explanation of this phenomenon in terms of the oxygen theory. Priestley thus regarded it as more prudent to stay with the imperfect, though well-tried and highly fertile 'hypothesis' of phlogiston than to replace it with an unknown, untried hypothesis of restricted scope and anomalous content.[55]

But Priestley's complacency was short-lived. The relevant anomalies in the oxygen theory were removed by the 'discovery' in 1783–4 that water was a compound of oxygen and a specific inflammable gas called '*hydrogène*'. This discovery enabled Lavoisier to explain all instances of the production of inflammable air in terms of the decomposition of water. Where the Phlogistians saw the release and absorption of phlogiston, the Antiphlogistians posited the analysis and synthesis of water. But Priestley's exploration of the analogical implications of his 1783 'principal fact' – which involved heating a variety of metallic calces with phlogiston-rich substance – generated a serious anomaly even for the mature oxygen theory, which was unable to explain the production of inflammable air (subsequently identified as carbon monoxide) in the reaction between finery cinder and charcoal, neither of which were thought to contain water. Still Priestley shifted the focus of his chemical inquiries and opposition to the Antiphlogistians from the calces of metals to the nature of water, sharing with Cavendish, between 1783 and 1786, a compromise position which treated water as a compound of dephlogisticated air and inflammable air. Between 1786 and 1788, however, Priestley treated the reaction between dephlogisticated and inflammable air to form water as the new 'principal fact' of his inquires. He also reverted to the traditional doctrine of the simplicity of water, seeing it as deposited, rather than formed, in this reaction. Between 1788 and 1796, Priestley developed a phlogistic account of the composition of the known gases based on the analogical extension of the principal fact that an acid was every bit as much a product as water of the reaction between dephlogisticated and inflammable air. Incorporating Lavoisier's oxygen-theory of acidity into his phlogistic speculations, Priestley argued that all the known airs were compounds of (simple) water, the principle of acidity and phlogiston, or the principle of inflammability and alkalinity. Priestley reduced the Chemical Revolution to a choice between two fundamental facts: either phlogiston was real and water was simple, or water was composed of oxygen and hydrogen and phlogiston did not exist. From 1796 until the end of this life, Priestley focused his chemical energies on the proliferation of anomalies for the oxygen theory, which, given his view of their interconnectedness, functioned as confirming instances for the reality of phlogiston.[56]

Priestley related his view of a close conceptual link between Lavoisier's concept of the composite nature of water and the denial of the reality of phlogiston to the possibility of chemistry as an intelligible mode of discourse and inquiry. He argued that if it is supposed that water is a simple substance, and not a compound of inflammable and vital air, as Lavoisier believed, and 'if there is no such thing as a principle of phlogiston' common to all combustibles, then it 'must be admitted that' inflammable air is a compound of water and the particular combustible used in combination with steam to produce it. Furthermore, according to such a view, the revivification of a metallic calx in inflammable air procured by means of another metal implies the 'interconvertibility, or transmutation of metals', and 'iron made of inflammable air from sulphur, ought ... to ... have the properties of *sulphurated iron*, which undoubtedly it would not have'. A 'hypothesis loaded with these difficulties must be inadmissible', since it undermines the very possibility of chemistry, which seeks to determine and differentiate the composition of substances through an examination of their observed effects.[57] Such an enterprise was possible for Priestley only if the inflammable air procured from water and a combustible was traced to either a source in the water, as Lavoisier suggested, or in the combustible, as Priestley maintained. According to this line of reasoning, eighteenth-century chemists were compelled, by considerations relating to the possibility of chemistry as an explanatory system, to accept the doctrine of the 'composition of water' once they had ceased to believe in the existence of phlogiston. In Priestley's mind at least, the Chemical Revolution was as much about the traditional doctrine of elementary water as it was about the reality of phlogiston. Fire and Water were inseparable components of a world in decline. The dialectic at the empirical and theoretical heart of the Chemical Revolution articulated a broader sociocultural transformation, in which the unity and coherence of the premodern world, exemplified in the doctrine of the four contraries, gave way, slowly but surely, to the analytical multiplicities associated with the Enlightenment and the modern world.

The Institutions and Practice (Instruments) of Chemistry

The institutional, material and instrumental dimensions of the contrast between Priestley's empiricism and Lavoisier's rationalism is captured in the juxtaposition of two historical images: Whereas English thought and practice in the eighteenth century was concrete, practical and individualistic, French thought and practice was abstract, principled and corporate. Whereas the English linked the liberal and analytical principles of the Enlightenment to the unfettered experience of the individual, the French associated rationality with the organization and administration of society. As a leading representative of the English Enlightenment, Priestley mingled liberal, Dissenting and Baconian motifs in a vision of epistemological progress based on the cooperation of individual experimentalists, united by a network of communications encompassing diverse views and interests and unencumbered by the institutional accretions of established theory and practice.[58] On this view, science could support the forces of progress, rationality and liberalism only if it avoided all dogmatic modes of thought and encouraged the intellectual and moral development of the individual. Priestley, accordingly, placed more epistemic value on the observations and judgements of individual experimentalists than on a shared body of theory, such as the 'French system', promulgated by an individual genius and perpetrated more by the techniques of indoctrination than by critical discourse. Criticizing John MacLean for basing his opinions on what he read rather than on what he observed, Priestley commented: 'I speak from my own experiments, and I only wish that Dr. MacLean would speak from his'.[59] Priestley's rejection of the oxygen theory was rooted in his refusal to take on faith results he could not duplicate in his own laboratory.

Priestley's individualistic notion of speaking 'from his own observations' contrasts vividly with Lavoisier's sense of participating in a 'sort of community of opinions, in which it is often difficult for everyone to know his own'.[60] Although Lavoisier was the undisputed leader of this hierarchical community, the cooperative division of theoretical and empirical labour among its members, and their readiness to close ranks against external criticism and dissent, betokened a collectivist enterprise at odds with Priestley's individualistic preference for thinking, reading, working and publishing on his own. A similar divergence in epistemological sensibilities underscored Priestley's Lockean view that language expresses, or communicates, the prelinguistic experiences and meanings of isolated individuals and the notion favoured by Lavoisier and Condillac that the thoughts and ideas of the individual are shaped by the communal tool of language. Yet again, whereas Priestley struggled, in the name of the civil and intellectual liberties of the individual, to minimize the power and influence of central governments and established institutions, Lavoisier was influenced by Turgot's programme for using the methods of science to reform, rationalize and, thereby, strengthen the administrative structure of the nation.[61]

The difference between Lavoisier's corporate view of knowledge and Priestley's individualistic epistemology highlights differences between the institutional organization of French and British science and society in the late eighteenth century. Whereas the highly integrated community of state-subsidized French theoreticians provided fertile ground for the flowering of paradigmatic conformity during the Chemical Revolution, the dissemination of Lavoisier's theory in England met with a more varied resistance, arising out of the individualistic mode of inquiry characteristic of British natural philosophy and its entrepreneurial context. The broader intellectual context within which these contrasting sensibilities were located is evident in the different agendas for the development of a science of man inherent in the English doctrine of associationism, which sought to construct human nature out of the experiences of the individual, and the programme adopted by the French *philosophes* for the construction of a social science based on actuarial considerations and the mathematical theory of probability.[62] It is symptomatic of the diversity within the unitary movement of the Enlightenment that Lavoisier found the centralization of power and authority associated with the dissemination of his theoretical doctrines perfectly compatible with the liberal and analytical principles that led Priestley to oppose him.[63]

Sociologically minded historians of science related the social, institutional and organizational dimensions of eighteenth-century chemistry to the uses and development of its apparatus and instruments. They described the role of instruments in the replication of experiments, the transmission of manipulative skills, the winning of assent and creation of audiences, the articulation of a new Enlightenment discourse for chemistry, and the dissemination of wider cultural and political values and interests. In keeping with Althusser's sense of the relative autonomy of the different levels of a social totality, some of these commentators encouraged the notion of an instrumental dimension of science following developmental pathways not necessarily in phase with experimental and theoretical transformations in the discipline. Of even greater relevance to the themes of dissolution and autonomization emphasized in this chapter are the accounts developed by Roberts and Klein of a deep-seated movement in eighteenth-century chemistry, which transformed it from a 'manipulative art', tied to the 'workshop tradition', to an independent science, in which the integrated commercial and technical activities and results of apothecaries, metallurgists and artisans were used in the service of purely 'epistemic' ends. This process involved the 'death of the sensuous chemist', as new theoretical objects and standardized forms of manipulation and instrumentation replaced the 'workshop' chemists' embodied practices and reliance on the evidence of their own refined senses. The rise of pneumatic chemistry, which called for new and sensitive instruments to handle and manipulate its insensible and rarefied objects, accelerated this development, though once again Lavoisier took it further than Priestley. By and large, Priestley pursued his observational goals with simplified modifications and improvements of the common or garden household apparatus Hales, Brownrigg and Cavendish deployed, relying mainly on 'readily available objects', such as ordinary jars, wooden or earthen troughs, gun barrels and clay pipe stems, in his domestic laboratory. Lavoisier, on the other hand, developed more robust and specialized instruments in a systematic instrumental space separated from the uncertain contingencies of workshops, kitchens, garden sheds and other local and colloquial places of chemical practice. Whereas Lavoisier designed apparatus to further the standardization of chemical theory and practice, Priestley developed apparatus and processes adapted to his own 'peculiar views'.[64]

Interest in the materials of eighteenth-century chemistry is still a minority one among historians of chemistry, and it falls to future scholars, conversant as much with the material as with the literary culture of chemistry, to explore in more detail the changes and development in instrumentation associated with the dissolution of the integrated workshop tradition and the resultant emergence of chemistry as an autonomous theoretical science. In their recent study *Materials in Eighteenth-Century Science: A Historical Ontology*, Ursula Klein and Wolfgang Lefèvre delineate a likely trajectory for these future studies.[65] Their concept of eighteenth-century 'chemical substances' as 'multidimensional objects of inquiry', formed and transformed by the intersecting and interacting practices of early modern chemists, artisans and workers certainly supports the historiography of complexity advocated in this chapter.

Summary and Conclusion

Robust contextualism offers a dialectical account of the integrated moments of continuity and discontinuity in the Chemical Revolution which transcends the inadequacies and partialities of existing philosophical and sociological reconstructions of the history of science. It projects an image of the ineluctable complexity and autonomy of history which nullifies the reductive extremes of philosophy and sociology by positing a dynamic, decentred totality, in which the uniformity of modernist temporality and the diversity of postmodernist spatiality are incorporated into a model of the Chemical Revolution as a complex conjunction of distinct, but interrelated, events or levels. Robust contextualism replaces the focus of the sociology of scientific knowledge on local practices, concrete skills and particular situations with a concern for the place of the Chemical Revolution in the broader intellectual, cultural, social and economic life of the eighteenth century. In place of the attenuated contextualism of the historiography of disciplinarity, which contrasts the unchanging 'intrinsic' identity of the discipline of eighteenth-century chemistry with its variable 'extrinsic' contexts, robust contextualism emphasizes the mutability of eighteenth-century chemistry and the internality of its relations to so-called 'extrinsic' factors. Robust contextualism offers a dynamic account of the emergence and unfolding of the Chemical Revolution rooted in a dialectical interaction between mutable historical structures and purposive agents, individual as well as collective. 'Complexity' is an important feature of this account, serving to emphasize unitary, but non-reductive, relations of dependence and interaction between the different levels of the totality it depicts, as well as the specificity and autonomy that accrues to it as a historical totality.

Robust contextualism adopts the Marxist sense of the historicity and temporality of human existence, rejecting its obfuscation by the closed ontologies of realism (universalism) and nominalism. Opposed to both the spurious homogeneity of philosophical realism and the fragmented plurality of sociological nominalism, robust contextualism highlights conceptions of the specificity and autonomy of history and the methods used to understand it. Adopting Althusser's Marxist notion of the 'relative autonomy' of the different levels or domains of a 'decentred totality', this model elucidates the way in which the socioeconomic transition from feudalism to capitalism generated, through a long drawn-out process of simultaneous dissolution and autonomization, 'the Enlightenment' as a dominant ideological apparatus which stabilized and reproduced the emerging conditions of production of eighteenth-century capitalism. The Enlightenment thus articulated a 'hierarchy of effectivity', which included the Chemical Revolution as a dynamic pattern of 'multiple existences', or distinct disciplinary interests and practices, with a determinate shape and duration.

This interpretive model shows how the different articulations of a core set of ideological principles, grounded ultimately in the differential unfolding of an underlying economic structure and expressed succinctly in the Enlightenment notion of the self-defining subject, shaped the Chemical Revolution as a coherent and dynamic conjunction of distinct, yet interrelated, developments in the ontological, epistemological, methodological, linguistic, theoretical, empirical, instrumental, institutional and material dimensions of eighteenth-century chemistry. But as it stands, this account of the Chemical Revolution is incomplete and one-sided. It focuses almost exclusively on the dialectic between Priestley and Lavoisier, ignores the positive forward-looking thrust of the Enlightenment, and stresses the moment of Priestley's 'refusal' of the new chemistry while saying nothing about its almost complete integration into the British chemical community by the early nineteenth century. What this study does suggest, however, is that future attempts to balance these deficiencies with accounts of the critical and constructive role of the Enlightenment, or its demise, in the assimilation of the new chemistry in the European community as a whole must pay due attention to the inherent complexity and multidimensionality of the Chemical Revolution in particular and scientific change in general.

Robust contextualism offers a more balanced account than is currently available of the moments of continuity and discontinuity in scientific change and development. While sociologists of scientific knowledge eschewed the problematic of scientific change entirely, positivist philosophers of science identified *empirical* continuity with *scientific* continuity and postpositivist philosophers of science identified *theoretical* discontinuity with *scientific* discontinuity because they mistook a partial monomial aspect or single level of science for its complex, polynomial identity. This identity implies the complexity of scientific change, which can involve continuity and cumulativity at any one level, such as the empirical or ontological, and discontinuity at any other level, such as the theoretical or methodological. This analysis undercuts the relativist account of scientific development inherent in the fashionable concept of incommensurability, according to which competing (global) theories are sufficiently incongruous to rule out the possibility of comparison on a shared set of criteria. This image of scientific change has left the lasting image of a yawing gap between the two great protagonists of the Chemical Revolution, Priestley and Lavoisier. But, as Larry Holmes recently argued, Priestley and Lavoisier had more in common than they had differences. The dispute between them was not that between an entrenched defender of the 'reigning paradigm' of chemistry and its usurper, but a competition between 'competing research programmes' equally, but in different ways, at odds with the reigning paradigm. But whereas Holmes used his analysis of the conflict between Priestley and Lavoisier to support his decentred view of the Chemical Revolution as a set of specific events in local contexts, robust contextualism links the differences between Priestley and Lavoisier to divergent articulations of their shared Enlightenment heritage. In stressing the dynamic unity-in-diversity that constituted the Chemical Revolution, robust contextualism projects an image of scientific change as a complex phenomenon, constituted by the intermingling and interconnection of contrary moments and differential temporalities, such as the continuous and discontinuous, the gradual and revolutionary, and the progressive and retrogressive.

NOTES

Introduction: The Philosophical and Historiographical Terrain

1. C. Perrin, 'Research Traditions, Lavoisier, and the Chemical Revolution', in A. Donovan (ed.), *The Chemical Revolution: Essays in Reinterpretation, Osiris, 2nd Series, Volume 4* (Philadelphia, PA: History of Science Society, 1988), pp. 53–81, pp. 79–80.

2. See A. Thackray, *Atoms and Powers: An Essay on Newtonian Matter Theory and the Development of Chemistry* (Cambridge, MA: Harvard University Press, 1970), p. vii.

3. The transition from defensive empiricism to greater reflexive sophistication is considered in L. Laudan, *Progress and Its Problems: Toward a Theory of Scientific Growth* (Berkeley, CA: University of California Press, 1977), pp. 164–7; S. Shapin, 'History of Science and Its Sociological Reconstructions', *History of Science*, 20 (1982), pp. 157–211, pp. 157–177.

4. See A. Thackray, 'Science: Has Its Present Past a Future?', in R. H. Steuwer (ed.), *Historical and Philosophical Perspectives of Science* (1970; New York: Gordon and Breach, 1989), pp. 122–33 for a discussion of how the discipline of the history of science functioned as a legitimating activity of science itself.

5. See J. R. R. Christie, 'The Development of the Historiography of Science', in R.C. Olby, G. N. Cantor, J. R. R. Christie and M. J. S. Hodge (eds), *Companion to the History of Modern Science* (London: Routledge, 1990), pp. 5–22; D. Forbes, 'Scientific Whiggism: Adam Smith and John Miller', *Cambridge Journal*, 7 (1954), pp. 643–70.

6. G. Motzkin, 'The Catholic Response to Secularization and the Rise of the History of Science as a Discipline', *Science in Context*, 3 (1989), pp. 203–22, p. 203. See also H. Kragh, *An Introduction to the Historiography of Science* (Cambridge: Cambridge University Press, 1987), pp. 110–11.

7. See G. Sarton, 'The History of Science', *Monist*, 26 (1916), pp. 321–65, pp. 344–65; T. Frängsmyr, 'Science or History: George Sarton and the Positivist Tradition in the History of Science', *Lychnos*, 74 (1973–4), pp. 104–44, pp. 123–8.

8. W. H. Brock, *The Fontana History of Chemistry* (London: Fontana, 1992), pp. xxii–xxiii.

9. J. M. Maienschein, J. M. Laubichler and A. Loettgers, 'How Can the History of Science Matter to Scientists?', *Isis*, 99 (2008), pp. 341–49, pp. 341 and 349.

10. See, respectively, J. Rouse, 'Philosophy of Science and the Persistent Narratives of Modernity', *Studies in History and Philosophy of Science*, 22 (1991), pp. 141–62; R. Illife, 'Rhetorical Vices: Outlines of a Feyerabendian History of Science', *History of Science*, 30 (1992), pp. 199–219.

11. See, e.g. L. Marx, 'Does Improved Technology Mean Progress?', in A. H. Teich (ed.), *Technology and the Future*, 9th edn (Belmont, CA: Thompson Wadsworth, 2003), pp. 3–12; Christie, 'Development of the Historiography of Science', pp. 10–11.

12. S. Shapin, 'Discipline and Bounding. The History of Science as Seen Through the Externalism/Internalism Debate', *History of Science*, 30 (1992), pp. 333–69, p. 339.

13. S. Toulmin, *Foresight and Understanding: An Inquiry into the Aims of Science* (New York: Harper and Rowe, 1977), p. 150.

14. I. B. Cohen, 'The Many Faces of the History of Science. A Font of Examples for Philosophers, a Scientific Type of History, an Archaeology of Discovery, a Branch of Sociology, a Variant of Intellectual History – or *What?*', in F. Delzell (ed.), *The Future of History: Essays in the Vanderbilt University Centennial Symposium* (Nashville, TN: Vanderbilt University Press, 1977), pp. 65–110.

15. For conflicting views of the hegemonic relationship between philosophy, sociology and history see S. Fuller, 'Is History and Philosophy of Science Withering on the Vine?', *Philosophy of the Social Sciences*, 21 (1991), pp. 149–74; J. Habermas, *On the Logic of the Social Sciences*, trans. S. W. Nicholsen and J. A. Stark (Cambridge, MA: MIT Press, 1988), pp. 16–22.

16. R. G. Collingwood, *An Autobiography* (Oxford: Oxford University Press, 1939), p. 132.

17. R. S. Westman and D. C. Lindberg, 'Introduction', in R. S. Westman and D. C. Lindberg (eds), *Reappraisals of the Scientific Revolution* (Cambridge: Cambridge University Press, 1990), pp. xvii–xxvii, p. xiv.

18. See, e.g. A. Richardson, 'Scientific Philosophy as a Topic for the History of Science', *Isis*, 99 (2008), pp. 88–96; M. Friedman, 'History and Philosophy of Science in a New Key', *Isis*, 99 (2008), pp. 125–34.

19. Perrin, 'Research Traditions', pp. 78–80.

20. T. S. Kuhn, *The Structure of Scientific Revolutions*, 2nd edn (1962; Chicago, IL: University of Chicago Press, 1970), p. 3.

21. A. R. Hall, 'On Whiggism', *History of Science*, 21 (1983), pp. 45–59, p. 57.

22. H. Butterfield, *The Whig Interpretation of History* (1931; New York: W. W. Norton, 1965), p. 30.

23. H. Butterfield, *The Englishman and His History* (Hamden, CT: Archon Books, 1970), pp. 3–4.

24. B. Bensaude-Vincent and I. Stengers, *A History of Chemistry*, trans. D. van Dam (Cambridge, MA: Harvard University Press, 1996), p. 4.

25. C. Hartfoot, 'The Missing Synthesis in the Historiography of Science', *History of Science*, 24 (1991), pp. 207–16.

26. See M. Gordy, 'Reading Althusser: Time and the Social Whole', *History and Theory* 22 (1983), pp. 1–21.

27. See Habermas, *On the Logic of the Social Sciences*, pp. 155–7.

28. See B. Bensaude-Vincent, 'Between History and Memory: Centennial and Bicentennial Images of Lavoisier', *Isis*, 87 (1996), pp. 217–37.

29. For this view of the problem field of historiography see J. R. R. Christie, 'Narrative and Rhetoric in Hélène Metzger's Historiography of Eighteenth-Century Chemistry', *History of Science*, 25 (1987), pp. 99–109; P. Anderson, *Arguments within English Marxism* (London: New Left Books, 1980), pp. 1–15.

30. Christie, 'Narrative and Rhetoric', p. 99.

31. L. Laudan, 'The Pseudo-Science of Science', *Philosophy of the Social Sciences*, 11 (1981), pp. 173–98, pp. 18.

32. I. Lakatos, 'History of Science and Its Rational Reconstructions', in C. Howson (ed.), *Method and Appraisal in the Physical Sciences: The Critical Backgroudn to Modern Science, 1800–1905* (Cambridge: Cambridge University Press, 1976), pp. 1–39, pp. 21 and 32.

33. D. Bloor, *Knowledge and Social Imagery*, 2nd edn (1976; Chicago, IL: Chicago University Press, 1991), p. 7.

34. Laudan, *Progress and Its Problems*, p. 201.

35. Ibid.

36. S. Woolgar and M. Ashmore, 'The Next Step: An Introduction to the Reflexive Project', in S. Woolgar (ed.), *Knowledge and Reflexivity: New Frontiers in the Sociology of Knowledge* (Beverly Hills, CA and London: Sage Publications, 1988), pp. 1–11, p. 2.

37. For a critical exploration of the hermeneutic approach to history see Habermas, *On the Logic of the Social Sciences*, pp. 162–75.

38. See E. J. Hobsbawm, *The Age of Revolution, Europe 1789–1948* (London: Abacus, 1977).

39. I. B. Cohen, 'The Eighteenth-Century Origins of the Concept of Scientific Revolution', *Journal of the History of Ideas*, 37 (1976), pp. 257–88, pp. 257–8.

40. I. B. Cohen, *Revolution in Science* (Cambridge, MA: Harvard University Press, 1985), p. 230.

41. J. Priestley, *Experiments and Observations on Different Kinds of Air, and Other Branches of Natural Philosophy Concerned with the Subject, in Three Volumes Being the Former Six Volumes Abridged and Methodized, with Many Additions,* 3 vols (Birmingham: J. Johnson, 1790), vol. 1, p. xxv.

42. J. Priestley, *Experiments and Observations Relating to the Analysis of Atmospherical Air ... to which are added Considerations on the Doctrine of Phlogiston and the Decomposition of Water* (London: J. Johnson, 1796), p. 36.

43. A. Donovan, 'Introduction', in A. Donovan (ed.), *The Chemical Revolution: Essays in Reinterpretation, Osiris, 2nd Series, Volume 4* (Philadelphia, PA: History of Science Society, 1988), pp. 5–12, p. 5; Cohen, *Revolution in Science*, p. 236.

44. Priestley, *Experiments and Observations Relating to the Analysis of Atmospherical Air*, p. 35.

45. See, e.g. A. Donovan, 'The Chemical Revolution Revisited', in S. H. Sutcliffe (ed.), *Science and Technology in the Eighteenth-Century: Essays of the Lawrence Gipson Institute for Eighteenth-Century Studies* (Bethlehem, PA: Gipson Institute, Lehigh University, 1984), pp. 1–15, pp. 6–9.

46. B. Bensaude-Vincent, 'Introductory Essay: A Geographical History of Eighteenth-Century Chemistry', in B. Bensaude-Vincent and F. Abbri (eds), *Lavoisier in European Context: A New Language for Chemistry* (Cambridge, MA: Science History Publications, 1995), pp. 1–17, pp. 8 and 13.

1 Positivism, Whiggism and the Chemical Revolution

1. F. Suppe, 'The Search for Philosophic Understanding of Scientific Theories', in F. Suppe (ed.), *The Structure of Scientific Theories* (Urbana, IL: University of Illinois Press, 1974), pp. 3–232, p. 3. A more complete documentation of the issues covered in this chapter can be found in J. G. McEvoy, 'Positivism, Whiggism and the Chemical Revolution: A Study in the Historiography of Science', *History of Science,* 35 (1997), pp. 1–33.

2. See J. W. Burrow, *A Liberal Descent: Victorian Historians and the English Past* (Cambridge: Cambridge University Press, 1981), p. 296; C. G. Jones, 'The Pathology of English History', *New Left Review*, 46 (1967), pp. 29–43, pp. 30–2.

3. See K Löwith, *Meaning in History: Theological Implications of the Philosophy of History* (Chicago, IL: Chicago University Press, 1949), p. 73; J. B. Bury, *The Idea of Progress: An Inquiry into Its Origin and Growth* (London: Macmillan, 1924), p. 236; Forbes, 'Scientific Whiggism', pp. 643–70.

4. R. Jann, *The Art and Science of Victorian History* (Columbus, OH: Ohio University Press, 1985), p. 208.

5. Butterfield, *The Whig Interpretation of History*, p. v. See also Jones, 'The Pathology of English History', pp. 30–1.

6. Hall, 'On Whiggism', p. 50.

7. Butterfield, *The Whig Interpretation of History*, pp. 19–20, 28–39, 34–9 and 46–7 and 96.

8. J. Agassi, *Towards an Historiography of Science* ('s-Gravenhage: Mouton, 1963), p. 32; G. Gutting, 'Continental Philosophy of Science', in P. D. Asquith and H. E. Kyburg (eds), *Current Research in Philosophy of Science: Proceedings of the PSA Cultural Research Problems Conference* (East Lansing, MI: Philosophy of Science Association, 1979), pp. 94–117, pp. 95–7. For a discussion and evaluation of scientism see J. Habermas, *Knowledge and Human Interests*, trans. J. J. Shapiro (Boston, MA: Beacon Press, 1971), pp. 67–9.

9. On the history and philosophy of positivism see Habermas, *Knowledge and Human Interests*, ch. 4–6. On the contrast between Comtean positivism and Logical Positivism see L. Laudan, *Science and Hypothesis: Historical Essays on Scientific Method* (Dordrecht: D. Reidel), ch. 9 and 13.

10. Habermas, *Knowledge and Human Interests*, p. 71. See also ibid., pp. 67–74.

11. A. Comte, 'The Positive Philosophy of August Comte (Selections)', in P. Gardiner (ed.), *Theories of History: From Classical to Contemporary Sources* (1893; Glenco, IL: Free Press, 1959), pp. 73–82, p. 76.

12. M. Pickering, *Auguste Comte: An Intellectual Biography* (Cambridge: Cambridge University Press, 1983), pp. 277–89 and 683–4.

13. P. Galison, 'History, Philosophy and the Central Metaphor', *Science in Context*, 2 (1988), pp. 197–212.

14. On Logical Positivism as 'a synthesis of traditional empiricism and traditional rationalism' see M. Friedman, *Reconsidering Logical Positivism* (Cambridge: Cambridge University Press, 1999), pp. 5 and 9.

15. Habermas, *Knowledge and Human Interests*, pp. 73–81.

16. See C. G. Hempel, *Aspects of Scientific Explanation and Other Essays in the Philosophy of Science* (New York: Free Press, 1965).

17. See Laudan, *Science and Hypothesis*, ch. 9; Pickering, *Auguste Comte*, pp. 213–4, 294–6 and 567–70.

18. Richardson, 'Scientific Philosophy as a Topic for History of Science', pp. 90–1.

19. L. Laudan, 'The History of Science and the Philosophy of Science', in R. C. Olby, G. N. Cantor, J. R. R. Christie and M. J. S. Hodge (eds), *Companion to the History of Modern Science* (London: Routledge, 1990), pp. 47–59, pp. 42–6.

20. Ibid., pp. 42 and 49.

21. Frängsmyr, 'Science or History', p. 107.

22. G. Buchdahl, *The Image of Newton and Locke in the Age of Reason* (London: Sheed and Ward, 1961), p. 2; Frängsmyr, 'Science or History', pp. 121–3.

23. S. Schaffer, 'Scientific Discoveries and the End of Natural Philosophy', *Social Studies of Science*, 16 (1986), pp. 387–420, pp. 387–90 and 413.

24. G. Sarton, *Introduction to the History of Science*, 3 vols (Baltimore, MD: Williams and Wilkins, 1927–48), vol. 1, p. 32.

25. Frängsmyr, 'Science or History', p. 124.

26. Agassi, *Towards an Historiography of Science*, p. 12.

27. J. A. Schuster and R. R. Yeo, 'Introduction', in J. A. Schuster and R. R. Yeo (eds), *The Politics and Rhetoric of Scientific Method: Historical Studies* (Dordrecht: D. Reidel, 1986), pp. ix–xxvii, pp. vii–ix.

28. J. B. Conant, 'Introduction', in J. B Conant and L. K. Nash (eds), *Harvard Case Histories in Experimental Science* (Cambridge, MA: Harvard University Press, 1957), pp. vii–xvi, p. ix.

29. See, e.g. A. K. Mayer, 'Moralizing Science: The Uses of Science's Past in National Education in the 1920s', *British Journal for the History of Science*, 30 (1997), pp. 51–70.

30. On identity philosophy see A. Callinicos, *Althusser's Marxism* (London: Pluto Press, 1976), p. 19 and ch. 3; P. Dews, *Logics of Disintegration: Poststructuralist Thought and the Claims of Critical Theory* (London: Verso, 1987), ch. 1 and 3.

31. Sarton, *Introduction to the History of Science*, vol. 3, p. 15; Frängsmyr, 'Science or History', pp. 110–13.

32. See Agassi, *Towards an Historiography of Science*, pp. 1–3.

33. Sarton, *Introduction to the History of Science*, vol. 3, p. 15.

34. On the concept of 'expressive causality' see Callinicos, *Althusser's Marxism*, pp. 40–1.

35. See T. Benton, *The Rise and Fall of Structural Marxism: Althusser and His Influence* (London: Macmillan, 1984), pp. 59–61.

36. Butterfield, *The Whig Interpretation of History*, p. 8.

37. See Gordy, 'Reading Althusser. Time and the Social Whole', pp. 1–21, pp. 1–4.

38. See K. R. Popper, *The Poverty of Historicism* (Boston, MA: Beacon Press, 1957).

39. Kuhn, *Structure of Scientific Revolutions*, pp. 1–2; Agassi, *Towards an Historiography of Science*, pp. 2, 7 and 25.

40. H. Butterfield, *The Origins of Modern Science, 1300–1800* (New York: Free Press, 1965), ch. 11.

41. See L. Laudan, *Science and Values: An Essay on the Aims of Science and their Role in Scientific Debate* (Berkeley, CA: University of California Press, 1984), pp. 1–13.

42. See Lakatos, 'History of Science', pp. 2–9; Laudan, *Science and Hypothesis*, pp. 91–3.

43. See I. Lakatos, 'Falsification and the Methodology of Scientific Research Programmes', in I. Lakatos and A. Musgrave (eds), *Criticism and the Growth of Knowledge* (Cambridge: Cambridge University Press, 1970), pp. 91–196, pp. 97–103.

44. S. Toulmin, 'Crucial Experiments: Priestley and Lavoisier', *Journal of the History of Ideas*, 18 (1957), pp. 205–22, p. 206.

45. Butterfield, *The Whig Interpretation of History*, pp. 64–5.

46. J. R. Partington, *A Short History of Chemistry* (London: Macmillan, 1937), p. 129; Butterfield, *The Origins of Modern Science, 1300–1800*, p. 204.

47. Butterfield, *The Origins of Modern Science, 1300–1800*, p. 208; C. C. Gillispie, *The Edge of Objectivity: An Essay in the History of Scientific Ideas* (Princeton, NJ: Princeton University Press, 1960), p. 205; See B. Bensaude-Vincent, 'A Founder Myth in the History of Science? – the Lavoisier Case', in L. Graham, P. Lepinies and W. Weingart (eds),

Functions and Uses of Disciplinary Histories (Sociology of Sciences Yearbook, Volume vii) (Dordrecht: D. Reidel, 1983), pp. 53–78, p. 53.

48. R. P. Multhauf, 'On the Use of the Balance in Chemistry', *Proceedings of the American Philosophical Society*, 106 (1965), pp. 210–18, pp. 210, 213 and 218.

49. H. Hartley, *Studies in the History of Chemistry* (Oxford: Clarendon Press, 1971), pp. 27–8.

50. See, e.g. H. Kopp, *Beiträge zur Geschichte der Chemie* (Braunschweig: F. Vieweg, 1878), pp. 70–7; P. Duhem, *Le Mixte et le Combinasion Chimique: Essay sur l'Evolution d'une Idée* (Paris: Fayard, 1902), p. 17; T. L. Davis, 'The First Edition of the *Skeptical Chymist*', *Isis*, 8 (1926), pp. 71–6, p. 71.

51. See, e.g. G. F. Rodwell, 'On the Theory of Phlogiston', *Philosophical Magazine*, 35 (1868), pp. 1–32, pp. 26–32; J. H. White, *The History of the Phlogiston Theory* (London: E. Arnold, 1932), pp. 1–14; D. McKie, *Antoine Lavoisier: The Father of Modern Chemistry* (Philadelphia, PA: J. B. Lippincott, 1935), pp. 61–4.

52. Bensaude-Vincent, 'A Founder Myth in the History of Science?', pp. 55 and 69–76; Donovan, 'The Chemical Revolution Revisited', p. 9.

53. See C. Meinel, 'Demarcation Debates: Lavoisier in Germany', in M. Beretta (ed.), *Lavoisier in Perspective* (Munich: Deutsches Museum, 2005), pp. 153–66.

54. See, e.g. M. Muir, *Heroes of Science: Chemists* (New York: E. and J. B. Young, 1883), p. 76. See also McEvoy, 'Positivism, Whiggism and the Chemical Revolution', p. 30, n. 58.

55. H. Guerlac, *Lavoisier – the Crucial Year: The Background and Origin of His First Experiments in Combustion in 1772* (Ithaca, NY: Cornell University Press, 1961), p. xiv.

56. F. Engels, 'Preface', in K. Marx, *Capital*, 3 vols (1867–94; New York International Publishers, 1967), vol. 2, pp. 1–19, pp. 14–15. See also A. N. Whitehead, *Science and the Modern World* (New York: Macmillan, 1925), p. 60.

57. Butterfield, *The Origins of Modern Science 1300–1800*, p. 20.

58. White, *The History of the Phlogiston Theory*, pp. 11–12; Butterfield, *The Origins of Modern Science 1300–1800*, ch. 11; M. P. Crosland, 'Chemistry and the Chemical Revolution', in G. S. Rousseau and R. Porter (eds), *The Ferment of Knowledge: Studies in the Historiography of Eighteenth-Century Science* (Cambridge: Cambridge University Press, 1980), pp. 389–416, p. 392.

59. See Donovan, 'Introduction', p. 11; Schaffer, 'Scientific Discoveries', pp. 406–13.

60. See Bensaude-Vincent, 'A Founder Myth in the History of Science?', pp. 54–60; M. Beretta, *The Enlightenment of Matter: The Definition of Chemistry from Agricola to Lavoisier* (Canton, MA: Science History Publications, 1993), pp. 197–206.

61. W. C. Dampier, *A History of Science and its Relation with Philosophy and Religion* (Cambridge: Cambridge University Press, 1929), p. 184; Crosland, 'Chemistry and the Chemical Revolution', p. 393.

62. On the historiography of Boyle's role in the Chemical Revolution see especially T. S. Kuhn, 'Robert Boyle and Structural Chemistry in the Seventeenth Century', *Isis*, 43 (1952), pp. 1–23.

63. Davis, 'The First Edition of the *Skeptical Chymist*', p. 79; Agassi, *Towards an Historiography of Science*, p. 44. See also Cochrane, *Lavoisier*, pp. 76–7; McEvoy, 'Positivism, Whiggism and the Chemical Revolution', p. 31, n. 63.

64. Davis, 'The First Edition of the *Skeptical Chymist*', pp. 71 and 91; Kuhn, 'Robert Boyle and Structural Chemistry', p. 14.

65. For a helpful summary of this view of chemistry in the early eighteenth century see Butterfield, *The Origins of Modern Science, 1300–1800*, pp. 202–13.

66. See, e.g. Rodwell, 'On the Theory of Phlogiston', pp. 26–32; White, *The History of the Phlogiston Theory*, pp. 11–14 and 88; Gillispie, *The Edge of Objectivity*, pp. 204–5.
67. J. R. Partington and D. McKie, 'Historical Studies on the Phlogiston Theory – IV Last Phases of the Theory', *Annals of Science*, 4 (1939), pp. 113–49, p. 149; Butterfield, *The Origins of Modern Science, 1300–1800*, pp. 210–11. See also McEvoy, 'Positivism, Whiggism and the Chemical Revolution', p. 31, n. 64.
68. See McEvoy, 'Positivism, Whiggism and the Chemical Revolution', p. 31, n. 65.
69. Gillispie *The Edge of Objectivity*, pp. 204–5 and 231; M. P. Crosland, 'The Development of Chemistry in the Eighteenth Century', *Studies on Voltaire and the Eighteenth Century*, 34 (1963), pp. 369–442, p. 397; Multhauf, 'On the Use of the Balance in Chemistry', pp. 213–15. See also McEvoy, 'Positivism, Whiggism and the Chemical Revolution', p. 31, n. 66.
70. Hartley, *Studies in the History of Chemistry* p. 27; McKie, *Antoine Lavoisier,* pp. 209–11.
71. See A. R. Hall, *The Scientific Revolution, 1500–1800: The Formation of the Modern Scientific Attitude* (London: Langmans, 1962), pp. 334–7.
72. See Hartley, *Studies in the History of Chemistry*, p. 7; Rodwell, 'On the Theory of Phlogiston', p. 31.
73. J. B. Conant, 'The Overthrow of the Phlogiston Theory: The Chemical Revolution of 1775–1789', in J. B. Conant and L. K. Nash (eds), *Harvard Case Studies in Experimental Science*, 2 vols (Cambridge, MA: Harvard University Press, 1957), vol. 1, pp. 65–116, p. 111; J. C. Gregory, *Combustion from Heraclitus to Lavoisier* (London: Edward Arnold, 1934), p. 21. See also F. W. Gibbs, *Joseph Priestley: Adventure in Science and Champion of Truth* (London: Nelson, 1965), p. 83; A. Musgrave, 'Why did Oxygen Supplant Phlogiston? Research Programmes in the Chemical Revolution', in C. Howson (ed.), *Method and Appraisal in the Physical Sciences: The Critical Background to Modern Science* (Cambridge: Cambridge University Press, 1976), p. 186.
74. See Rodwell, 'On the Theory of Phlogiston', p. 29. See also McEvoy, 'Positivism, Whiggism and the Chemical Revolution', p. 32. n. 71.
75. Gillispie, *The Edge of Objectivity*, pp. 212–13.
76. Butterfield, *The Origins of Modern Science,* p. 211; Gillispie, *The Edge of Objectivity*, p. 218; Partington, *A Short History of Chemistry*, p. 150; Rodwell, 'On the Theory of Phlogiston', p. 3.
77. See J. G. McEvoy, 'Joseph Priestley, "Aerial Philosopher": Metaphysics and Methodology in Priestley's Chemical Thought, 1772–1781, Part 1', *Ambix*, 25 (1978), pp. 1–55, p. 2. See also ibid., pp. 1–4 for a discussion of the historiography of Priestley's role in the Chemical Revolution.
78. See S. Schaffer, 'Priestley Questions: An Historiographic Survey', *History of Science*, 22 (1984), pp. 151–83, pp. 152–4.
79. Gillispie, *The Edge of Objectivity*, p. 211.
80. See, e.g. Gibbs, *Joseph Priestley*, pp. 117–18; P. Hartog, 'The Newer Views of Priestley and Lavoisier', *Annals of Science*, 5 (1941), pp. 1–56, p. 27.

2 Postpositivism and the Historiography of Science

1. Guerlac, *Lavoisier – the Crucial Year*, p. xvi.
2. Perrin, 'Research Traditions', p. 54.
3. Kuhn, 'Robert Boyle and Structural Chemistry', pp. 14–36.

4. Ibid., p. 36.

5. J. G. McEvoy, 'Continuity and Discontinuity in the Chemical Revolution', in A. Donovan (ed.), *The Chemical Revolution: Essays in Reinterpretation, Osiris, 2nd Series, Volume 4* (Philadelphia, PA: History of Science Society, 1988), pp. 195–213, pp. 203–9.

6. R. Siegfried and B. J. Dobbs, 'Composition: A Neglected Aspect of the Chemical Revolution', *Annals of Science*, 24 (1968), pp. 275–93.

7. Toulmin, 'Crucial Experiments'.

8. Musgrave, 'Why did Oxygen Supplant Phlogiston?', pp. 182–7.

9. McEvoy, 'Continuity and Discontinuity in the Chemical Revolution', pp. 199–208.

10. See R. E. Schofield, 'The Scientific Background of Joseph Priestley', *Annals of Science*, 13 (1957), pp. 148–63; J. G. McEvoy, 'Enlightenment and Dissent in Science: Joseph Priestley and the Limits of Theoretical Reasoning', *Enlightenment and Dissent*, 1 (1983), pp. 47–67, pp. 56–9.

11. See note 10 to the Introduction, above.

12. See, e.g. M. Schlick, 'The Foundations of Knowledge', in A. J. Ayer (ed.), *Logical Positivism* (1934; New York: Free Press, 1959), pp. 209–27, p. 226.

13. See Benton, *The Rise and Fall of Structural Marxism*, pp. 10–14; T Eagleton, *Literary Theory: An Introduction* (Minneapolis, MN: University of Minnesota Press, 1983), pp. 91–126; J. Golinski, *Making Natural Knowledge: Constructivism and the History of Science* (Cambridge: Cambridge University Press, 1998), ch. 1 and 5.

14. A. Callinicos, *Against Postmodernism: A Marxist Critique* (New York: St. Martin's Press, 1989), pp. 107–111.

15. K. R. Popper, *The Logic of Scientific Discovery* (1935; London: Hutchinson, 1968), pp. 107–111.

16. T. W. Adorno, 'Why Philosophy', in D. Ingram and J. Simon-Ingram (eds), *Critical Theory: The Essential Readings* (1963; New York: Paragon House, 1992), pp. 20–30, p. 24; K. R. Popper, *Conjectures and Refutations: The Growth of Scientific Knowledge* (New York: Harper Torchbooks, 1968), pp. 33–6 and 215–20.

17. W. V. O. Quine, 'Two Dogmas of Empiricism', in M. Curd and J. A. Clover (eds), *Philosophy of Science. The Central Issues* (1953; London: Macmillan, 1998), pp. 280–302, pp, 296–7.

18. K. R, Popper, *Objective Knowledge: An Evolutionary Approach* (Oxford: Oxford University Press, 1972), pp. 106–52.

19. Popper, *The Logic of Scientific Discovery*, p. 105.

20. Popper used the term 'logic of discovery' to denote the 'logic of science', by which he meant the (logical) context of justification, not the (psychological) context of discovery.

21. Popper, *The Logic of Scientific Discovery*, p. 42.

22. See ibid., pp. 106–11.

23. Schaffer, 'Scientific Discoveries', p. 413.

24. See, e.g. Popper, *The Logic of Scientific Discovery*, pp. 272–8.

25. Ibid., pp. 79–92.

26. Ibid., p. 107.

27. T. S. Kuhn, 'Logic of Discovery or Psychology of Research', in I. Lakatos and A. Musgrave (eds), *Criticism and the Growth of Knowledge* (Cambridge: Cambridge University Press, 1970), pp. 1–23, p. 15. Kuhn's integration of semantic considerations into Popper's syntactical theory of meaning served to distinguish his idealism from Popper's realism.

28. Ibid., p. 16.

29. See N. R. Hanson, *Patterns of Discovery: An Inquiry into the Conceptual Foundations of Science* (London: The Scientific Book Guild, 1962), pp. 4–30. See also L. Wittgenstein, *Philosophical Investigations*, trans. G. E. M. Anscombe (New York: Macmillan, 1953), p. 5.

30. Hanson, *Patterns of Discovery*, pp. 18–19 and 72; Kuhn, *The Structure of Scientific Revolutions*, pp. 111–38.

31. See S. Toulmin, *The Philosophy of Science* (1953; London: Arrow Books, 1962), pp. 88–9 and 93–4; S, Toulmin, *Human Understanding: The Collective Use and Evolution of Concepts* (Princeton, NJ: Princeton University Press, 1972), pp. 37, 67, 106 and 133–8.

32. See M. Polanyi, *Personal Knowledge: Towards a Post-Critical Philosophy* (Chicago, IL: University of Chicago Press, 1958), p. 60.

33. Kuhn, *The Structure of Scientific Revolutions*, pp. 44–6, 91 and 191. See also Wittgenstein, *Philosophical Investigations*, pp. 32–3; Golinski, *Making Natural Knowledge*, pp. 15–17.

34. P. K. Feyerabend, 'Reply to Criticism: Some Comments on Smart, Sellars and Putnam', in R. S. Cohen and M. Wartofsky (eds), *Boston Studies in the Philosophy of Science* (New York: Humanities Press, 1965), pp. 223–61, p. 223.

35. Lakatos, 'History of Science', p. 91.

36. See Musgrave, 'Why did Oxygen Supplant Phlogiston?'

37. Lakatos, 'History of Science', p. 18.

38. See Laudan, *Progress and Its Problems*, pp. 196–222 for a critical discussion of the arationality assumption and the standoff between philosophers of science and sociologists of scientific knowledge.

39. Wittgenstein, *Philosophical Investigations*, p. 49.

40. Toulmin, *Human Understanding*, pp. 85–96 and 478–84.

41. Kuhn, 'Logic of Discovery or Psychology of Research', p. 20; Kuhn, *The Structure of Scientific Revolutions*, pp. 8–9. On the prescriptive/descriptive distinction see T. Nickles, 'Remarks on the Use of History as Evidence', *Synthese*, 69 (1986), pp. 253–66.

42. T. S. Kuhn, *The Essential Tension: Selected Studies in Scientific Tradition and Change* (Chicago, IL: University of Chicago Press, 1977), p. 18.

43. Kuhn, 'Logic of Discovery or Psychology of Research', p. 21; Kuhn, *The Structure of Scientific Revolutions*, pp. 178–81.

44. See, e.g. Golinski, *Making Natural Knowledge*, pp. 19–21.

45. Kuhn, *The Essential Tension*, pp. 118–20.

46. Kuhn, *The Structure of Scientific Revolutions*, pp. 8–9; Laudan, *Progress and Its Problems*, p. 156.

47. See Laudan, 'The History of Science and the Philosophy of Science', pp. 47–9.

48. Lakatos, 'History of Science ', p. 7; Agassi, *Towards an Historiography of Science*, pp. 64–74.

49. Laudan, *Progress and Its Problems*, p. 72.

50. R. M. Burian, 'More than a Marriage of Convenience: On the Inextricability of History and Philosophy of Science', Philosophy of Science, 44 (1977), pp. 1–42, p. 1.

51. See M. Masterman, 'The Nature of a Paradigm', in I. Lakatos and A. Musgrave (eds), *Criticism and the Growth of Knowledge* (Cambridge: Cambridge University Press, 1970), pp. 59–89. Masterman incorrectly assimilated Kuhn's concept of a paradigm to Mary Hesse's account of the essential role of concrete models in the interpretation of abstract calculi in science. But Kuhn used the term 'paradigm' to highlight the 'tacit dimension' of meaning in both concrete models and abstract calculi.

52. See Kuhn, The Structure of Scientific Revolutions, pp. 176–98. See also Golinski, Making Natural Knowledge, pp. 14–27.

53. T. S. Kuhn, 'Reflections on My Critics', in I. Lakatos and A. Musgrave (eds), *Criticism and the Growth of Knowledge* (Cambridge: Cambridge University Press), pp. 231–77, pp. 241–59.

54. On the ambiguities in Kuhn's concept of incommensurability see G. Doppelt, 'Kuhn's Epistemological Relativism: An Interpretation and Defense', *Inquiry*, 21 (1978), pp. 33–86.

55. Kuhn, *The Structure of Scientific Revolutions*, pp. 118, 169, 193 and 206.

56. Lakatos, 'Falsification and the Methodology of Scientific Research Programmes', p. 115.

57. Ibid., pp. 119 and 132–8.

58. Ibid., pp. 102–5, 154–9; Lakatos, 'History of Science', pp. 8–12.

59. P. K. Feyerabend, 'Consolations for the Specialist', in I. Lakatos and M. Musgrave (eds), *Criticism and the Growth of Knowledge* (Cambridge: Cambridge University Press, 1970), pp. 197–230, pp. 214–19.

60. On the developmental trajectory of Feyerabend's philosophical career see J. G. McEvoy, 'A Revolutionary Philosophy of Science: Paul K. Feyerabend and the Degeneration of Critical Rationalism into Sceptical Fallibilism', *Philosophy of Science*, 47 (1975), pp. 49–66.

61. Laudan, *Progress and Its Problems*, p. 6.

62. Ibid., pp. 170 and 210.

63. Laudan, *Progress and Its Problems*, pp. 75, 95–100 and 106–19.

64. Ibid., ch.2; Toulmin, *Human Understanding*, pp. 173–91; G. Buchdahl, 'History of Science and Criteria of Choice', in R. H. Stuewer (ed.), Historical and Philosophical Perspectives of Science (1970; New York: Gordon and Breach, 1989), pp. 204–30, pp. 207–13.

65. See G. Doppelt, 'Review Discussion: Laudan's Pragmatic Alternative to Relativist and Holist Theories of Science', *Inquiry*, 24 (1981), pp. 253–71, p. 269; H. LeGrand, 'Theory and Application: The Early Chemical Works of J. A. C. Chaptal', *British Journal for the History of Science*, 17 (1984), pp. 31–46, p. 39; Toulmin, *Human Understanding*, pp. 41–130; McEvoy, 'Continuity and Discontinuity in the Chemical Revolution'.

66. Laudan, *Progress and Its Problems*, pp. 139–43; Laudan, *Science and Values*, ch. 1, 2 and 3.

67. Ibid., pp. 50 and 62–6.

68. See G. Gutting, *Michel Foucault's Archaeology of Scientific Reason* (Cambridge: Cambridge University Press, 1989), p. 4. See also ibid., pp. 9–13.

69. See R. Bhaskar, 'Feyerabend and Bachelard: Two Philosophies of Science', *New Left Review*, 94 (1974), pp. 31–55.

70. Gutting, *Michel Foucault's Archaeology of Scientific Reason*, p. 20.

71. Ibid., pp. 25–6; Bhaskar, 'Feyerabend and Bachelard', p. 5; P. Dews, 'Althusser, Structuralism and the French Epistemological Tradition', in G. Elliot (ed.), *Althusser: A Critical Reader* (London: Blackwell, 1994), pp. 104–41, p. 123.

72. Gutting, *Michel Foucault's Archaeology of Scientific Reason*, p. 31.

73. On Canguilheim's philosophy of science see ibid., pp. 32–54.

74. See L. Althusser, *For Marx*, trans. B. Brewster (New York: Vintage Books, 1971), pp. 219–42. See also Benton, *The Rise and Fall of Structural Marxism*, pp. 10–14.

75. See, e.g. Althusser, *For Marx*, pp. 21–34; L. Althusser and É. Balibar, *Reading Capital*, trans. B. Brewster (London: New Left Books, 1970), pp. 13–198.

76. J. Cronin, *Foucault's Antihumanist Historiography* (Lewiston, NY: Edwin Mellen Press, 2001), p. 1; Dews, 'Althusser, Structuralism, and the French Epistemological Tradition', p. 123.

77. See M. Foucault, *Mental Illness and Psychology* (1954), trans. A. M. Sheridan-Smith (Berkeley, CA: University of California Press, 1987), p. 69.

78. See M. Foucault, *The Birth of the Clinic: An Archaeology of Medical Perception* (1963), trans. A. M. Sheridan-Smith (New York: Vintage Books, 1973).

79. M. Foucault, *The Order of Things: An Archaeology of the Human Sciences* (1966), trans A. M. Sheridan-Smith (New York: Vintage Books, 1973), p. xxii.

80. See M. Foucault, *The Archaeology of Knowledge and the Discourse on Language*, trans. A. M. Sheridan-Smith (New York, Pantheon Books, 1972).

81. See M. Biagioli, 'Meyerson: Science and the "Irrational"', *Studies in History and Philosophy of Science*, 94 (1988), pp. 5–42, pp. 15–16 and 34–76; Christie, 'The Development of the Historiography of Science', pp. 16–18.

82. A. Koyré, 'Commentary', in A. C. Crombie (ed.), *Scientific Change: Historical Studies on the Intellectual Social, and Technical Conditions for Scientific Discovery and Technical Invention from Antiquity to the Present* (Ithaca, NY: Cornell University Press, 1963), pp. 847–57, p. 856.

83. D. C. Lindberg, 'Conceptions of the Scientific Revolution from Bacon to Butterfield', in D. C. Lindberg and R. S. Westman (eds), *Reappraisals of the Scientific Revolution* (Cambridge: Cambridge University Press, 1990), pp. 1–26, p. 19.

84. See e.g. Christie, 'The Development of the Historiography of Science', pp. 17–18.

85. For the influence of Cold War perspectives on British historians of science after World War 2 see A.-K. Mayer, 'Setting Up a Discipline, II: British History of Science and the "End of Ideology"', *Studies in History and Philosophy of Science*, 35 (2004), pp. 41–72, pp. 56–64. pp. 56–64.

86. R. M. Young, 'Marxism and the History of Science', in R. C. Olby, G. N. Cantor, J. R. R. Christie and M. J. S. Hodge (eds), *Companion to the History of Modern Science* (London: Routledge, 1990), pp. 77–86, pp. 83–4.

87. Koyré, 'Commentary', p. 856.

88. H. Guerlac, 'Discussion', in A. C. Crombie (ed.), *Scientific Change: Historical Studies on the Intellectual Social, and Technical Conditions for Scientific Discovery and Technical Invention from Antiquity to the Present* (Ithaca, NY: Cornell University Press, 1963), pp. 875–6, p. 875.

89. R. S. Westfall, 'Marxism and the History of Science: Reflections on Ravetz's Essay', *Isis*, 72 (1951), pp. 402–5, p. 405.

90. See B. Bensaude-Vincent, 'Chemistry in the French Tradition of the Philosophy of Science: Duhem, Meyerson, Metzger, and Bachelard', *Studies in History and Philosophy of Science*, 36 (2005), pp. 627–48.

91. Christie, 'Narrative and Rhetoric', pp. 102–5.

92. B. Bensaude-Vincent, 'Hélène Metzger's *La Chimie*: A Popular Treatise', *History of Science*, 25 (1987), pp. 71–84, p. 79.

3 Postpositivist Interpretations of the Chemical Revolution

1. H. Guerlac, 'Some French Antecedents of the Chemical Revolution', in H. Guerlac, *Essays and Papers in the History of Modern Science* (1959; Baltimore, MD: Johns Hopkins University Press, 1977), pp. 34–74, p. 361.

2. C. E. Perrin, 'Revolution or Reform: The Chemical Revolution and Eighteenth-Century Concepts of Scientific Change', *History of Science*, 25 (1987), pp. 395–423, pp. 417–19.

3. Guerlac, 'Some French Antecedents of the Chemical Revolution', pp. 362–3.

4. M. Beretta, 'Chemists in the Storm: Lavoisier, Priestley and the Chemical Revolution', *Nuncius*, 8 (1993), pp. 75–104, p. 99.

5. F. Verbruggen, 'How to Explain Priestley's Defense of Phlogiston', *Janus*, 54 (1972), pp. 47–89, p. 66.

6. Toulmin, 'Crucial Experiments', pp. 214 and 220.

7. Ibid., p. 218.

8. Musgrave, 'Why did Oxygen Supplant Phlogiston?', pp. 205–7.

9. Kuhn, *The Structure of Scientific Revolutions*, pp. 69–72 and 107.

10. Ibid., p. 107; Doppelt, 'Kuhn's Epistemological Relativism', pp. 42–5.

11. K. Hufbauer, *The Formation of the German Chemical Community (1720–1795)* (Berkeley, CA: University of California Press, 1982), p. 3.

12. H. G. McCann, *Chemistry Transformed: The Paradigmatic Shift from Phlogiston to Oxygen* (Norwood, NJ: Ablex Publishing, 1978), p. 124.

13. Crosland, 'Chemistry and the Chemical Revolution', pp. 402 and 406.

14. R. E. Schofield, *Mechanism and Materialism: British Natural Philosophy in an Age of Reason* (Princeton, NJ: Princeton University Press, 1970), pp. 230–1 and 272–3.

15. J. B. Gough, 'Lavoisier and the Fulfillment of the Stahlian Revolution', in A. Donovan (ed.), *The Chemical Revolution: Essays in Reinterpretation, Osiris, 2nd Series, Volume 4* (Philadelphia, PA: History of Science Society, 1988), pp. 15–33, p. 20; R. Siegfried, 'Lavoisier and the Phlogistic Connection', *Ambix*, 36 (1989), pp. 31–40, pp. 33 and 37.

16. M. P. Crosland, 'Priestley Memorial Lecture: A Practical Perspective on Joseph Priestley as Natural Philosopher', *British Journal for the History of Science*, 16 (1983), p. 223–38, p. 237.

17. J. H. Brooke, '"A Sower Went Forth": Joseph Priestley and the Ministry of Reform', in T. A. Schwartz and J. G. McEvoy (eds), *Motion Towards Perfection: The Achievement of Joseph Priestley* (Boston, MA: Skinner House Books, 1990), pp. 21–56, p. 24. A Donovan, *Antoine Lavoisier: Science, Administration and Revolution* (Oxford: Blackwell, 1993), p. 139.

18. Toulmin, 'Crucial Experiments', p. 220; Kuhn, *The Structure of Scientific Revolutions*, p. 151.

19. Kuhn, *The Structure of Scientific Revolutions*, p. 159.

20. McCann, *Chemistry Transformed*, pp. 56–7.

21. Musgrave, 'Why did Oxygen Supplant Phlogiston?', pp. 201–3.

22. See R. E. Schofield, *A Scientific Autobiography of Joseph Priestley (1733–1804); Selected Scientific Correspondence: Edited with Commentary by Robert E. Schofield* (Cambridge, MA: MIT Press, 1966), p. 271.

23. See P. Kitcher, *The Advancement of Science: Science Without Legend, Objectivity Without Illusions* (Oxford: Oxford University Press, 1983), ch. 3, 4 and 5.

24. Ibid., pp. 78, 97–103 and 272–90.

25. Kitcher, *The Advancement of Science*, p. 100.

26. Guerlac, *Lavoisier – the Crucial Year*, pp. 5–8, 96 and 192–6.

27. Ibid., pp. xviii–xix.

28. See e.g. Gough, 'Lavoisier and the Fulfillment of the Stahlian Revolution'.

29. M. A. Finocchiaro, *History of Science as Explanation* (Detroit, MI: Wayne State University Press, 1973), p. 83.

30. R. E. Kohler, 'The Origins of Lavoisier's First Experiments on Combustion', *Isis*, 63 (1972), pp. 349–66.

31. M. P. Crosland, 'Lavoisier's Theory of Acidity', *Isis*, 64 (1973), pp. 306–25.

32. H. E. LeGrand, 'Lavoisier's Oxygen Theory of Acidity', *Annals of Science*, 29 (1972), pp. 1–18.

33. J. B. Gough, 'Lavoisier's Early Career in Science: An Examination of Some New Evidence', *British Journal for the History of Science*, 4 (1968), pp. 52–7.

34. M. Fichman, 'French Stahlism and Chemical Studies of Air', *Ambix*, 18 (1971), pp. 94–122.

35. See J. A. Gough, 'The Origins of Lavoisier's Theory of the Gaseous State', in H. Wolf (ed.), *The Analytic Spirit: Essays in the History of Science in Honor of Henry Guerlac* (Ithaca, NY: Cornell University Press, 1981), pp. 15–39.

36. Fichman, 'French Stahlism and Chemical Studies of Air', p. 122.

37. R. Siegfried, 'Lavoisier's View of the Gaseous State and Its Early Application to Pneumatic Chemistry', *Isis*, 63 (1972), pp. 59–78, p. 78.

38. S. H. Mauskopf, 'Gunpowder and the Chemical Revolution', in A. Donovan, *The Chemical Revolution: Essays in Reinterpretation, Osiris, 2nd Series, Volume 4)*, (Philadelphia, PA: History of Science Society, 1988), pp. 93–118, p. 93.

39. R. J. Morris, 'Lavoisier and the Caloric Theory', *British Journal for the History of Science*, 6 (1972), pp. 1–38, pp. 1 and 38.

40. Gough, 'The Origins of Lavoisier's Theory of the Gaseous State', p. 18.

41. Siegfried, 'Lavoisier's View of the Gaseous State', pp. 7–8 and 59.

42. J. R. R. Christie, 'Ether and the Science of Chemistry: 1740–1790', in G. Cantor and M. J. S. Hodge (eds), *Conceptions of the Ether: Studies in the History of Ether Theories, 1740–1790* (Cambridge: Cambridge University Press, 1981), pp. 85–110.

43. Gough, 'The Origins of Lavoisier's Theory of the Gaseous State', p. 57.

44. C. E. Perrin, 'Document, Text and Myth: Lavoisier's Crucial Year Revisited', *British Journal for the History of Science*, 22 (1989), pp. 3–25, p. 23.

45. Perrin, 'Research Traditions', pp. 55–9.

46. Ibid., p. 157.

47. Ibid., pp. 59–64.

48. Ibid., pp. 64–9 and 73–4.

49. Ibid., pp. 77–8.

50. Perrin, 'Document, Text and Myth', p. 24.

51. F. L. Holmes, 'Lavoisier's Conceptual Passage', in A. Donovan *The Chemical Revolution: Essays in Reinterpretation, Osiris, 2nd series, Volume 4* (Philadelphia, PA: History of Science Society, 1988), pp. 82–92, p. 82.

52. F. L. Holmes, *Lavoisier and the Chemistry of Life: An Exploration of Scientific Creativity* (Madison, WI: University of Wisconsin Press, 1985), p. 497.

53. F. L. Holmes, *Eighteenth-Century Chemistry as an Investigative Enterprise* (Berkeley, CA: Office for History of Science and Technology, University of California, 1989), p. 6; Holmes, *Lavoisier and the Chemistry of Life*, p. 268.

54. Holmes, *Eighteenth-Century Chemistry as an Investigative Enterprise*, p. 107.

55. See F. L. Holmes, 'The "Revolution in Chemistry and Physics": Overthrow of a Reigning Paradigm or Competition Between Competing Research Programmes', *Isis*, 91 (2000), pp. 735–53.

56. Holmes, *Eighteenth-Century Chemistry as an Investigative Enterprise*, pp. 102 and 126.

57. Holmes, *Lavoisier and the Chemistry of Life*, pp. 501–2; Holmes, *Eighteenth-Century Chemistry as an Investigative Enterprise*, p. 126.

58. Donovan, 'The Chemical Revolution Revisited', p. 2; A. Donovan, 'Lavoisier and the Origins of Modern Chemistry', in A. Donovan *The Chemical Revolution: Essays in Reinterpretation, Osiris, 2 series, volume 4* (Philadelphia, PA: History of Science Society, 1988), pp. 214–31, pp. 214, 223 and 226–8.

59. Donovan, 'Lavoisier and the Origins of Modern Chemistry', p. 231.

60. A. Donovan, 'Buffon, Lavoisier and the Transformation of French Chemistry', in J. Guyon (ed.), *Buffon 88: Actes du Colloque International pour le Bicentennaire de le Mort de Buffon* (Paris: Vrin, 1992), pp. 387–95, pp. 387–8; A Donovan, 'British Chemistry and the Concept of Science in the Eighteenth Century', *Albion*, 7 (1975), pp. 131–44, pp. 143–4.

61. E. Melhado, 'Toward an Understanding of the Chemical Revolution', *Knowledge and Society – Studies in the Sociology of Culture Past and Present*, 8 (1989), pp. 123–37, pp. 123–4.

62. E. Melhado, 'Chemistry, Physics and the Chemical Revolution', *Isis*, 76 (1985), pp. 195–211, p. 195; E. Melhado, 'Metzger, Kuhn and Eighteenth-Century Disciplinary History', in G. Freudenthal (ed.), *Etudes Sur/Studies On Hélène Metzger* (Leiden: E. J. Brill, 1990), pp. 111–35, pp. 121–6.

63. Melhado, 'Chemistry, Physics and the Chemical Revolution', pp. 196 and 200.

64. Ibid., pp. 204 and 207–8.

65. Melhado, 'Toward an Understanding of the Chemical Revolution', pp. 127–8.

66. C. E. Perrin, 'Chemistry as Peer of Physics: A Response to Donovan and Melhado on Lavoisier', *Isis*, 81 (1990), pp. 259–70, pp. 260–1, 262 and 269.

67. A. Donovan, 'Lavoisier as Chemist and Experimental Physicist: A Reply to Perrin', *Isis*, 81 (1990), pp. 270–2, p. 271.

68. Ibid., p. 272.

69. E. Melhado, 'On the Historiography of Science: A Reply to Perrin', *Isis,* 81 (1990), pp. 273–6.

70. W. R. Albury, 'The Logic of Condillac and the Structure of French Chemical and Biological Thought, 1780–1801' (PhD dissertation, Johns Hopkins University, Baltimore, MD, 1972), pp. 64–72 and 128–34.

71. T. H. Levere, 'Lavoisier, Language, Instruments and the Chemical Revolution', in T. H. Levere and W. R. Shea (eds), *Nature, Experiment and the Sciences: Essays on Galileo and the History of Science in Honour of Stillman Drake* (Dordrecht: Kluwer Academic Publishers, 990), pp. 207–23, p. 214.

72. Albury, 'The Logic of Condillac', p. 185.

73. Beretta, *The Enlightenment of Matter*, pp. xiv–xvi.

74. Kuhn, 'Robert Boyle and Structural Chemistry', pp. 32 and 36.

75. Thackray, *Atoms and Powers*, pp. 5–6 and 193–8.

76. R. E. Schofield, 'The Counter-Reformation in Eighteenth-Century Science – Last Phase', in D. H. D. Roller (ed.), *Perspectives in the History of Science and Technology* (Normal, OK: Oklahoma University Press, 1971), pp. 39–54, p. 40.

77. Ibid., pp. 40–5.

78. Ibid., pp. 46–7.

79. Ibid., pp. 49–51.

80. See C. E. Perrin, 'Lavoisier's Table of the Elements: A Reappraisal', *Ambix*, 20 (1973), pp. 128–44; Morris, 'Lavoisier and the Caloric Theory', p. 34; M. Daumas, *Lavoisier, Théore-cien et Expérimentateur* (Paris: Presses Universitaires de France, 1955), pp. 157–78.

81. M. (Boas) Hall, 'Structure of Matter and Chemical Theory in the Seventeenth and Eighteenth Centuries', in M. Clagett (ed.), *Critical Problems in the History of Science* (Madison, WI: University of Wisconsin Press, 1969), pp. 499–514, pp. 499 and 501–4.

82. Ibid., pp. 511–12.

83. Ibid., p. 505.

84. M. Duncan, 'The Function of Affinity Tables in Lavoisier's List of Elements', *Ambix*, 17 (1970), pp. 28–42, pp. 40–1.

85. J. W. Llana, 'A Contribution of Natural History to the Chemical Revolution in France', *Ambix*, 32 (1985), 71–91, p. 85; T. M. Porter, 'The Promotion of Mining and the Advancement of Science: The Chemical Revolution of Mineralogy', *Annals of Science*, 38 (1981), pp. 543–70, p. 543; D. Oldroyd, 'Mineralogy and the Chemical Revolution', *Centaurus* 21 (1975), pp. 54–71.

86. D. Oldroyd, 'An Examination of G. E. Stahl's *Philosophical Problems of Universal Chemistry*', *Ambix*, 20 (1973), pp. 36–52, p. 52. See also D. Oldroyd, 'The Doctrine of Property-Conferring Principles in Chemistry: Origins and Antecedents', *Organon*, 12–13 (1976–7), pp. 441–62.

87. Siegfried and Dobbs, 'Composition: A Neglected Aspect of the Chemical Revolution', p. 278.

88. Ibid., p. 276.

89. See ibid., pp. 281–5.

90. R. Siegfried, 'Lavoisier's Table of Simple Substances: Its Origins and Interpretation', *Ambix* 29 (1982), pp. 29–48.

91. F. Abbri, 'Romanticism versus Enlightenment: Sir Humphry Davy's Idea of Chemical Philosophy', in S. Poggi and M. Bossi (eds), *Romanticism in Science: Science in Europe, 1790–1840* (Boston, MA: Kluwer Academic Publishers, 1994), pp. 31–45, p. 41.

92. For postpositivist accounts of responses to Lavoisier's chemistry in Italy and Germany see Abbri, 'Romanticism versus Enlightenment', p. 32 and Hufbauer, *The Formation of the German Chemical Community*. For the situation in Sweden, Spain and the Netherlands see the papers by A. Lundgren, R. Gago and H. A. M. Smelders, in A. Donovan *The Chemical Revolution: Essays in Reinterpretation, Osiris, 2nd Series, Volume 4* (Philadelphia, PA: History of Science Society, 1988), pp. 146–68, 169–92 and 121–45.

93. C. E. Perrin, 'The Chemical Revolution: Shifts in Guiding Assumptions', in A. Donovan, L. Laudan and R. Laudan (eds), *Scrutinizing Science: Empirical Studies of Scientific Change* (Dordrecht: Kluwer Academic Publishers, 1988), pp. 104–25, p. 121.

94. D. Allchin, 'Phlogiston After Oxygen', *Ambix*, 39 (1992), pp. 110–16, p. 113.

95. Perrin, 'The Chemical Revolution', pp. 107–8 and 114–22; C. E. Perrin, 'Early Opposition to the Phlogiston Theory: Two Anonymous Attacks', *British Journal for the History of Science*, 5 (1970), pp. 128–44, p. 142.

96. C. E. Perrin, 'A Reluctant Catalyst: Joseph Black and the Edinburgh Reception of Lavoisier's Chemistry', *Ambix*, 29 (1982), pp. 141–76, pp. 153–7 and 168–70.

97. A. Donovan, 'The New Nomenclature among the Scots: Assessing Novel Chemical Claims in a Culture Under Strain', in B. Bensaude-Vincent and F. Abbri (eds), *Lavoisier in European Context: Negotiating a New Language for Chemistry* (Canton, MA: Science History Publications), pp. 113–22, p. 116.

98. A Donovan, 'Scottish Responses to the New Chemistry of Lavoisier', *Studies in Eighteenth-Century Culture*, 9 (1979), pp. 237–49, pp. 246–7.

99. J. R. R. Christie, 'Joseph Black and John Robison', in A. D. Simpson (ed.), *Joseph Black 1728–1799: A Commemorative Symposium* (Edinburgh: Royal Scottish Museum, 1982), pp. 47–52, p. 51.

100. See, See J. G. McEvoy, 'Joseph Priestley: "Aerial Philosopher"'; McEvoy, 'Continuity and Discontinuity in the Chemical Revolution', pp. 199–204.

101. See McEvoy, 'Continuity and Discontinuity in the Chemical Revolution', p. 198.

102. See J. Golinski, 'The Chemical Revolution and the Politics of Language', *Eighteenth Century: Theory and Interpretation*, 33 (1992), pp. 238–51; J. G. McEvoy, 'Priestley Responds to Lavoisier's Nomenclature: Language, Liberty and Chemistry in the English Enlightenment', in B. Bensaude-Vincent and F. Abbri (eds), *Lavoisier in European Context: Negotiating a New Language for Chemistry*, (Canton, MA: Science History Publications, 1995), pp. 123–42.

103. M. Conlin, 'Joseph Priestley's American Defense of Phlogiston Reconsidered', *Ambix*, 43 (1996), pp. 129–45.

104. See J. G. McEvoy, 'Joseph Priestley, Natural Philosopher: Some Comments on Professor Schofield's Views', *Ambix*, 15 (1968), pp. 115–23; J. G. McEvoy, 'Causes and Laws, Powers and Principles: The Metaphysical Foundations of Priestley's Concept of Phlogiston', in R. G. W. Anderson and C. Lawrence (eds), *Science, Medicine and Dissent: Joseph Priestley (1733–1804)* (London: Wellcome Trust and the Science Museum, 1987), pp. 55–71, pp. 60–2.

105. See J. G. McEvoy and J. E. McGuire, 'God and Nature: Priestley's Way of Rational Dissent', *Historical Studies in the Physical Sciences*, 6 (1975), pp. 325–404; McEvoy, 'Joseph Priestley: "Aerial Philosopher"'; J. G. McEvoy, 'Joseph Priestley and the Chemical Revolution: A Thematic Overview', in A. T. Schwartz and J. G. McEvoy (eds), *Motion Toward Perfection: The Achievement of Joseph Priestley* (Boston, MA: Skinner House Books, 1990), pp. 129–60.

106. F. L. Holmes, 'Beyond the Boundaries: Concluding Remarks on the Workshop', in B. Bensaude-Vincent and F. Abbri (eds), *Lavoisier in European Context: Negotiating a New Language for Chemistry*, (Canton, MA: Science History Publications, 1995), pp. 267–78, p. 273.

4 From Modernism to Postmodernism: Changing Philosophical Images of Science

1. J. R. R. Christie and J. Golinski, 'The Spreading of the Word: New Directions in the Historiography of Chemistry', *History of Science*, 20 (1982), pp. 235–66, pp. 235–75.

2. Bloor, *Knowledge and Social Imagery*, pp. 1–5, 13 and 163–5.

3. See ibid., pp. 2–16; M. Foucault, 'On the Archaeology of the Sciences: Response to the Epistemology Circle', in F. D. Faubion (ed.), *Aesthetics, Method and Epistemology: Essential Works of Michel Foucault 1954–1984* (1968; New York: The New Press, 1998), pp. 297–333, p. 308. See also Golinski, *Making Natural Knowledge*, pp. 7–8.

4. Laudan, *Progress and Its Problems*, p. 201; M. Mulkay, *Science and the Sociology of Knowledge* (Bloomington, IN: Indiana University Press, 1991), p. 62; B. Latour, *Science in Action: How to Follow Engineers through Society* (Cambridge, MA: Harvard University

Press, 1988), p. 99; H. M. Collins, *Changing Order: Replication and Induction in Scientific Practice* (1985; Chicago, IL: University of Chicago Press, 1992), pp. 16 and 74.

5. B Barnes, 'On the Causal Explanation of Scientific Judgment', *Social Science Information*, 19 (1980), pp. 685–95, p. 686; See D. Bloor, 'Anti-Latour', *Studies in History and Philosophy of Science*, 30 (1999), pp. 81–112, pp. 87–94.

6. R. Rorty, 'Solidarity or Objectivity', in L. Cahoune (ed.), *From Modernism to Postmodernism: An Anthology* (1985; Oxford: Blackwell Publishers, 2003), pp. 447–56, p. 449. See also Bloor, 'Anti–Latour', p. 102; Golinski, *Making Natural Knowledge*, pp. ix and 8–9.

7. K. Knorr-Cetina and M. Mulkay, 'Introduction: Emerging Principles in the Social Studies of Science', in K. Knorr-Cetina and M. Mulkay (eds), *Science Observed: Perspectives in the Social Studies of Science* (Beverly Hills, CA and London: Sage Publications, 1983), pp. 1–17, p. 6.

8. Callinicos, *Against Postmodernism*, p. 2.

9. J.-F. Lyotard, *The Postmodern Condition: A Report on Knowledge*, trans. G. Bennington (1979; Minneapolis, MN: University of Minnesota Press, 1984), pp. xxiv and 31–47.

10. F. Jameson, 'Postmodernism, or the Cultural Logic of Late Capitalism', *New Left Review*, 146 (1984), pp. 53–94, pp. 77–8.

11. M. Teich, 'Afterword', in R. Porter and M. Teich (eds), *The Enlightenment in National Context* (Cambridge: Cambridge University Press, 1981), pp. 215–17, p. 216; D. Ingram, *Habermas and the Dialectic of Reason* (New Haven, CT and London: Yale University Press, 1987), pp. 80 and 95.

12. R. B. Pippin, *Modernism as a Philosophical Problem: On the Dissatisfactions of European High Culture* (Oxford: Blackwell, 1991), pp. 1–8, 51–79 and 120–2; M. Foucault, 'What is Enlightenment?' (1977), in P. Rabinow (ed.), *The Foucault Reader* (New York: Pantheon Books, 1984), pp. 32–50, pp. 32–42.

13. Dews, *Logics of Disintegration*, p. 103; J. Habermas, 'Modernity versus Postmodernity', *New German Critique*, 22 (1981), pp. 3–16, pp. 8–9.

14. J. Habermas, 'Modernity: An Unfinished Project', in M. P. d'Entrèves and S. Benhabib (eds), *Habermas and the Unfinished Project of Modernity: Critical Essays on the Philosophical Discourse of Modernity* (Cambridge: Polity Press, 1996), pp. 38–58, p. 45.

15. C. Dawson, *Christianity and the New Age (Essays in Order)* (London: Sheed and Ward, 1931), p. 66, quoted in B. Willey, *The Seventeenth-Century Background: Studies in the Thought of the Age in Relation to Poetry and Religion* (1934; Hammondsworth: Penguin Books, 1964), p. 16.

16. See Pippin, *Modernism as a Philosophical Problem*, pp. 51–61 and 115–16; Rouse, 'Philosophy of Science and the Persistent Narratives of Modernity', p. 146.

17. Foucault, *The Order of Things*, p. 55. See also ibid., pp. 17–77.

18. J. Habermas, *The Philosophical Discourse of Modernity: Twelve Lectures* (1981; Cambridge: Polity Press, 1987), p. 7; R. Rorty, 'Habermas and Lyotard on Modernity', in R. J. Bernstein (ed.), *Habermas and Modernity* (Cambridge, MA: MIT Press, 1985), pp. 161–75, pp. 167–8.

19. Feyerabend, 'Reply to Criticism', p. 223.

20. R. Rorty, *Philosophy and the Mirror of Nature* (Princeton, NJ: Princeton University Press, 1979), pp. 45–51; R. W. Rosemann, 'Heidegger's Transcendental History', *Journal of the History of Philosophy*, 40 (2002), pp. 501–23, pp. 511–12.

21. T. Eagleton, 'Capitalism, Modernism and Postmodernism', *New Left Review*, 152 (1985), pp. 60–73, p. 62.

22. B. Thomas, 'The New Historicism and Other Old-Fashioned Topics', in A. H. Veeser (ed.), *The New Historicism* (New York: Routledge, 1989), pp. 182–203, pp. 189–93; S. Lash and J. Friedman, 'Introduction: Subjectivity and Modernity's Other', in S. Lash and J. Friedman (eds), *Modernity and Identity* (Oxford: Blackwell, 1992), pp. 1–30, p. 10.

23. B. Latour, *We Have Never Been Modern*, trans. C. Porter (Cambridge, MA: Harvard University Press, 1993), p. 68; Thomas, 'The New Historicism', pp. 189–93.

24. Rouse, 'Philosophy of Science and the Persistent Narratives of Modernity', p. 148.

25. Galison, 'History, Philosophy and the Central Metaphor', p. 207; Rouse, 'Philosophy of Science and the Persistent Narratives of Modernity', pp. 156–60.

26. Feuerbach used the contrast between 'monarchical' time and the 'liberalism of space' to distinguish Hegel's historicism from his own materialism (see Z. Hanfi (ed.), *The Fiery Brook: Selected Writing of Ludwig Feuerbach* (New York: Anchor Books, 1972), p. 54. See also M. Foucault, 'Of Other Spaces', *Diacritics*, 16 (1986), pp. 22–7, p. 23.

27. See A. Ophir and S. Shapin, 'The Places of Knowledge: A Methodological Survey', *Science in Context*, 4 (1991), pp. 3–21.

28. See, e.g., Pippin, *Modernism as a Philosophical Problem*, p. 156.

29. Rorty, 'Habermas and Lyotard on Postmodernity', p. 175. On 'affirmative' and 'sceptical' postmodernists see P. M. Rosenau, *Postmodernism and the Social Sciences: Insights, Inroads and Intrusions* (Princeton, NJ: Princeton University Press, 1992), pp. 14–20.

30. See P. Starr, *Logics of Failed Revolts: French Thought after May '68* (Stanford, CA: Stanford University Press, 1995), pp. 27–8.; C. Lemert, 'General Social Theory, Irony, Postmodernism', in C. Seidman and D. Wagner (eds), *Postmodernism and Social Theory: The Debate over General Theory* (Cambridge, MA: Blackwell, 1992), pp. 17–46, pp. 31–2.

31. Ingram, *Habermas and the Dialectic of Reason*, p. 77.

32. R. Wolin, 'Modernism versus Postmodernism in Debates in Contemporary Culture', *Telos*, 62 (1984/5), pp. 9–29, p. 18.

33. P. Anderson, *The Origins of Postmodernity* (London: Verso, 1998), p. 59; F. Jameson, 'Postmodernism, or the Cultural Logic of Late Capitalism', *New Left Review*, 146 (1984), pp. 59–92, p. 65; Lash and Friedman, 'Introduction: Subjectivity and Modernity's Other', pp. 1–2.

34. F. Jameson, 'Marxism and Postmodernism', *New Left Review*, 176 (1989), pp. 31–45, p. 36. For Foucault's 'spatialization of reason' see H. L. Dreyfus and Rabinow, *Michel Foucault: Beyond Structuralism and Hermeneutics*, 2nd edn (Chicago, IL: University of Chicago Press, 1983), pp. 104–7.

35. B. Barnes, D. Bloor and J. Henry, *Scientific Knowledge: A Sociological Analysis* (Chicago, IL: University of Chicago Press, 1996), p. 85.

36. Eagleton, *Literary Theory*, pp. 106–34 and 172–5; Callinicos, *Against Postmodernism*, pp. 2–3, 71–3 and 87.

37. R. Young, 'Poststructuralism: An Introduction', in R. Young (ed.), *Untying the Text: A Poststructuralist Reader* (Boston, MA: Routledge and Kegan Paul, 1981), pp. 1–28, p. 9. See also Eagleton, *Literary Theory*, p. 115.

38. P. Dews, 'Adorno, Poststructuralism and the Critique of Identity', *New Left Review*, 157 (1986), pp. 28–44, p. 31.

39. Eagleton, *Literary Theory*, p. 131. See also ibid., pp. 127–50.

40. J. Derrida, 'But Beyond … (Open Letter to Anne McLintock and Rob Nixon)', *Critical Inquiry*, 13 (1986), pp. 155–70, pp. 167–8.

41. J. Derrida, *Of Grammatology*, trans. G. Chakravorty Spirak (1967; Baltimore, MD: Johns Hopkins University Press, 1976), p. 158. See also T. Lenoir, 'Inscription Practices and Materialities of Communication', in T. Lenoir and H. U. Gumbrecht (eds), *Inscribing Science: Scientific Texts and the Materialities of Communication* (Stanford, CA: Stanford University Press, 1998), pp. 4–8.

42. T. Lenoir, 'Practice, Reason, Context: The Dialectic Between Theory and Practice', *Science in Context*, pp. 3–22, p. 28. See also Lenoir, 'Inscription Practices', pp. 16–19; Golinski, *Making Natural Knowledge*, pp. 103–19.

43. Golinski, *Making Natural Knowledge*, p. 119.

44. See Eagleton, *Literary Theory*, pp. 66–74. See H.-G. Gadamer, *Truth and Method* (1960), trans. J. Weinsheimer and D. G. Marshall (New York: Crossroads, 1975).

45. Habermas, *On the Logic of the Social Sciences,* pp. 155–6.

46. Bensaude-Vincent, 'Between History and Memory', p. 499.

47. See M. G. Kim, 'Lavoisier, the Father of Modern Chemistry?', in M. Beretta (ed.), *Lavoisier in Perspective* (Munich: Deutsches Museum, 2005), pp. 167–91, p. 167; Meinel, 'Demarcation Debates: Lavoisier in Germany'.

48. See Eagleton, *Literary Theory*, pp. 74–90; Rosenau, *Postmodernism and the Social Sciences*, pp. 25–41.

49. N. Jardine, 'A Dip into the Future', *Studies in History and Philosophy of Science*, 20 (1989), pp. 15–18, p. 17. On sociological analyses of scientific language see Golinski, *Making Natural Knowledge* 103–32.

50. M. Foucault, 'Truth and Power' (1977), in P. Rabinow (ed.), *The Foucault Reader* (New York: Pantheon Books, 1984), pp. 51–5, p. 55; M. Foucault, 'Nietzsche, Genealogy, History' (1971), in P. Rabinow (ed.), *The Foucault Reader* (New York: Pantheon Books, 1984), pp. 76–100, p. 76; Foucault, 'What is Enlightenment?', p. 45.

51. For Nietzsche's influence on Foucault see Cronin, *Foucault's Antihumanist Historiography*, pp. 31–55. For Merleau-Ponty's influence on Foucault see Dreyfus and Rabinow, *Michel Foucault*, pp. 33–4, 111–12 and 166–7.

52. M. Foucault, *Discipline and Punish: The Birth of the Prison* (1975), trans. A. S. Smith (New York: Pantheon Books, 1978), pp. 27–8.

53. J. Rouse, 'Power/Knowledge', in G. Gutting (ed.), *The Cambridge Companion to Foucault* (Cambridge: Cambridge University Press, 1994), pp. 92–114, p. 102. See also Foucault, *Discipline and Punish*, p. 27.

54. Foucault, *Discipline and Punish*, pp. 205 and 216. See also Dreyfus and Rabinow, *Michel Foucault*, pp. 143–67.

55. Foucault, *Discipline and Punish*, p. 143. See also Rouse, 'Power/Knowledge', pp. 92–9.

56. J. Weeks, 'Foucault for Historians', *History Workshop*, 14 (1982), pp. 106–19, p. 111.

57. See Dreyfus and Rabinow, *Michel Foucault*, pp. 116–7, 133–4, 160–2 and 197–200.

58. J. Rouse, 'Foucault and the Natural Sciences', in J. Caputo and M. Yount (eds), *Foucault and the Critique of Institutions* (University Park, PA: Pennsylvania University Press, 1993), pp. 137–62, pp. 149–57.

59. See J. Rouse, *Knowledge and Power: Toward a Political Philosophy of Science* (Ithaca, NY: Cornell University Press, 1987), pp. 220–36; Golinski, *Making Natural Knowledge*, pp. 47–78.

60. See M. Lynch, *Scientific Practice and Ordinary Action: Ethnomethodology and the Social Studies of Science* (Cambridge: Cambridge University Press, 1993), pp. 125–41.

61. A. Schutz, 'Concept and Theory Formation in the Social Sciences', in F. Dallmayr and T. McCarthy (eds), *Understanding and Social Inquiry* (1954; Notre Dame, IN and London: University of Notre Dame Press, 1977), pp. 225–39, p. 236.

62. See F. Dallmayr and T. McCarthy, 'Introduction', in F. Dallmayr and T. McCarthy (eds), *Understanding and Social Inquiry* (Notre Dame, IN and London: University of Notre Dame Press, 1977), pp. 219–24; Habermas, *On the Logic of the Social Sciences*, pp. 109–16; Lynch, *Scientific Practice*, pp. 133–7.

63. Lynch, *Scientific Practice*, p. 25. See also ibid., pp. 1–25; Dallmayr and McCarthy, 'Introduction', pp. 222–4. See H. Garfinkel, 'What Is Ethnomethodology?', in F. Dallmayr and T. McCarthy (eds), *Understanding and Social Inquiry* (1967; Notre Dame, IN and London: University of Notre Dame Press, 1977), pp. 240–61.

64. M. Lynch, E. Livingston and H. Garfinkel, 'Temporal Order in Laboratory Work', in K. Knorr-Cetina and M. Mulkay (eds), *Science Observed: Perspectives on the Social Studies of Science* (Beverly Hills, CA and London: Sage Publications, 1983), pp. 205–38, pp.207–8.

65. K. Knorr-Cetina, 'The Ethnographic Study of Scientific Work: Toward a Constructivist Interpretation of Science', in K. Knorr-Cetina and M. Mulkay (eds), *Science Observed: Perspectives on the Social Studies of Science* (Beverly Hills, CA and London: Sage Publications, 1983), pp. 115–39, pp. 134–6.

66. See Lynch, Livingston and Garfinkel, 'Temporal Order in Laboratory Work', pp. 224–31; Lynch, *Scientific Practice*, pp. 114–5 and 134–41.

67. Lynch, *Scientific Practice*, p. 128. See also ibid., pp. 127–33; Rouse, *Knowledge and Power*, pp. 73–80.

68. See Rouse, *Knowledge and Power*, pp. 58–68; M. Lynch, *Art and Artifact in Laboratory Science* (London: Routledge and Kegan Paul, 1985), p. 10.

69. Rouse, *Knowledge and Power*, p. 64 . See ibid., pp. 69–126.

70. Ibid., pp. 125–6.

71. See the collection of papers in H. Veeser (ed.), *The New Historicism* (New York: Routledge, 1989).

72. L. A. Montrose, 'Professing the Renaissance: The Poetics and Politics of Culture', in H. Veeser (ed.), *The New Historicism* (New York: Routledge, 1989), pp. 15–36, pp. 20–1; Thomas, 'The New Historicism', pp. 192–203.

73. See R. Johnson, 'Edward Thompson, Eugene Geonevese, and Socialist-Humanist Historians', *History Workshop*, 6 (1978), pp. 79–100, p. 80.

74. E. P. Thompson, 'The Poverty of Theory', in E. P. Thompson, *The Poverty of Theory and Other Essays* (New York: Monthly Review Press, 1978), pp. 1–210, pp. 83–4 and 87. See also ibid., pp. 10–16, 45–6; Benton, *The Rise and Fall of Structural Marxism*, p. 210.

75. K. Knorr-Cetina, 'The Couch, the Cathedral, and the Laboratory: On the Relationship Between Experiment and Laboratory in Science', in A. Pickering (ed.), *Science as Practice and Culture* (Chicago, IL: University of Chicago Press, 1992), pp. 113–38, p. 119; L. Roberts, 'Understanding Science: Beyond the Antics of Ontics (Symposium on the Possibility of a Postmodern Philosophy of Science)', *Social Epistemology*, 5 (1991), pp. 247–55, p. 248; Latour, *Science in Action*; S. Shapin, 'Here and Everywhere: Sociology of Scientific Knowledge', *Annual Review of Sociology*, 12 (1995), pp. 289–321, pp. 303 and 312; G. Bowker and B. Latour, 'A Booming Discipline Short of Discipline: (Social) Studies of Science in France', *Social Studies of Science*, 17 (1987), pp. 715–48, p. 729; P. Galison, *How Experiments End* (Chicago, IL: University of Chicago Press, 1987); A. Pickering, *The Mangle of Practice: Time, Agency, and Science* (Chicago, IL: University of

Chicago Press, 1995); J. Golinski, 'The Theory of Practice and the Practice of Theory: Sociological Approaches to the History of Science', *Isis*, 81 (1990), pp. 492–505, pp. 499–505.

76. Jardine, 'A Dip into the Future', p. 18.

77. B. Barnes, *T. S. Kuhn and Social Science* (New York: Columbia University Press, 1982), pp. 31 and 87; D. Bloor, *Wittgenstein: A Sociological Theory of Knowledge* (New York: Columbia University Press, 1983), p. 183; Barnes, *T. S. Kuhn and Social Science*, p. 87.

78. B. Barnes and D. Bloor, 'Relativism, Rationalism and the Sociology of Scientific Knowledge', in M. Hollis and S. Lukes (eds), *Rationality and Relativism* (Oxford: Blackwell, 1982), pp. 21–47, pp. 35–43; D. Bloor, 'The Sociology of Reasons: Or Why 'Epistemic Factors' are Social Factors', in D. R. Brown (ed.), *The Rational and the Social* (Dordrecht: D. Reidel, 1984), pp. 295–324, pp. 303–5.

79. Barnes and Bloor, 'Relativism, Rationalism and the Sociology of Knowledge', pp. 44–5; Bloor, *Wittgenstein*, p. 183; Barnes, *T. S. Kuhn and Social Science*, pp. 30–1; Bloor, 'The Sociology of Reasons', pp. 302–3.

80. Shapin, 'Here and Everywhere', p. 302; Barnes, *T. S. Kuhn and Social Science*, pp. 25–7, 45–53 and 122.

81. S. Shapin, *The Scientific Revolution* (Chicago, IL: University of Chicago Press, 1996), p. 95; S. Schaffer, 'Making Certain (Essay Review of J. Shapiro, *Probability and Certainty in the Seventeenth Century*), *Social Studies of Science*, 14 (1984), pp. 137–52, p. 150.

82. See, e.g., Bloor, *Knowledge and Social Imagery*, pp. 163–73. See also S. Shapin, 'Social Uses of Science', in G. S. Rousseau and R. Porter (eds), *The Ferment of Knowledge: Studies in the Historiography of Eighteenth-Century Science* (Cambridge: Cambridge University Press, 1980), pp. 93–139.

83. Rorty, 'Solidarity or Objectivity', pp. 448–51; Rouse, *Knowledge and Power*, p. 25; Bernstein, *Praxis and Action*, pp. 173–87.

84. Shapin, 'Here and Everywhere', p. 304. See also ibid., pp. 304–9; Ophir and Shapin, 'The Place of Knowledge'.

85. Shapin, 'Here and Everywhere', pp. 305–6.

86. S. Schaffer, 'Natural Philosophy', in G. S. Rousseau and R. Porter (eds), *The Ferment of Knowledge: Studies in the Historiography of Eighteenth-Century Science* (Cambridge: Cambridge University Press, 1980), pp. 55–91, pp. 55–6 and 72–3.

87. See the papers by Nancy Cartwright, Bas van Fraassen and Arthur Fine in M. Curd and J. A. Cover (eds), *Philosophy of Science: The Central Issues* (New York: W. Norton, 1998), pp. 865–78, 1064–87, and 1186–1208. See also Rouse, *Knowledge and Power*, pp. 9–10.

88. I. Hacking, *Representing and Intervening: Introductory Topics in the Philosophy of Natural Science* (Cambridge: Cambridge University Press, 1983), pp. 150 and 220–32; I. Hacking, 'The Self-Validation of the Laboratory Sciences', in A. Pickering, *Science as Practice and Culture* (Chicago, IL: University of Chicago Press, 1992), pp. 29–64, p. 36.

89. Hacking, *Representing and Intervening*, pp. 149–60; Rouse, *Knowledge and Power*, pp. 87–9 and 99–101.

90. See, e.g., T. Nickles, 'Reconstructing Science: Discovery and Experiment', in D. Batens and J. P. van Bendegem (eds), *Theory and Experiment: Recent Insights and Perspective on Their Relation* (Dordrecht: D. Reidel, 1988), pp. 33–53, pp. 39–41.

91. See, e.g., A. van Helden and T. L. Hankins (eds), 'Introduction: Instruments in the History of Science', in *Instruments, Osiris, 2nd Series, volume 9* (Chicago, IL: University of Chicago Press, 1994), pp. 1–6.

92. L. Fleck, *Genesis and Development of a Scientific Fact* (1935), trans. F. Bradley and T. J. Trenn (Chicago, IL: University of Chicago Press, 1979), p. 98. See also Golinski, *Making Natural Knowledge*, pp. 32–7.

93. B. Barnes, 'Thomas Kuhn', in Q. skinner (ed.), *The Return of Grand Theory in the Human Sciences* (Cambridge: Cambridge University Press, 1985), pp. 83–100, p. 85.

94. See, Golinski, *Making Natural Knowledge*, pp. 16–21 and 27–9.

95. See P. Galison and D. J. Stump (eds), *The Disunity of Science: Boundaries, Contexts and Power* (Stanford, CA: Stanford University Press, 1996).

96. P. Anderson, *In the Tracks of Historical Materialism* (London: Verso, 1983), p. 48.

5 The Sociology of Scientific Knowledge and the History of Science

1. Shapin, 'Here and Everywhere', p. 302.

2. B. Barnes, 'On the Conventional Character of Knowledge and Cognition', in K. Knorr-Cetina and M. Mulkay (eds), *Science Observed: Perspectives on the Social Studies of Science* (Beverly Hills, CA and London: Sage Publications, 1983), pp. 19–51, p. 49.

3. M Foucault, *The Archaeology of Knowledge*, pp. 162–5. On the 'ostensive' and 'performative' models of society see S. S. Strum and B. Latour, 'Redefining the Social Link: From Baboons to Humans', *Information sur les Sciences Sociales*, 26 (1987), pp. 783–802.

4. See e.g. Barnes, 'On the Conventional Character of Knowledge and Cognition', pp. 23–37.

5. Golinski, *Making Natural Knowledge*, p. 26.

6. See D. Oldroyd, 'Grid/Group Analysis for Historians of Science', *History of Science*, 24 (1986), pp. 145–171.

7. Shapin, 'History of Science and Its Sociological Reconstructions', p. 164. See also D. MacKenzie and B. Barnes, 'The Biometry-Mendelism Controversy', in B. Barnes and S. Shapin (eds), *Natural Order: Historical Studies of Scientific Culture* (Beverly Hills, CA and London: Sage Publications, 1979), pp. 191–210; A. Pickering, *Constructing Quarks: A Sociological History of Particle Physics* (Edinburgh: Edinburgh University Press, 1984).

8. Bloor, *Knowledge and Social Imagery*, pp. 165–70.

9. L. Roberts, 'Going Dutch: Situating Science in the New Enlightenment', in W. Clark, J. Golinski and S. Schaffer (eds), *The Sciences in Enlightened Europe* (Chicago, IL: University of Chicago Press, 1999), pp. 350–88, p. 351.

10. Shapin, 'History of Science and Its Sociological Reconstructions', p. 197.

11. MacKenzie and Barnes, 'The Biometry-Mendelism Controversy', p. 205. See also Shapin, 'Social Uses of Science', p. 106.

12. Shapin, 'History of Science and Its Sociological Reconstructions', p. 198; Bloor, *Knowledge and Social Imagery*, pp. 172–3.

13. Bloor, *Knowledge and Social Imagery*, pp. 103–6 and 165–70. See also Shapin, 'Social Uses of Science', pp. 66–8.

14. See, e.g. ibid., p. 5; Collins, *Changing Order*.

15. Lynch, *Scientific Practice*, p. 86. See, e.g. H. M. Collins, 'The Role of the "Core Set" in Modern Science; Social Contingency with Methodological Propriety in Science', *History of Science*, 19 (1981), pp. 6–19.

16. H. M. Collins, 'An Empirical-Relativist Programme in the Sociology of Scientific Knowledge', in K. Knorr-Cetina and M. Mulkay (eds), *Science Observed: Perspectives on*

the Social Studies of Science (Beverly Hills, CA and London: Sage Publications, 1983), pp. 85–113, p. 92.

17. Ibid., p. 93. See P. Winch, *The Idea of a Social Science and Its Relation to Philosophy* (1958; London: Routledge, Kegan and Paul, 1970).

18. P. Dear, 'Cultural History of Science: An Overview with Reflections', *Science, Technology, and Human Values*, 20 (1995), pp. 150–70, p. 164. See C. Geertz, 'Thick Description: Toward an Interpretive Theory of Culture', *The Interpretation of Culture* (New York: Basic Books, 1973), pp. 3–30. See also Golinski, 'The Theory of Practice and the Practice of Theory', p. 494.

19. Lynch, *Scientific Practice*, p. 92.

20. Knorr-Cetina, 'The Ethnographic Study of Scientific Work', p. 219. See also Lynch, Livingston and Garfinkel, 'Temporal Order in Laboratory Work', p. 212.

21. Lynch, Livingston, and Garfinkel, 'Temporal Order in Laboratory Work', pp. 205–9 and 214.

22. H. M. Collins, 'The Sociology of Scientific Knowledge: Studies of Contemporary Science', *Annual Review of Sociology*, 9 (1983), pp. 265–85, p. 276.

23. B. Latour and S. Woolgar, *Laboratory Life: The Social Construction of Scientific Facts*, 2nd edn (Princeton, NJ: Princeton University Press, 1986), p. 64.

24. Ibid., pp. 105, 147, 177, 240 and 285.

25. Lynch, *Scientific Practice*, p. 95.

26. Golinski, *Making Natural Knowledge*, p. 92. See Lenoir, 'Inscription Practices', pp. 5 and 9.

27. See, e.g. the papers in T. Lenoir and H. U. Gumbrecht (eds), *Inscribing Science: Scientific Texts and the Materiality of Communication* (Stanford, CA: Stanford University Press, 1998); P. Dear (ed.), *The Literary Structure of Scientific Arguments: Historical Studies* (Philadelphia, PA: University of Pennsylvania Press, 1991).

28. G. N. Cantor, 'Weighing Light: The Role of Metaphor in Eighteenth-Century Optical Discourse', in A. E. Benjamin, G. N. Cantor and J. R. R. Christie (eds), *The Figural and the Literal: Problems of Language in the History of Science and Philosophy* (Manchester: Manchester University Press, 1987), pp. 124–46, p. 142. See Lenoir, 'Inscription Practices', p. 7; Golinski, 'The Theory of Practice and the Practice of Theory', pp. 497–8; Golinski, *Making Natural Knowledge*, pp. 103–32.

29. Lynch, *Scientific Practice*, pp. 96–8.

30. On Discourse Analysis in the history of science see S. Shapin, 'Talking History: Reflections on Discourse Analysis', *Isis*, 75 (1984), pp. 125–30.

31. See A. Brannigan, *The Social Basis of Scientific Discoveries* (Cambridge: Cambridge University Press, 1981); D. P. Miller, *Discovering Water: James Watt, Henry Cavendish and the Nineteenth-Century 'Water Controversy'* (Burlington, VT and Aldershot: Ashgate, 2004).

32. S. C. Ward, *Reconfiguring Truth: Postmodernism, Science Studies and the Search for a New Model of Knowledge* (London: Rowman and Littlefield, 1996), p. 86. See also ibid., pp. 73–9; Lynch, *Scientific Practice*, pp. 97–100.

33. Lynch, *Scientific Practice*, pp. 106–7. See e.g. Woolgar and Ashmore, 'The Next Step' and other papers in the anthology S. Woolgar (ed.), *Knowledge and Reflexivity: New Frontiers in the Sociology of Knowledge* (Beverly Hills, CA and London: Sage Publications, 1988).

34. Golinski, *Making Natural Knowledge*, p. 37. See also Lynch, *Scientific Practice*, pp. 104.

35. See A. Comte, *Introduction to Positive Philosophy* (1830), trans. F. Ferré (Cambridge, MA: Hacket, 1988), pp. 1–33; Shapin, 'Discipline and Bounding'; Golinski, *Making Natural Knowledge,* pp. 67–8.

36. Schaffer, 'Scientific Discoveries ', p. 387; Christie and Golinski, 'The Spreading of the Word', p. 261.

37. L. Roberts, 'Setting the Table: The Disciplinary Development of Eighteenth-Century Chemistry as Read through the Changing Structure of Its Tables', in P. Dear (ed.), *The Literary Structure of Scientific Arguments: Historical Studies* (Philadelphia, PA: University of Pennsylvania Press, 1991), pp. 99–132, p. 99.

38. Latour, *Science in Action*, p. 62.

39. Strum and Latour, 'Redefining the Social Link', p. 794; Latour, *Science in Action*, p. 201.

40. See B. Latour, 'One More Turn after the Social Turn', in E. McMullin (ed.), *The Social Dimensions of Science* (Notre Dame, IN: University of Notre Dame Press, 1972), pp. 272–94.

41. See Golinski, *Making Natural Knowledge*, p. 39.

42. Pickering, *The Mangle of Practice*, p. 11.

43. Lynch, *Scientific Practice*, p. 109.

44. Shapin, 'Here and Everywhere', p. 309.

45. Latour, *We Have Never Been Modern*, p. 55. See Golinski, *Making Natural Knowledge*, pp. 171–4; Shapin, 'Here and Everywhere', pp. 307–8.

46. B. Bensaude-Vincent, 'The Balance: Between Chemistry and Politics', *The Eighteenth Century: Theory and Interpretation*, 33 (1992), pp. 213–37, p. 234; Latour, *We Have Never Been Modern*, pp. 51–5 and 82–5.

47. Golinski, *Making Natural Knowledge*, pp. 172–85.

48. See e.g. Pickering, *The Mangle of Practice*, pp. 5–10 and 301–26; Golinski, *Making Natural Knowledge*, pp. 41–3.

49. Pickering, *The Mangle of Practice*, p. 22. See also Galison, *How Experiments End*, pp. 234–41.

50. A. Pickering, 'Knowledge, Practice and Mere Construction', *Social Studies of Science*, 20 (1990), pp. 682–729, p. 692; A. Pickering, 'From Science as Knowledge to Science as Practice', in A. Pickering (ed.), *Science as Practice and Culture* (Chicago, IL: University of Chicago Press, 1992), pp. 1–26, p. 17; Pickering, *The Mangle of Practice*, p. 14.

51. Pickering, *The Mangle of Practice*, pp. 5–7 and 15.

52. Ibid., pp. 3–5 and 20–2; Pickering, 'Knowledge, Practice and Mere Construction', pp. 686–7 and 689–709.

53. Pickering, 'Knowledge, Practice and Mere Construction', pp. 701–6.

54. Pickering, *The Mangle of Practice*, pp. 4, 15–20 and 24–6.

55. S. Shapin and A. Thackray, 'Prosopography as a Research Tool in the History of Science', *History of Science*, 12 (1974), pp. 1–28, p. 21.

56. See, e.g. S. Shapin, 'Homo Phrenologicus: Anthropological Perspectives on an Historical Problem', in B. Barnes and S. Shapin (eds), *Natural Order: Historical Studies of Scientific Culture* (Beverly Hills, CA and London: Sage Publications, 1979), pp. 41–72; S. Shapin, 'Of Gods and Kings: Natural Philosophy and Politics in the Leibniz-Clarke Correspondence', *Isis*, 72 (1981), pp. 187–215.

57. Schaffer, 'Natural Philosophy', p. 56; S. Schaffer, 'Machine Philosophy: Demonstration Devices in Georgian Mechanics', in A. van Helden and T. L. Hankins (eds), *Instruments, Osiris, 2 series; volume 9* (Chicago, IL: University of Chicago Press, 1994), pp. 157–82,

p. 157. See also S. Schaffer, 'Natural Philosophy and Public Spectacle in the Eighteenth Century', *History of Science*, 21 (1983), pp. 1–43.

58. S. Shapin and S. Schaffer, *Leviathan and the Air-Pump: Hobbes, Boyle and the Experimental Life* (Princeton, NJ: Princeton University Press, 1985), pp. 3, 6–7 and 13.

59. Ibid., pp. 15–16. See also Golinski, *Making Natural Knowledge*, pp. 11–12 and 21–2.

60. S. Shapin, 'Pump and Circumstance: Robert Boyle's Literary Technology', *Social Studies of Science*, 14 (1984), pp. 481–520, p. 484.

61. B. Latour, 'Essay Review: Postmodern? No, Simply Amodern!: Steps Towards an Anthropology of Science', *Studies in History and Philosophy of Science*, 21 (1990), pp. 145–71.

62. Golinski, *Making Natural Knowledge*, p. 30. See S. Shapin, *A Social History of Truth: Civility and Society in the Seventeenth Century* (Chicago, IL: University of Chicago Press, 1994).

63. Golinski, *Making Natural Knowledge*, pp. 53–4. See Shapin and Schaffer, *Leviathan and the Air-Pump*, pp. 341–2.

64. S. Shapin, 'Science and the Public', in R. C. Olby, G. N. Cantor, J. R. R. Christie and M. J. S. Hodge (eds), *Companion to the History of Modern Science* (London: Routledge, 1990), pp. 989–1007, p. 999.

65. S. Shapin, 'Placing the View From Nowhere: Historical and Sociological Problems in the Location of Science', *Transactions of the Institute of British Geographers*, 23 (1998), pp. 5–12, p. 5. See Ophir and Shapin, 'The Place of Knowledge', pp. 9–15.

66. Shapin, 'Here and Everywhere', pp. 313–15.

67. Schaffer, 'Making Certain', pp. 140 and 150.

68. See Latour, 'Essay Review: Postmodern?', p. 152; S. Schaffer, 'Measuring Virtue: Eudiometry, Enlightenment and Pneumatic Medicine', in A. Cunningham and R. French (eds), *The Medical Enlightenment of the Eighteenth Century* (Cambridge: Cambridge University Press, 1990), pp. 281–318, pp. 289–90.

69. Schaffer, 'Machine Philosophy', p. 182; S. Schaffer, 'Enlightened Automata', in W. Clark, J. Golinski and S. Schaffer (eds), *The Sciences in Enlightened Europe* (Chicago, IL: University of Chicago Press, 1999), pp. 126–65, p. 182.

6 Postmodernist and Sociological Interpretations of the Chemical Revolution

1. Schaffer, 'Natural Philosophy', pp. 55–6 and 72–3.

2. See B. Bensaude-Vincent, *Lavoisier: Mémoires d'une Révolution* (Paris: Flammarion, 1993).

3. See Thackray, *Atoms and Powers*, pp. 6–7; McEvoy, 'Continuity and Discontinuity in the Chemical Revolution', pp. 198–9.

4. A Lavoisier, 'Reflexions sur le Phlogistique', in J. B. Dumas and G. Grimaux (eds), *Ouevres de Lavoisier*, 6 vols (Paris: Impremerie Impériele, 1862–93), vol. 3, pp. 623–55, p. 640.

5. Holmes, 'The Revolution in Chemistry and Physics', pp. 735–7.

6. Ibid., pp. 738–46 and 750–3.

7. F. Abbri, 'Some Ingenious Systems: Lavoisier and the Northern Chemists', in M. Beretta (ed.), *Lavoisier in Perspective* (Munich: Deutsches Museum, 2005), pp. 96–108, pp. 96 and 108. See also Bensaude-Vincent, 'Introductory Essay'.

8. J. Perkins, 'Creating Chemistry in Provincial France before the Revolution: The Example of Nancy and Metz. Part I, Nancy', *Ambix*, 50 (2003), pp. 145–81, pp. 145–6; J. Perkins,

'Creating Chemistry in Provincial France before the Revolution: The Example of Nancy and Metz. Part II, Metz', *Ambix*, 51 (2004), pp. 43–76, p. 73.

9. Holmes, 'Beyond the Boundaries', p. 277.

10. P. Brett, 'Power, Sociability and Dissemination of Science: Lavoisier and the Learned Societies', in M. Beretta (ed.), *Lavoisier in Perspective* (Munich: Deutsche Museum, 2005), pp. 129–52, p. 151.

11. Golinski, *Science as Public Culture: Chemistry and the Enlightenment in Britain, 1760–1820* (Cambridge: Cambridge University Press, 1992), p. 65; Christie and Golinski, 'The Spreading of the Word', pp. 255–7.

12. W. Clark, J. Golinski and S. Schaffer, 'Introduction', in W. Clark, J. Golinski and S. Schaffer (eds), *The Sciences in Enlightened Europe* (Chicago, IL: University of Chicago Press, 1999), pp. 3–31, p. 29. See ibid., on pp. 5–10 and 16–19.

13. A. Nordmann, 'The Passion for Truth: Lavoisier's and Lichtenberg's Enlightenments', in M Beretta (ed.), *Lavoisier in Perspective* (Munich: Deutsche Museum, 2005), pp. 109–28, pp. 117–8 and 127.

14. Ibid., pp. 108 and 126.

15. R. Chartier, *The Cultural Origins of the French Revolution* (Durham, NC: Duke University Press, 1991), p. 17.

16. Bensaude-Vincent, 'Introductory Essay', pp. 6–8 and 13;.

17. Crosland, 'Priestley's Memorial Lecture', pp. 234 and 237.

18. C. Bazerman, 'How Natural Philosophers Can Cooperate: The Literary Technology of Coordinated Investigation in Joseph Priestley's *History and Present State of Electricity* (1767)', in T. C. Kynell and M. G. Moran (eds), *Three Keys to the Past: The History of Technological Communication (Contemporary Studies in Technological Communication, volume 7)* (Stamford, CT: Ablex Publishers, 1999), pp. 21–48, pp. 22–24. See also Golinski, *Making Natural Knowledge*, pp. 112–4.

19. W. C. Anderson, *Between the Library and the Laboratory: The Language of Chemistry in Eighteenth-Century France* (Baltimore, MD: Johns Hopkins University Press, 1984), pp. 2–3.

20. Anderson, *Between the Library and the Laboratory*, pp. 5, 123 and 151.

21. Ibid., p. 3.

22. Schaffer, 'Priestley Questions', pp. 157, 170 and 174–5; S. Schaffer, 'Priestley and the Politics of Spirit', in R. G. W. Anderson and C. Lawrence (eds), *Science, Medicine and Dissent Joseph Priestley (1733–1804)* (London: Wellcome Trust and the Science Museum, 1987), pp. 39–53, pp. 50–1.

23. Schaffer, 'Priestley and the Politics of Spirit', pp. 39–42 and 50.

24. Schaffer, 'Priestley Questions', pp. 166–7; Schaffer, 'Measuring Virtue', pp. 282–318.

25. Schaffer, 'Priestley Questions', p. 165.

26. Schaffer, 'Measuring Virtue', pp. 288–90 and 313.

27. Golinski, *Making Natural Knowledge,* pp. 3–6 and 166–7.

28. Golinski, *Science as Public Culture*, p. 5.

29. Christie and Golinski, 'The Spreading of the Word', pp. 237 and 243.

30. See, e.g. McEvoy, 'Continuity and Discontinuity in the Chemical Revolution', pp. 314–22.

31. J. Golinski, 'Precision Instruments and the Demonstrative Order of Proof in Lavoisier's Chemistry', in A. van Helden and T. L. Hankin (eds), *Instruments (Osiris, 2 series, volume 9)* (Chicago, IL: University of Chicago Press, 1994), pp. 30–47, p. 31.

32. Golinski, *Science as Public Culture*, pp. 130–7.

33. Ibid., p. 137.

34. Ibid., p. 148.

35. L. Stewart, 'Putting on Airs; Science, Medicine, and Polity in the Late Eighteenth Century', in T. H. Levere and G. Turner (eds), *Discussing Chemistry and Steam: The Minutes of the Coffee-House Philosophical Society, 1760–1783* (Oxford: Oxford University Press, 2002), pp. 207–55, pp. 208 and 215.

36. Golinski, *Science as Public Culture*, pp. 10 and 148–52.

37. L. Roberts, 'A Word and the World: The Significance of Naming the Calorimeter', *Isis*, 82 (1991a), pp. 199–222, p. 199.

38. Ibid., pp. 200–3.

39. Ibid., pp. 201, 214–18 and 221–2; Golinski, *Science as Public Culture*, pp. 140–2.

40. See Bensaude-Vincent, *Lavoisier*, pp. 15–82; A Levin, 'Venel, Lavoisier, Fourcroy, Cabanis, and the Idea of Scientific Revolution: The French Political Context and the General Patterns of Conceptualization of Scientific Change', *History of Science*, 22 (1984), pp. 303–20.

41. Bensaude-Vincent, *Lavoisier*, pp. 35–6 and 83–165.

42. Bensaude-Vincent, 'The Balance', pp. 218 and 234.

43. Bensaude-Vincent, 'A View of the Chemical Revolution through Contemporary Textbooks: Lavoisier, Fourcroy and Chaptal', *British Journal for the History of Science*, 23 (1990), pp. 434–60, pp. 437–43.

44. Bensaude-Vincent, *Lavoisier*, pp. 343–417.

45. Meinel, 'Demarcation Debates', pp. 154 and 165; Kim, 'Lavoisier, the Father of Modern Chemistry', pp. 169–70 and 175.

46. Christie and Golinski, 'The Spreading of the Word', pp. 235–6.

47. O. Hannaway, *The Chemists and the Word: The Didactic Origins of Chemistry* (Baltimore, MD: Johns Hopkins University Press, 1975), pp. 153–6.

48. Christie and Golinski, 'The Spreading of the Word', p. 260.

49. Ibid., pp. 245–56 and 259–261.

50. F. L. Holmes, *Antoine Lavoisier – The Next Crucial Year, or the Sources of His Quantitative Method in Chemistry* (Princeton, NJ: Princeton University Press, 1998), p. 125.

51. Ibid., pp. 125 and 149. See also pp. 103–5, above.

52. Roberts, 'Setting the Table', pp. 99–100 and 131.

53. Ibid., p. 101; L. Roberts, 'Filling the Space of Possibilities: Eighteenth-Century Chemistry's Transition from Art to Science', *Science in Context*, 6 (1993), pp. 511–53, pp. 518–19.

54. Ibid., pp. 535–9; L. Roberts, 'Condillac, Lavoisier and the Instrumentalization of Science', *Eighteenth Century: Theory and Interpretation*, 33 (1992), pp. 252–71, p. 265; Roberts, 'Filling the Space of Possibilities', p. 550.

55. Roberts, 'Condillac, Lavoisier and the Instrumentalization of Science', pp. 252–4 and 259.

56. L. Roberts, 'The Death of the Sensuous Chemist: The "New" Chemistry and the Transformation of Sensuous Technology', *Studies in History and Philosophy of Science*, 26 (1995), pp. 503–29, pp. 506, 515–6 and 522–9.

57. U. Klein, 'The Chemical Workshop Tradition and the Experimental Practice: Discontinuities within Continuities', *Science in Context*, 9:3 (1996), pp. 251–87, pp. 251–2; U. Klein, 'Origin of the Concept of Chemical Compound', *Science in Context*, 7:2 (1994), pp. 163–204, p. 163.

58. Klein, 'The Chemical Workshop Tradition', pp. 255, 263 and 268–9.

59. Ibid., pp. 257–9.
60. U. Klein, 'E. F. Geoffroy's Table of Different "Rapports" between Different Chemical Substances – A Reinterpretation', *Ambix*, 7:2 (1995), pp. 79–100, pp. 91–3; Klein, 'The Chemical Workshop Tradition', p. 279.
61. J. Simon, 'The Chemical Revolution and Pharmacy: A Disciplinary Perspective', *Ambix*, 45 (1998), pp. 1–13, pp. 2–3 and 4–7. See also J. Simon, *Chemistry, Pharmacy and Revolution in France, 1777–1809* (Burlington, VT and Aldershot: Ashgate, 2005), ch. 3.
62. Quoted at Simon, 'The Chemical Revolution and Pharmacy', p. 10.
63. Ibid., p. 6; J. Simon, 'Analysis and Hierarchy of Nature in Eighteenth-Century Chemistry', *British Journal for the History of Science*, 35 (2002), pp. 1–16, pp. 10 and 12–16.
64. Simon, 'The Chemical Revolution and Pharmacy', p. 1; J. Simon, 'Authority and Authorship in the Method of Chemical Nomenclature', *Ambix*, 44 (2002), pp. 206–26, pp. 207–9 and 222.
65. Simon, 'The Chemical Revolution and Pharmacy', pp. 9–12. See also Simon, *Chemistry, Pharmacy and Revolution in France*, ch. 4 and 5.
66. M. G. Kim, *Affinity, That Elusive Dream: A Genealogy of the Chemical Revolution* (Cambridge, MA: MIT Press, 2003), pp. 2–3 and 440–3.
67. Ibid., pp. 447–8.
68. Ibid., pp. 8–11.
69. Ibid., pp. 4–5.
70. Ibid., pp. 6–8, 13–14.
71. Ibid., p. 449–51 and 454–4.
72. T. H. Levere and F. Holmes, 'Introduction: A Practical Science', in F. L. Holmes and T. H. Levere (eds), *Instruments and Experimentation in the History of Chemistry* (Cambridge, MA: MIT Press, 2000), pp. vii–xvii, pp. vii–ix.
73. Van Helden and Hankins, 'Introduction: Instruments in the History of Science', pp. 1–3 and 6.
74. See the collection of papers in F. L. Holmes and T. L. Hankins (eds), *Instruments and Experiments in the History of Chemistry* (Cambridge, MA: MIT Press, 2000). See also T. H. Levere, 'Lavoisier's Gasometers and Others: Research, Control and Dissemination', in M. Beretta (ed.), *Lavoisier in Perspective* (Munich: Deutsches Museum, 2005), pp. 53–69.
75. See J. P. Prinz, 'Lavoisier's Experimental Method of His Research on Respiration', in M. Beretta (ed.), *Lavoisier in Perspective* (Munich: Deutsches Museum, 2005), pp. 43–51, p. 43; P. Heering, 'Weighing the Heat: The Replication of Experiments with the Ice Calorimeter of Lavoisier and Laplace', in M. Beretta (ed.), *Lavoisier in Perspective* (Munich: Deutsche Museum, 2005), pp. 27–41.
76. M. Daumas, 'Precision Measurement and Chemical Research in the Eighteenth Century', in A. C. Crombie (ed.), *Scientific Change: Historical Studies in the Intellectual, Social and Technical Conditions for Scientific Discovery and Technical Invention, from Antiquity to the Present* (New York: Basic Books, 1963), pp. 418–30, pp. 428–30.
77. M. Beretta and A. Scotti, 'Panopticon Lavoisier: A Presentation', in M. Beretta (ed.), *Lavoisier in Perspective* (Munich: Deutsches Museum, 2005), pp. 193–207, pp. 193–6.
78. For a discussion of Sarton's pedagogical vision for the history of science see A.-K. Mayer, 'When Things Don't Talk: Knowledge and Belief in the Inter-War Humanism of Charles Singer', *British Journal for the History of Science*, 38 (2005), pp. 325–47, pp. 330–1.
79. Heering, 'Weighing the Heat', pp. 28–30.

7 The Chemical Revolution as History

1. Shapin, 'Social Uses of Science', pp. 111–12.
2. G. Freudenthal, 'The Role of Shared Knowledge in Science: The Failure of the Constructivist Programme in the Sociology of Science', *Social Studies of Science*, 14 (1984), pp. 285–95, pp. 286–8.
3. J. Golinski, 'Science *in* the Enlightenment' (Review of T. Hankins, *Science and the Enlightenment*)', *History of Science*, 24 (1986), pp. 411–24, p. 419.
4. Ibid., pp. 415 and 419–20; Golinski, *Science as Public Culture*, p. 6.
5. Thackray, *Atoms and Powers*, pp. viii and 280–1; Donovan, 'Scottish Responses to the New Chemistry of Lavoisier', pp. 237–8.
6. See, e.g., 'Continuity and Discontinuity in the Chemical Revolution', pp. 197–9.
7. J. Priestley, *Considerations on the Doctrine of Phlogiston and the Decomposition of Water, Part 2* (Philadelphia, PA; Thomas Dobson, 1797), p. 29; A. Lavoisier, *Elements of Chemistry in a New Systematic Order: Containing All the Modern Discoveries* (1790), trans. R. Kerr (New York: Dover 1965), pp. xxxiii–xxxiv.
8. Montrose, 'Professing the Renaissance', p. 20; Benton, *The Rise and Fall of Structural Marxism*, p. 21. See also Anderson, *In the Tracks of Historical Materialism*, pp. 53–5; Laudan, *Progress and Its Problems*, pp. 98–100.
9. Benton, *The Rise and Fall of Structural Marxism*, p. 214; Anderson, *In the Tracks of Historical Materialism*, p. 54.
10. Holmes, *Eighteenth-Century Chemistry as an Investigative Enterprise*, p. 114.
11. Simon, 'The Chemical Revolution and Pharmacy', pp. 1 and 11.
12. Lash and Friedman, 'Introduction: Subjectivity and Modernity's Other', pp. 1–2.
13. Latour, *We Have Never Been Modern*, p. 81.
14. See H. White, 'New Historicism: A Commentary', in H. Veeser (ed.), *The New Historicism* (New York: Routledge, 1989), pp. 293–302, p. 295; Thompson, 'The Poverty of Theory', pp. 78–84.
15. Latour, *We Have Never Been Modern*, pp. 72–4.
16. Ibid., pp. 81–2.
17. K. Marx and F. Engels, *Karl Marx and Frederick Engels: Collected Works Volume 5, Marx and Engels 1845–1847* (New York: International Publishers, 1976), p. 28.
18. Marx, *Capital*, 3 vols, vol. 1, p. 565.
19. G. Lukács, *History and Class Consciousness. Studies in Marxist Dialectics* (Cambridge, MA: MIT Press, 1971), p. 13.
20. Foucault, *The Order of Things*, pp. 370–1.
21. Dews, 'Althusser, Structuralism and the French Epistemological Tradition', p. 112.
22. Gutting, *Michel Foucault's Archaeology of Scientific Reason*, pp. 240–1.
23. Dews, 'Althusser, Structuralism and the French Epistemological Tradition', p. 125. See Foucault, *The Archaeology of Knowledge*, pp. 45–9, 75–9, 117 and 191–2.
24. See Althusser and Balibar, *Reading Capital*, pp. 91–118; Althusser, *For Marx*, pp. 87–128 and 161–218; Gordy, 'Reading Althusser', pp. 8–13.
25. P. Anderson, 'Modernity and Revolution', *New Left Review*, 144 (1984), pp. 96–133.
26. Gordy, 'Reading Althusser', p. 12; Althusser and Balibar, *Reading Capital*, pp. 100 and 105. See also Latour, *We Have Never Been Modern*, p. 74.
27. F. Jameson, *The Political Unconscious: Narrative as a Symbolic Act* (New York: Cornell University Press, 1981), p. 36; Althusser and Balibar, *Reading Capital*, p. 100; Callinicos, *Althusser's Marxism*, p. 52.

28. Friedman, 'History and Philosophy of Science in a New Key', p. 133.

29. Teich, 'Afterword', p. 216;

30. L. Althusser, *Lenin and Philosophy and Other Essays* (New York: Monthly Review Press, 1971), p. 149; Benton, *The Rise and Fall of Structural Marxism*, p. 101.

31. Althusser, *Lenin and Philosophy and Other Essays*, p. 149; Shapin, *The Scientific Revolution*, p. 124.

32. For documentation of the claims made in this paragraph see J. G. McEvoy, 'The Chemical Revolution in Context', *The Eighteenth Century: Theory and Interpretation*, 33 (1992), pp. 192–216, p. 216, nn. 38, 39 and 40.

33. Donovan, 'Scottish Responses to the New Chemistry of Lavoisier', p. 247.

34. E. Burke, 'Letter to a Noble Lord'; quoted in Schaffer, 'Priestley and the Politics of Spirit', pp. 39–40.

35. See M. Crosland, 'The Image of Science as a Threat: Burke versus Priestley and the "Philosophical Revolution', *British Journal for the History of Science*, 20 (1987), pp. 277–307, pp. 285–7.

36. See E. P. Thompson, 'The Peculiarities of the English', in *The Poverty of Theory and Other Essays* (1965; New York: Monthly Review Press, 1978), pp. 245–301. See H. Heller, *The Bourgeois Revolution in France (1789–1815)* (New York and Oxford: Berghahn Books, 2006), for an account of the more centralized 'statist' situation in France.

37. See I. Kramnick, 'Eighteenth-Century Science and Radical Social Theory: The Case of Joseph Priestley's Scientific Liberalism', in T. L. Schwartz and J. G. McEvoy (eds), *Motion Toward Perfection: The Achievement of Joseph Priestley* (Boston, MA: Skinner House Books, 1990), pp. 57–92, pp. 73–86; R. Porter, 'Science, Provincial Culture and Public Opinion in Eighteenth-Century England', *British Journal for Eighteenth-Century Studies* 3 (1980), pp. 20–45, pp. 27–31.

38. R. L. Heilbroner, *The Nature and Logic of Capitalism* (New York: Norton, 1985), pp. 86. See also ibid., pp. 85–95. Marx discussed the connection between the rise of capitalism and the dissolution of feudalism in 'On the Jewish Question' and *The German Ideology*.

39. K. Marx and F. Engels, *The German Ideology, Part 1* (New York: International Publishers, 1970), p. 57; Heilbroner, *The Nature and Logic of Capitalism*, p. 89. See also ibid., pp. 89–91.

40. Clark, Golinski and Schaffer, 'Introduction', pp. 23–6 and 38–9.

41. For a fuller account and documentation of the Enlightenment conception of the self-defining subject outlined here see J. G. McEvoy, 'The Enlightenment and the Chemical Revolution', in R. S. Woolhouse (ed.), *Metaphysics and Philosophy of Science in the Seventeenth and Eighteenth Centuries: Essays in Honour of Gerd Buchdahl* (Dordrecht: D. Reidel, 1988), pp. 307–25, pp. 307–11.

42. See Thompson, 'The Peculiarities of the English', pp. 269–9; N. Hampson, 'The Enlightenment in France', in R. Porter and M. Teich (eds), *The Enlightenment in National Context* (Cambridge; Cambridge University Press, 1981), pp. 41–53, p. 45; R. Porter, 'The Enlightenment in England', in R. Porter and M. Teich (eds), *The Enlightenment in National Context* (Cambridge: Cambridge University Press, 1981), pp. 1–19, pp. 5–6.

43. Lavoisier, *Elements of Chemistry in a New Systematic Order*, p. xxiv. For a fuller account and documentation of the ontological issues discussed in this section see McEvoy, 'The Enlightenment and the Chemical Revolution', pp. 311–14; McEvoy, 'Continuity and Discontinuity in the Chemical Revolution', pp. 199–203.

44. J. Priestley, *Heads of Lectures on a Course of Experimental Philosophy, Particularly Including Chemistry, Delivered at the New College at Hackney* (London: J. Johnson, 1794), p. 4.

45. J. Priestley, 'Experiments on the Production of Air by the Freezing of Water', *New York Medical Repository*, 4 (1801), pp. 17–21, p. 21.

46. Lavoisier, *Elements of Chemistry in a New Systematic Order*, pp. xxii–xxiv.

47. For a fuller account and documentation of the epistemological issues covered in this section see McEvoy, 'The Enlightenment and the Chemical Revolution', pp. 314–17; McEvoy, 'Continuity and Discontinuity in the Chemical Revolution', pp. 203–9.

48. See Lavoisier, *Elements of Chemistry in a New Systematic Order*, pp. xv–xvi.

49. L. B. Guyton de Morveau, A. Lavoisier, C.-L. Berthollet and A. F. Fourcroy, *Méthode de Nomenclature Cimique* (Paris: Cuchet, 1787), p. 298.

50. J. Priestley, 'Further Experiments Relating to the Decomposition of Dephlogisticated and Inflammable Air', *Philosophical Transactions*, 81 (1791), pp. 213–22. See J. Priestley, *The Doctrine of Phlogiston Established and that of the Decomposition of Water Refuted* (Philadelphia, PA: P. Byrne, 1803), p. 103.

51. See Foucault, *The Order of Things*, pp. 17–63. See also Hannaway, *The Chemists and the Word*, pp. 62–72. For a fuller discussion and documentation of the issues covered in this section see McEvoy, 'Priestley Responds to Lavoisier's Nomenclature'. See also McEvoy, 'The Enlightenment and the Chemical Revolution', pp. 317–19.

52. See McEvoy and McGuire, 'God and Nature', pp. 354–7.

53. See Bensaude-Vincent, 'The Balance'; J. Barrell, *English Literature in History 1730–1780: An Equal Wide Survey* (London: Hutchinson, 1883), pp. 21–5.

54. See Holmes, *Lavoisier and the Chemistry of Life*, pp. 1–147.

55. See J. Priestley, 'Experiments and Observations Relating to Phlogiston and the Seeming Conversion of Water into Air', *Philosophical Transactions*, 73 (1983), pp. 398–434; McEvoy, 'Enlightenment and Dissent in Science', pp. 56–9.

56. See, e.g., Priestley, *Heads of Lectures*, pp. 8–9.

57. J. Priestley, 'Experiments and Observations Relating to the Principle of Acidity, the Composition of Water and Phlogiston', *Philosophical Transactions*, 78 (1788), pp. 147–57, pp. 155–6.

58. See Porter, 'The Enlightenment in England'; McEvoy, 'Enlightenment and Dissent in Science'.

59. Priestley, 'Considerations on the Doctrine of Phlogiston', p. 29.

60. Lavoisier, *Elements of Chemistry in a New Systematic Order*, pp. xxxiii–xxxiv;

61. See Donovan, *Antoine Lavoisier*, pp. 244–50; Eagleton, *Literary Theory*, pp. 60–1.

62. See McEvoy and McGuire, 'God and Nature', pp. 348–57; T. L. Hankins, *Science and the Enlightenment* (Cambridge University Press, 1985), pp. 179–90.

63. Guerlac, *Antoine-Laurent Lavoisier: Chemist and Revolutionary* (New York: Scribners, 1975), pp. 124–31.

64. See F. L. Holmes, 'The Evolution of Lavoisier's Chemical Apparatus', in F. L. Holmes and T. H. Levere (eds), *Instruments and Experimentation in the History of Chemistry* (Cambridge, MA: MIT Press, 2000), pp. 137-52, p. 138; J. Parascandola and J. H. Ihde, History of the Pneumatic Trough', *Isis*, 55 (1969), pp. 351–61; L. Badash, 'Joseph Priestley's Apparatus for Pneumatic Chemistry', *Journal of the History of Medicine*, 19 (1964), pp. 139–155, p. 139.

65. See U. Klein and W. Lefèvre, *Materials in Eighteenth-Century Science: A Historical Ontology* (Cambridge, MA: MIT Press, 2007).

WORKS CITED

Abbri, F., 'Romanticism versus Enlightenment: Sir Humphry Davy's Idea of Chemical Philosophy', in S. Poggi and M. Bossi (eds), *Romanticism in Science: Science in Europe, 1790–1840* (Boston, MA: Kluwer Academic Publishers, 1994), pp. 31–45.

—, 'Some Ingenious Systems: Lavoisier and the Northern Chemists', in M. Beretta (ed.), *Lavoisier in Perspective* (Munich: Deutsches Museum, 2005), pp. 95–108.

Abbri, F., and M. Segala, (eds), *The Routes of Learning: Italy and Europe in the Modern Age* (Florence: Olschki, 2003).

Adorno, T. W., 'Why Philosophy', in D. Ingram and J. Simon-Ingram (eds), *Critical Theory: The Essential Readings* (1963; New York: Paragon House, 1992), pp. 20–30.

Agassi, J., *Towards an Historiography of Science* (Gravenhage: Mouton, 1963).

Albury, W. R., 'The Logic of Condillac and the Structure of French Chemical and Biological Theory, 1780–1801' (PhD dissertation, Johns Hopkins University, Baltimore, MD, 1972).

Allchin, D., 'Phlogiston after Oxygen', *Ambix*, 39 (1992), pp. 110–6.

Althusser, L., *For Marx*, trans. B. Brewster (New York: Vintage Books, 1970).

—, *Lenin and Philosophy and Other Essays*, trans. B. Brewster (New York: Monthly Review Press, 1971).

Althusser, L., and É. Balibar, *Reading Capital*, trans. B. Brewster (London: New Left Books, 1970).

Anderson, P., *Arguments within English Marxism* (London: New Left Books, 1980).

—, *In the Tracks of Historical Materialism* (London: Verso, 1983).

—, 'Modernity and Revolution', *New Left Review*, 144 (1984), pp. 96–133.

—, *The Origins of Postmodernity* (London: Verso, 1998).

Anderson, W. C., *Between the Library and the Laboratory: The Language of Chemistry in Eighteenth-Century France* (Baltimore, MD: Johns Hopkins University Press, 1984).

Badash, L., 'Joseph Priestley's Apparatus for Pneumatic Chemistry', *Journal of the History of Medicine*, 19 (1964), pp. 139–55.

Barnes, B., 'On the Causal Explanation of Scientific Judgment', *Social Science Information*, 19 (1980), pp. 685–95.

—, *T. S. Kuhn and Social Science* (New York: Columbia University Press, 1982).

—, 'On the Conventional Character of Knowledge and Cognition', in K. D. Knorr-Cetina and M. Mulkay (eds), *Science Observed: Perspectives on the Social Studies of Science* (Beverly Hills, CA and London: Sage Publications, 1983), pp. 19–51.

—, 'Thomas Kuhn', in Q. Skinner (ed.), *The Return of Grand Theory in the Human Sciences* (Cambridge: Cambridge University Press, 1985), pp. 83–100.

Barnes, B., and D. Bloor, 'Relativism, Rationalism, and the Sociology of Knowledge', in M. Hollis and S. Lukes (eds), *Rationality and Relativism* (Oxford: Blackwell, 1982), pp. 21–47.

Barnes, B., D. Bloor and J. Henry, *Scientific Knowledge: A Sociological Analysis* (Chicago, IL: University of Chicago Press, 1996).

Barrell, J., *English Literature in History 1730–80: An Equal Wide Survey* (London: Hutchinson, 1983).

Bazerman, C., 'How Natural Philosophers Can Cooperate: The Literary Technology of Coordinated Investigation in Joseph Priestley's *History and Present State of Electricity* (1767)', in T. C. Kynell and M. G. Moran (eds), *Three Keys to the Past: The History of Technological Communication (Contemporary Studies in Technological Communication Volume 7)* (Stamford, CT: Ablex Publishers, 1999), pp. 21–48.

Bensaude-Vincent, B., 'A Founder Myth in the History of Science? – the Lavoisier Case', in L. Graham, W. Lepinies and P. Weingart (eds), *Function and Uses of Disciplinary Histories (Sociology of the Sciences Yearbook, Volume Vii)* (Dordrecht: D. Reidel, 1983), pp. 53–78.

—, 'Hélène Metzger's *La Chimie*: A Popular Treatise', *History of Science*, 25 (1987), pp. 71–84.

—, 'A View of the Chemical Revolution through Contemporary Textbooks: Lavoisier, Fourcroy, and Chaptal', *British Journal for the History of Science*, 23 (1990), pp. 434–60.

—, 'The Balance: Between Chemistry and Politics', *The Eighteenth Century: Theory and Interpretation)*, 33 (1992), pp. 217–37.

—, *Lavoisier: Mémoires D'une Révolution* (Paris: Flammarion, 1993).

—, 'Introductory Essay: A Geographical History of Eighteenth-Century Chemistry', in B. Bensaude-Vincent and F. Abbri (eds), *Lavoisier in European Context: Negotiating a New Language for Chemistry* (Canton, MA: Science History Publications, 1995), pp. 1–17.

—, 'Between History and Memory: Centennial and Bicentennial Images of Lavoisier', *Isis*, 87 (1996), pp. 481–99.

—, 'Chemistry in the French Tradition of the Philosophy of Science: Duhem, Meyerson, Metzger and Bachelard', *Studies in History and Philosophy of Science*, 36 (2005), pp. 627–48.

Bensaude-Vincent, B., and I. Stengers, *A History of Chemistry*, trans. D. van Dam (Cambridge, MA: Harvard University Press, 1996).

Benton, T., *The Rise and Fall of Structural Marxism: Althusser and His Influence* (London: Macmillan, 1984).

Beretta, M., 'Chemists in the Storm: Lavoisier, Priestley and the Chemical Revolution', *Nuncius*, 8 (1993), pp. 75–104.

—, *The Enlightenment of Matter: The Definition of Chemistry from Agricola to Lavoisier* (Canton, MA: Science History Publications, 1993).

Beretta, M., and A. Scotti, 'Panopticon Lavoisier: A Presentation', in M. Beretta (ed.), *Lavoisier in Perspective* (Munich: Deutsches Museum, 2005), pp. 193–207.

Bhaskar, R., 'Feyerabend and Bachelard: Two Philosophies of Science', *New Left Review*, 94 (1975), pp. 31–55.

Biagioli, M., 'Meyerson: Science and the "Irrational"', *Studies in History and Philosophy of Science*, 19 (1988), pp. 31–55.

Bloor, D., *Wittgenstein: A Social Theory of Knowledge* (New York: Columbia University Press, 1983).

—, 'The Sociology of Reasons: Or Why "Epistemic Factors" Are Social Factors', in D. R. Brown (ed.), *The Rational and the Social* (Dordrecht: D. Reidel, 1984), pp. 295–324.

—, *Knowledge and Social Imagery*, 2nd edn (1976; Chicago, IL: University of Chicago Press, 1991).

—, 'Anti-Latour', *Studies in History and Philosophy of Science*, 30 (1999), pp. 81–112.

Bowker, G., and B. Latour, 'A Booming Discipline Short of Discipline: (Social) Studies of Science in France', *Social Studies of Science*, 17 (1987), pp. 715–48.

Brannigan, A., *The Social Basis of Scientific Discoveries* (Cambridge: Cambridge University Press, 1981).

Brett, P., 'Power, Sociability and Dissemination of Science: Lavoisier and the Learned Societies', in M. Beretta (ed.), *Lavoisier in Perspective* (Munich: Deutsches Museum, 2005), pp. 129–52.

Brock, W. H., *The Fontana History of Chemistry* (London: Fontana, 1992).

Brooke, J. H., '"A Sower Went Forth": Joseph Priestley and the Ministry of Reform', in T. A. Schwartz and J. G. McEvoy (eds), *Motion toward Perfection: The Achievement of Joseph Priestley* (Boston, MA: Skinner House Books, 1990), pp. 21–56.

Buchdahl, G., *The Image of Newton and Locke in the Age of Reason* (London: Sheed and Ward, 1961).

—, 'History of Science and Criteria of Choice', in R. H. Stuewer (ed.), *Historical and Philosophical Perspectives of Science* (1970; New York: Gordon and Breach, 1989), pp. 204–30.

Burian, R. M., 'More Than a Marriage of Convenience: On the Inextricability of History and Philosophy of Science', *Philosophy of Science*, 44 (1977), pp. 1–42.

Burrow, J. W., *A Liberal Descent: Victorian Historians and the English Past* (Cambridge: Cambridge University Press, 1981).

Bury, J. B., *The Idea of Progress: An Inquiry into Its Origin and Growth* (London: Macmillan, 1924).

Butterfield, H., *The Origins of Modern Science, 1300–1800*, revised edn (New York: Free Press, 1965).

—, *The Whig Interpretation of History* (New York: W. W. Norton, 1965 (1931)).

—, *The Englishman and His History* (Hamden, CT: Archon Books, 1970).

Cadzow, H., 'New Historicism', in M. Groden, M. Kreiswirth and I. Szeman (eds), *The Johns Hopkins Guide to Literary Theory and Criticism* (Baltiomre, MD: Johns Hopkins University Press, 1994), pp. 534–40.

Cahoune, L., 'Introduction', in L. Cahoune (ed.), *From Modernism to Postmodernism: An Anthology* (Oxford: Blackwell Publishers, 2003), pp. 1–13, 221–3.

Callinicos, A., *Althusser's Marxism* (London: Pluto Press, 1976).

—, *Against Postmodernism: A Marxist Critique* (New York: St Martin's Press, 1989).

Cantor, G. N., 'Weighing Light: The Role of Metaphor in Eighteenth-Century Optical Discourse', in A. E. Benjamin, G. N. Cantor and J. R. R. Christie (eds), *The Figural and the Literal: Problems of Language in the History of Science and Philosophy* (Manchester: Manchester University Press, 1987), pp. 124–46.

Chartier, R., *The Cultural Origins of the French Revolution* (Durham, NC: Duke University Press, 1991).

Christie, J. R. R., 'Ether and the Science of Chemistry: 1740–1790', in G. Cantor and M. J. S. Hodge (eds), *Conceptions of Ether: Studies in the History of Ether Theories 1740–1900* (Cambridge: Cambridge University Press, 1981), pp. 85–110.

—, 'Joseph Black and John Robison', in A. D. C. Simpson (ed.), *Joseph Black, 1728–1799: A Commemorative Symposium* (Edinburgh: Royal Scottish Museum, 1982), pp. 47–52.

—, 'Narrative and Rhetoric in Hélène Metzger's Historiography of Eighteenth Century Chemistry', *History of Science*, 25 (1987), pp. 99–109.

—, 'The Development of the Historiography of Science', in R. C. Olby, G. N. Cantor, J. R. R. Christie and M. J. S. Hodge (eds), *Companion to the History of Modern Science* (London: Routledge, 1990), pp. 5–22.

Christie, J. R. R., and J. Y. Golinski, 'The Spreading of the Word: New Directions in the Historiography of Chemistry, 1600–1800', *History of Science*, 20 (1982), pp. 235–66.

Clark, J. C. D., *Our Shadowed Present: Modernism, Postmodernism and History* (London: Atlantic Books, 2003).

Clark, W., J. Golinski and S. Schaffer, 'Introduction', in W. Clark, J. Golinski and S. Schaffer (eds), *The Sciences in Enlightened Europe* (Chicago, IL and London: University of Chicago Press, 1999), pp. 3–31.

Cohen, I. B., 'The Eighteenth-Century Origins of the Concept of Scientific Revolution', *Journal of the History of Ideas*, 37 (1976), pp. 257–88.

—, 'The Many Faces of the History of Science: A Font of Examples for Philosophers, a Scientific Type of History, an Archaeology of Discovery, a Branch of Sociology, a Variant of Intellectual or Social History – or *What?*' in C. F. Delzell (ed.), *The Future of History: Essays in the Vanderbilt University Centennial Symposium* (Nashville, TN: Vanderbilt University Press, 1977), pp. 65–110.

—, *Revolution in Science* (Cambridge, MA: Harvard University Press, 1985).

Collingwood, R. G., *An Autobiography* (Oxford: Oxford University Press, 1939).

Collins, H. M., 'The Role of The "Core Set" In Modern Science: Social Contingency with Methodological Propriety in Science', *History of Science*, 19 (1981), pp. 6–19.

—, 'An Empirical Relativist Programme in the Sociology of Scientific Knowledge', in K. D. Knorr-Cetina and M. Mulkay (eds), *Science Observed: Perspectives on the Social Studies of Science* (Beverly Hills, CA and London: Sage Publications, 1983), pp. 83–113.

—, 'The Sociology of Scientific Knowledge: Studies of Contemporary Science', *Annual Review of Sociology*, 9 (1983), pp. 265–85.

—, *Changing Order: Replication and Induction in Scientific Practice* (1985; Beverly Hills, CA and London: Sage Publications, 1992).

Comte, A., 'The Positive Philosophy of Auguste Comte (Selections)', in P. Gardiner (ed.), *Theories of History: From Classical to Contemporary Sources* (Glenco, IL: Free Press, 1959 (1993)), pp. 73–9.

—, *Introduction to Positive Philosophy* (1830), trans. F. Ferré (Cambridge, MA: Hackett, 1988).

Conant, J. B., 'Introduction', in J. B. Conant and L. K. Nash (eds), *Harvard Case Histories in Experimental Science* (Cambridge, MA: Harvard University Press, 1957), pp. vii–xvi.

—, 'The Overthrow of the Phlogiston Theory: The Chemical Revolution of 1775–1789', in J. B. Conant and L. K. Nash (eds), *Harvard Case Studies in Experimental Science* (Harvard, MA: Harvard University Press, 1957), pp. 65–116.

Conlin, M., 'Joseph Priestley's American Defense of Phlogiston Reconsidered', *Ambix*, 43 (1996), pp. 129–45.

Cronin, J., *Foucault's Antihumanist Historiography* (Lewiston, NY: Edwin Mellen Press, 2001).

Crosland, M. P., 'The Development of Chemistry in the Eighteenth Century', *Studies on Voltaire and the Eighteenth Century*, 24 (1963), pp. 369–441.

—, 'Lavoisier's Theory of Acidity', *Isis*, 64 (1973), pp. 306–25.

—, 'Chemistry and the Chemical Revolution', in G. S. Rousseau and R. Porter (eds), *The Ferment of Knowledge: Studies in the Historiography of Eighteenth-Century Science* (Cambridge: Cambridge University Press, 1980), pp. 389–416.

—, 'Priestley Memorial Lecture: A Practical Perspective on Joseph Priestley as a Natural Philosopher', *British Journal for the History of Science*, 16 (1983), pp. 223–38.

—, 'The Image of Science as a Threat: Burke Versus Priestley and the "Philosophical Revolution"', *British Journal for the History of Science*, 20 (1987), pp. 277–307.

Curd, M., and J. A. Cover, (eds), *Philosophy of Science: The Central Issues* (New York: W. W. Norton, 1998).

Dallmayr, F. D., and T. A. McCarthy, 'Introduction', in F. D. Dallmayr and T. A. McCarthy (eds), *Understanding and Social Inquiry* (Notre Dame, IN and London: University of Notre Dame Press, 1977), pp. 219–24.

Dampier, W. C., *A History of Science and its Relation with Philosophy and Religion* (Cambridge: Cambridge University Press, 1929).

Daumas, M., *Lavoisier: Théorecien Et Expérimentateur* (Paris: Presses Universitaires de France, 1955).

—, 'Precision of Measurement and Chemical Research in the Eighteenth Century', in A. C. Crombie (ed.), *Scientific Change: Historical Studeis in the Intellectual, Social, and Technical Conditions for Scientific Discovery and Technical Invention, from Antiquity to the Present* (New York: Basic Books, 1963), pp. 418–30.

Davis, T. L., 'The First Edition of the *Skeptical Chymist*', *Isis*, 8 (1926), pp. 71–6.

Dear, P., 'Cultural History of Science: An Overview with Reflections', *Science, Technology and Human Values*, 20 (1995), pp. 150–70.

—, *The Literary Structure of Scientific Arguments: Historical Studies* (Philadelphia, PA: University of Pennsylvania Press, 1991).

Derrida, J., *Of Grammatology*, trans. G. C. Spirak (Baltimore, MA: Johns Hopkins University Press, 1976 (1967)).

—, 'But, Beyond ... (Open Letter to Anne Mcclintock and Rob Nixon)', *Critical Inquiry*, 13 (1986), pp. 155–70.

Dews, P., 'Adorno, Post-Structuralism and the Critique of Identity', *New Left Review*, 157 (1986), pp. 28–44.

—, *Logics of Disintegration: Post-Structuralist Thought and the Claims of Critical Theory* (London: Verso, 1987).

—, 'Althusser, Structuralism, and the French Epistemological Tradition', in G. Elliot (ed.), *Althusser: A Critical Reader* (London: Blackwell, 1994), pp. 104–41.

Donovan, A., 'British Chemistry and the Concept of Science in the Eighteenth Century', *Albion*, 7 (1975), pp. 131–44.

—, 'Scottish Responses to the New Chemistry of Lavoisier', *Studies in Eighteenth-Century Culture*, 9 (1979), pp. 237–49.

—, 'The Chemical Revolution Revisited', in S. H. Cutcliffe (ed.), *Science and Technology in the Eighteenth Century: Essays of the Lawrence Gipson Institute for Eighteenth Century Studies* (Bethlehem, PA: Gipson Institute, Lehigh University, 1984), pp. 1–15.

—, 'Introduction', in A. Donovan (ed.), *The Chemical Revolution: Essays in Reinterpretation (Osiris, 2nd Series, Volume 4)* (Philadelphia, PA: History of Science Society, 1988), pp. 5–12.

—, 'Lavoisier and the Origins of Modern Chemistry', in A. Donovan (ed.), *The Chemical Revolution: Essays in Reinterpretation, Osiris, 2nd Series, Volume 4* (Philadelphia, PA: History of Science Society, 1988), pp. 214–31.

—, 'Lavoisier as Chemist and Experimental Physicist: A Reply to Perrin', *Isis*, 81 (1990), pp. 270–2.

—, 'Buffon, Lavoisier and the Transformation of French Chemistry', in J. Gayon (ed.), *Buffon 88: Actes Du Colloque International Pour Le Bicentennaire De La Mort De Buffon* (Paris: Vrin, 1992), pp. 397–95.

—, *Antoine Lavoisier: Science, Administration, and Revolution* (Oxford: Blackwell, 1993).

—, 'The New Nomenclature among the Scots: Assessing Novel Chemical Claims in a Culture under Strain', in B. Bensaude-Vincent and F. Abbri (eds), *Lavoisier in European Context: Negotiating a New Language for Chemistry* (Canton, MA: Science History Publications, 1995), pp. 113–22.

Doppelt, G., 'Kuhn's Epistemological Relativism: An Interpretation and Defense', *Inquiry*, 21 (1978), pp. 33–86.

—, 'Review Discussion: Laudan's Pragmatic Alternative to Reletivist and Historicist Theories of Science', *Inquiry*, 24 (1981), pp. 253–71.

Dreyfus, H. L., and P. Rabinow, *Michel Foucault: Beyond Structuralism and Hermeneutics*, 2nd edn (Chicago, IL: University of Chicago Press, 1983).

Duhem, P., *Le Mixte Et Le Combinasion Chimique: Essay Sur L'evolution D'une Idée* (Paris: Fayard, 1902).

Duncan, A. M., 'The Function of Affinity Tables in Lavoisier's List of Elements', *Ambix*, 17 (1970), pp. 28–42.

Eagleton, T., *Literary Theory: An Introduction* (Minneapolis, MN: University of Minnesota Press, 1983).

—, 'Capitalism, Modernism, and Postmodernism', *New Left Review*, 152 (1985), pp. 60–73.

Engels, F., 'Preface', in K. Marx (ed.), *Capital*, 1967 (1867–94)), pp. 1–19.

Feyerabend, P. K., 'Reply to Criticism: Comments on Smart, Sellars, and Putnam', in R. S. Cohen and M. W. Wartofsky (eds), *Boston Studies in the Philosophy of Science* (New York: Humanities Press, 1965), pp. 223–61.

—, 'Consolations for the Specialist', in I. Lakatos and A. Musgrave (eds), *Criticisms and the Growth of Knowledge* (Cambridge: Cambridge University Press, 1970), pp. 197–230.

Fichman, M., 'French Stahlism and Chemical Studies of Air, 1750–1770', *Ambix*, 18 (1971), pp. 94–122.

Finocchiaro, M. A., *History of Science as Explanation* (Detroit, MI: Wayne State University Press, 1973).

Fleck, L., *Genesis and Development of a Scientific Fact* (1935), trans. F. Bradley and T. J. Trenn (Chicago, IL: University of Chicago Press, 1979).

Forbes, D., 'Scientific Whiggism: Adam Smith and John Miller', *Cambridge Journal*, 7 (1954), pp. 643–70.

Foucault, M., *The Archaeology of Knowledge and the Discourse on Language*, trans. A. M. Sheridan Smith (New York: Harper and Row, 1972).

—, *The Birth of the Clinic: An Archaeology of Medical Perception* (1963), trans. A. Sheridan Smith (New York: Vintage Books, 1973).

—, *The Order of Things: An Archaeology of the Human Sciences* (1966), trans. A. Sheridan Smith (New York: Vintage Books, 1973).

—, *Discipline and Punish: The Birth of the Prison* (1975), trans. A. S. Smith (New York: Pantheon Books, 1978).

—, 'What Is Enlightenment' (1977), in P. Rabinow (ed.), *The Foucault Reader* (New York: Pantheon Books, 1984), pp. 32–50.

—, 'Nietzsche, Genealogy, History' (1971), in P. Rabinow (ed.), *The Foucault Reader* (New York: Pantheon Books, 1984), pp. 76–100.

—, 'Truth and Power' (1977), in P. Rabinow (ed.), *The Foucault Reader* (New York: Pantheon Books, 1984), pp. 51–75.

—, 'Of Other Spaces', *Diacritics*, 16 (1986), pp. 22–7.

—, *Mental Illness and Psychology* (1954), trans. A. M. Sheridan-Smith (Berkeley, CA: University of California Press, 1987).

—, 'On the Archaelogy of the Sciences: Response to the Epistemology Circle', in J. D. Faubion (ed.), *Aesthetics, Method, and Epistemology: Foucault* (1968; New York: The New Press, 1998), pp. 433–58.

Frängsmyr, T., 'Science or History: George Sarton and the Positivist Tradition in the History of Science', *Lychnos*, 74 (1973–4), pp. 104–44.

Freudenthal, G., 'The Role of Shared Knowledge in Science: The Failure of the Constructivist Progamme in the Sociology of Science', *Social Studies of Science*, 14 (1984), pp. 285–95.

Friedman, M., *Reconsidering Logical Positivism* (Cambridge: Cambridge University Press, 1999).

—, 'History and Philosophy of Science in a New Key', *Isis*, 99 (2008), pp. 125–34.

Fuller, S., 'Is the History and Philosophy of Science Withering on the Vine', *Philosophy of the Social Sciences*, 21 (1991), pp. 149–74.

Gadamer, H.-G., *Truth and Method* (1960), trans. J. Weinsheimer and D. G. Marshall (New York: Crossroad, 1975).

Galison, P., *How Experiments End* (Chicago, IL: University of Chicago Press, 1987).

—, 'History, Philosophy, and the Central Metaphor', *Science in Context*, 2 (1988), pp. 197–212.

Galison, P., and D. J. Stump, (eds), *The Disunity of Science: Boundaries, Contexts, and Power* (Stanford, CA: Stanford University Press, 1996).

Garfinkel, H., 'What Is Ethnomethodology?' in F. Dallmayr and T. McCarthy (eds), *Understanding and Scoical Inquiry* (Notre Dame, IN and London: University of Notre Dame Press, 1977 (1967)), pp. 240–61.

Geertz, C., 'Thick Description: Toward an Interpretive Theory of Culture', *The Interpretation of Culture* (New York: Basic Books, 1973), pp. 3–30.

Gibbs, F. W., *Joseph Priestley: Adventurer in Science and Champion of Truth* (London: Nelson, 1965).

Gillispie, C. C., *The Edge of Objectivity: An Essay in the History of Scientific Ideas* (Princeton, NJ: Princeton University Press, 1960).

Golinski, J., 'Science *in* the Enlightenment (Review of T. Hankins, *Science and the Enlightenment*)', *History of Science*, 24 (1986), pp. 411–24.

—, 'The Theory of Practice and the Practice of Theory: Sociological Approaches to the History of Science', *Isis*, 81 (1990), pp. 492–505.

—, *Science as Public Culture: Chemistry and the Enlightenment in Britain, 1760–1820* (Cambridge: Cambridge University Press, 1992).

—, 'The Chemical Revolution and the Politics of Language', *Eighteenth Century: Theory and Interpretation*, 33 (1992), pp. 238–51.

—, 'Prescision Instruments and the Demonstrative Order of Proof in Lavoisier's Chemistry', in A. van Helden and T. L. Hankins (eds), *Instruments (Osiris, 2nd Series, Volume 9)* (Philadelphia, PA: History of Science Society, 1994), pp. 30–47.

—, *Making Natural Knowledge: Constructivism and the History of Science* (Cambridge: Cambridge University Press, 1998).

Gordy, M., 'Reading Althusser: Time and the Social Whole', *History and Theory*, 22 (1983), pp. 1–21.

Gough, J. B., 'Lavoisier's Early Career in Science: An Examination of Some New Evidence', *British Journal for the History of Science*, 4 (1968), pp. 52–7.

—, 'The Origins of Lavoisier's Theory of the Gaseous State', in H. Woolf (ed.), *The Analytic Spirit: Essays in the History of Science in Honor of Henry Guerlac* (Ithaca, NY: Cornell University Press, 1981), pp. 15–39.

—, 'Lavoisier and the Fulfillment of the Stahlian Revolution', in A. Donovan (ed.), *The Chemical Revolution: Essays in Reinterpretation (Osiris, 2nd Series, Volume 4)* (Philadelphia, PA: History of Science Society, 1988), pp. 15–33.

Gregory, J. C., *Combustion from Hericlitus to Lavoisier* (London: Edward Arnold, 1934).

Guerlac, H., *Lavoisier – the Crucial Year: The Background and Origin of His First Experiments on Combustion in 1772* (Ithaca, NY: Cornell University Press, 1961).

—, 'Discussion', in A. C. Crombie (ed.), *Scientific Change: Historical Studies in the Intellectual, Social, and Technical Conditions for Scientific Discovery and Technical Invention from Antiquity to the Present* (New York: Basic Books, 1963), pp. 875–6.

—, *Antoine-Laurent Lavoisier: Chemist and Revolutionary* (New York: Scribners, 1975).

—, 'Some French Antecedents of the Chemical Revolution', in H. Guerlac, *Essays and Papers in the History of Modern Science* (1959; Baltimore, MD: Johns Hopkins University Press, 1977), pp. 340–74.

Gutting, G., 'Continental Philosophy of Science', in P. D. Asquith and H. E. Kyburg (eds), *Current Research in Philosophy of Science: Proceedings of the PSA Critical Research Problems Conference* (East Lansing, MI: Philosophy of Science Association, 1979), pp. 94–117.

—, *Michel Foucault's Archaeology of Scientific Reason* (Cambridge: Cambridge University Press, 1989).

Guyton de Morveau, L. B., A.-L. Lavoisier, C.-L. Berthollet, and A. F. Fourcroy, *Méthode De Nomenclature Chimique* (Paris: Cuchet, 1987).

Habermas, J., *Knowledge and Human Interests*, trans. J. J. Shapiro (Boston, MA: Beacon Press, 1971).

—, 'Modernity Versus Postmodernity', *New German Critique*, 22 (1981), pp. 3–14.

—, *The Philosophical Discourse of Modernity: Twelve Lectures*, trans. F. G. Lawrence (1981; Cambridge: Polity Press, 1987).

—, *On the Logic of the Social Sciences*, trans. S. W. Nicholsen and J. A. Stark (Cambridge, MA: MIT Press, 1988).

—, 'Modernity: An Unfinished Project', in M. P. d'Entrèves and S. Benhabib (eds), *Habermas and the Unfinished Project of Modernity: Critical Essays on the Philosophical Discourse of Modernity* (Cambridge: Polity Press, 1996), pp. 38–58.

Hacking, I., *Representing and Intervening: Introductory Topics in the Philosophy of Natural Science* (Cambridge: Cambridge University Press, 1983).

—, 'The Self-Validation of the Laboratory Sciences', in A. Pickering (ed.), *Science as Practice and Culture* (Chicago, IL: University of Chicago Press, 1992), pp. 29–64.

Hall (Boas), M., 'Structure of Matter and Chemical Theory in the Seventeenth and Eighteenth Centuries', in M. Clagett (ed.), *Critical Problems in the History of Science* (Madison, WI: University of Wisconsin Press, 1969), pp. 499–514.

Hall, A. R., *The Scientific Revolution, 1500–1800: The Formation of the Modern Scientific Attitude*, 2nd edn (London: Longmans, 1962).

—, 'On Whiggism', *History of Science*, 21 (1983), pp. 45–59.

Hampson, N., 'The Enlightenment in France', in R. Porter and T. Mikuláš (eds), *The Enlightenment in National Context* (Cambridge: Cambridge University Press, 1981), pp. 41–53.

Hanfi, Z., (ed.) *The Fiery Brook: Selected Writings of Ludwig Feuerbach: Tranlated with an Introduction by Zawar Hanfi* (New York: Anchor Books, 1972).

Hankins, T. L., *Science and the Enlightenment* (Cambridge: Cambridge University Press, 1985).

Hannaway, O., *The Chemists and the Word: The Didactic Origins of Chemistry* (Baltimore, MD: Johns Hopkins University Press, 1975).

Hanson, N. R., *Patterns of Discovery: An Inquiry into the Conceptual Foundations of Science* (London: The Scientific Book Guild, 1962).

Hartfoot, C., 'The Missing Synthesis in the Historiography of Science', *History of Science*, 24 (1991), pp. 207–16.

Hartley, H., *Studies in the History of Chemistry* (Oxford: Clarendon Press, 1971).

Hartog, P., 'The Newer Views of Priestley and Lavoisier', *Annals of Science*, 5 (1941), pp. 1–56.

Heering, P., 'Weighting the Heat: The Replication of Experiments with the Ice-Calorimeter of Lavoisier and Laplace', in M. Beretta (ed.), *Lavoisier in Perspective* (Munich: Deutsches Museum, 2005), pp. 27–41.

Heilbroner, R. L., *The Nature and Logic of Capitalism* (New York: Norton, 1985).

Heller, H., *The Bourgeois Revolution in France (1789–1815)* (New York and Oxford: Berghahn Books, 2006).

Hempel, C. G., *Aspects of Scientific Explanation and Other Essays in the Philosophy of Science* (New York: Free Press, 1965).

Hobsbawm, E. J., *The Age of Revolution, Europe 1789–1848* (London: Abacus, 1977).

Holmes, F. L., *Lavoisier and the Chemistry of Life: An Exploration of Scientific Creativity* (Madison, WI: University of Wisconsin Press, 1985).

—, 'Lavoisier's Conceptual Passage', in A. Donovan (ed.), *The Chemical Revolution: Essays in Reinterpretation, Osiris, 2nd Series, Volume 4* (Philadelphia, PA: History of Science Society, 1988), pp. 82–92.

—, *Eighteenth-Century Chemistry as an Investigative Enterprise* (Berkeley, CA: Office for History of Science and Technology, University of California, 1989).

—, 'Beyond the Boundaries: Concluding Remarks on the Workshop', in B. Bensaude-Vincent and F. Abbri (eds), *Lavoisier in European Context: Negotiating a New Language for Chemistry* (Canton, MA: Science Publications International, 1995), pp. 267–78.

—, *Antoine Lavoisier – the Next Crucial Year, or the Sources of His Quantitative Method in Chemistry* (Princeton, NJ: Princeton University Press, 1998).

—, 'The "Revolution in Chemistry and Physics": Overthrow of a Reigning Paradigm or Competition between Contemporary Research Programs', *Isis*, 91 (2000), pp. 735–53.

Holmes, F. L., and T. H. Levere, (eds), *Instruments and Experimentation in the History of Chemistry* (Cambridge, MA: MIT Press, 2000).

Hufbauer, K., *The Formation of the German Chemical Community (1720–1795)* (Berkeley, CA: University of California Press, 1982).

Illife, R., 'Rhetorical Vices: Outlines of a Feyerabendian History of Science', *History of Science*, 30 (1992), pp. 199–219.

Ingram, D., *Habermas and the Dialectic of Reason* (New Haven, CT and London: Yale University Press, 1987).

Jameson, F., *The Political Unconscious: Narrative as a Symbolic Act* (New York: Cornell University Press, 1981).

—, 'Postmodernism, or the Cultural Logic of Late Capitalism', *New Left Review*, 146 (1984), pp. 59–92.

—, 'Marxism and Postmodernism', *New Left Review*, 176 (1989), pp. 31–45.

Jann, R., *The Art and Science of Victorian History* (Columbus, OH: Ohio State University Press, 1985).

Jardine, N., 'A Dip into the Future', *Studies in History and Philosophy of Science*, 20 (1989), pp. 15–8.

Johnson, R., 'Edward Thompson, Eugene Genovese, and Socialist-Humanist History', *History Workshop*, 6 (1978), pp. 79–100.

Jones, G. S., 'The Pathology of English History', *New Left Review*, 46 (1967), pp. 29–43.

Kim, M. G., *Affinity, That Elusive Dream: A Genealogy of the Chemical Revolution* (Cambridge, MA: MIT Press, 2003).

—, 'Lavoisier, the Father of Modern Chemistry?', in M. Beretta (ed.), *Lavoisier in Perspective* (Munich: Deutsches Museum, 2005), pp. 167–91.

Kitcher, P., *The Advancement of Science: Science without Legend, Objectivity without Illusions* (Oxford: Oxford University Press, 1993).

Klein, U., 'Origin of the Concept of Chemical Compound', *Science in Context*, 7:2 (1994), pp. 163–204.

—, 'E. F. Geoffroy's Table of Different 'Rapports' Observed between Different Chemical Substances – a Reinterpretation', *Ambix*, 7:2 (1995), pp. 79–100.

—, 'The Chemical Workshop Tradition and the Experimental Practice: Discontinuities within Continuities', *Science in Context*, 9:3 (1996), pp. 251–87.

Klein, U., and W. Lefèvre, *Materials in Eighteenth-Century Science: A Historical Ontology* (Cambridge, MA: MIT Press, 2007).

Knorr-Cetina, K. D., 'The Ethnographic Study of Science Work: Toward a Constructivist Interpretation of Science', in K. Knorr-Cetina and M. Mulkay (eds), *Science Observed: Persepctives on the Social Study of Science* (Beverly Hills, CA and London: Sage Publications, 1983), pp. 115–39.

—, 'The Couch, the Cathedral, and the Laboratory: On the Relationship between Experiment and Laboratory in Science', in A. Pickering (ed.), *Science as Practice and Culture* (Chicago, IL: University of Chicago Press, 1992), pp. 113–38.

Knorr-Cetina, K. D., and M. Mulkay, 'Introduction: Emerging Principles in the Social Studies of Science', in K. D. Knorr-Cetina and M. Mulkay (eds), *Science Observed: Perspectives on the Social Studies of Science* (London and Beverly Hills, CA: Sage Publications, 1983), pp. 1–17.

Kohler, R. E., 'The Origin of Lavoisier's First Experiments on Combustion', *Isis*, 63 (1972), pp. 349–55.

Kopp, H., *Beiträge Zur Geschichte Der Chemie*, 2 vols(Braunschweig: F. Vieweg, 1875).

Koyré, A., 'Commentary', in A. C. Crombie (ed.), *Scientific Change: Historical Studies in the Intellectual, Social, and Technical Conditions for Scientific Discovery and Technical Invention, from Antiquity to the Present* (Ithaca, NY: Cornell University Press, 1963), pp. 847–57.

Kragh, H., *An Introduction to the Historiography of Science* (Cambridge: Cambridge University Press, 1987).

Kramnick, I., 'Eighteenth-Century Science and Radical Social Theory: The Case of Joseph Priestley's Scientific Liberalism', in A. T. Schwartz. and J. G. McEvoy (eds), *Motion toward Perfection: The Achievement of Joseph Priestley* (Boston, MA: Skinner House Books, 1990), pp. 57–92.

Kuhn, T. S., 'Robert Boyle and Structural Chemistry in the Seventeenth Century', *Isis*, 43 (1952), pp. 1–23.

—, *The Structure of Scientific Revolutions*, 2nd edn (1962; Chicago, IL: University of Chicago Press, 1970).

—, 'Logic of Discovery or Psychology of Research', in I. Lakatos and A. Musgrave (eds), *Criticism and the Growth of Knowledge* (Cambridge: Cambridge University Press, 1970), pp. 1–23.

—, 'Reflections on My Critics', in I. Lakatos and A. Musgrave (eds), *Criticism and the Growth of Knowledge* (Cambridge: Cambridge University Press, 1970), pp. 231–77.

—, *The Essential Tension: Selected Studies in Scientific Tradition and Change* (Chicago, IL: University of Chicago Press, 1977).

Lakatos, I., 'Falsification and the Methodology of Scientific Research Programmes', in I. Lakatos and A. Musgrave (eds), *Criticism and the Growth of Knowledge* (Cambridge: Cambridge University Press, 1970), pp. 91–196.

—, 'History of Science and Its Rational Reconstructions', in C. Howson (ed.), *Method and Appraisal in the Physical Sciences: The Critical Background to Modern Science, 1800–1905* (Cambridge: Cambridge University Press, 1976), pp. 1–39.

Lash, S., and J. Friedman, 'Introduction: Subjectivity and Modernity's Other', in S. Lash and J. Friedman (eds), *Modernity and Identity* (Oxford: Blackwell, 1992), pp. 1–30.

Latour, B., *Science in Action: How to Follow Scientists and Engineers through Society* (Cambridge, MA: Harvard University Press, 1987).

—, 'Essay Review: Postmodern? No, Simply Amodern! Steps Towards an Anthropology of Science', *Studies in History and Philosophy of Science*, 21 (1990), pp. 145–71.

—, 'One More Turn after the Social Turn ...' in E. McMullin (ed.), *The Social Dimensions of Science* (Notre Dame, IN: University of Notre Dame Press, 1992), pp. 272–94.

—, *We Have Never Been Modern*, trans. C. Porter (Cambridge, MA: Harvard University Press, 1993).

Latour, B., and S. Woolgar, *Laboratory Life: The Social Construction of Scientific Facts*, 2nd edn (Princeton, NJ: Princeton University Press, 1986).

Laudan, L., *Progress and Its Problems: Toward a Theory of Scientific Growth* (Berkeley, CA: University of California Press, 1977).

—, *Science and Hypothesis: Historical Essays on Scientific Method* (Dordrecht: D. Reidel, 1981).

—, 'The Pseudo-Science of Science', *Philosophy of the Social Sciences*, 11 (1981), pp. 173–98.

—, *Science and Values: An Essay on the Aims of Science and Their Role in Scientific Debate* (Berkeley, CA: University of California Press, 1984).

—, 'The History of Science and the Philosophy of Science', in R. C. Olby, G. N. Cantor, J. R. R. Christie and M. J. S. Hodge (eds), *Companion to the History of Modern Science* (London: Routledge, 1990), pp. 47–59.

Lavoisier, A., 'Reflexionsa Sur Le Phlogistique', in J. B. Dumas and E. Grimaux (eds), *Ouvres De Lavoisier*, 6 vols (Paris: Imprimerie Impériale, 1862–1893), vol. 3, pp. 623–55.

—, *Elements of Chemistry in a New Systematic Order, Continaing All the Modern Discoveries* (1790), trans. R. Kerr (New York: Dover, 1965).

LeGrand, H. E., 'Lavoisier's Oxygen Theory of Acidity', *Annals of Science*, 29 (1972), pp. 1–18.

—, 'Theory and Application: The Early Chemical Work of J.A.C. Chaptal', *British Journal for the History of Science*, 17 (1984), pp. 31–46.

Lemert, C., 'General Social Theory, Irony, Postmodernism', in S. Seidman and D. Wagner (eds), *Postmodernism and Social Theory: The Debate over General Theory* (Cambridge, MA: Blackwell, 1992), pp. 17–46.

Lenoir, T., 'Practice, Reason, Context: The Dialogue between Theory and Experiment', *Science in Context*, 2 (1988), pp. 3–22.

—, 'Inscription Practices and Materialities of Communication', in T. Lenoir and H. U. Gumbrecht (eds), *Inscribing Science: Scientific Texts and the Materiality of Communication* (Sanford, CA: Stanford University Press, 1998), pp. 1–19.

Lenoir, T., and H. U. Gumbrecht, (eds), *Inscribing Science: Scientific Texts and the Materiality of Communication* (Stanford, CA: Stanford University Press, 1988).

Levere, T. H., 'Lavoisier, Language, Instruments, and the Chemical Revolution', in T. H. Levere and W. R. Shea (eds), *Nature, Experiment, and the Sciences; Essays on Galileo and the History of Science in Honor of Stillman Drake* (Dordrecht: Kluwer Academic Publishers, 1990), pp. 207–23.

—, 'Lavoisier's Gasometers and Others: Research, Control, and Dissemination', in M. Beretta (ed.), *Lavoisier in Perspective* (Munich: Deutsches Museum, 2005), pp. 53–69.

Levere, T. H., and F. L. Holmes, 'Introduction: A Practical Science', in F. L. Holmes and T. H. Levere (eds), *Instruments and Experimentation in the History of Chemistry* (Cambridge, MA: MIT Press, 2000), pp. vii–xvii.

Levin, A., 'Venel, Lavoisier, Fourcroy, Cabanis, and the Idea of Scientific Revolutions: The French Political Context and General Patterns of Conceptualization of Scientific Change', *History of Science*, 22 (1984), pp. 303–20.

Lindberg, D. C., 'Conceptions of the Scientific Revolution from Bacon to Butterfield', in D. C. Lindberg and R. S. Westman (eds), *Reappraisals of the Scientific Revolution* (Cambridge: Cambridge University Press, 1990), pp. 1–26.

Llana, J. W., 'A Contribution of Natural History to the Chemical Revolution in France', *Ambix*, 32 (1985), pp. 71–91.

Löwith, K., *Meaning in History: Theological Implications of the Philosophy of History* (Chicago IL: University of Chicago Press, 1949).

Lukács, G., *History and Class Consciousness. Studies in Marxist Dialectics* (Cambridge, MA: MIT Press, 1971).

Lynch, M., *Art and Artifact in Laboratory Science* (London: Routledge and Kegan Paul, 1985).

—, *Scientific Practice and Ordinary Action: Ethnomethodology and Social Studies of Science* (Cambridge: Cambridge University Press, 1993).

Lynch, M., E. Livingstone and H. Garfinkel, 'Temporal Order in Laboratory Work', in K. Knorr-Cetina and M. Mulkay (eds), *Science Observed: Perspectives on the Social Study of Science* (Beverly Hills, CA and London: Sage Publications, 1983), pp. 205–38.

Lyotard, J.-F., *The Postmodern Condition: A Report on Knowledge*, trans. G. Bennington and B. Massumi (1979; Minneapolis, MN: University of Minnesota Press, 1984).

MacKenzie, D., and B. Barnes, 'Scientific Judgment: The Biometry-Mendelism Controversy', in B. Barnes and S. Shapin (eds), *Natural Order: Historical Studies of Scienfic Culture* (Beverly Hills, CA and London: Sage Publications, 1979), pp. 191–210.

Maienschein, J., M. Laubichler and A. Loettgers, 'How Can the History of Science Matter to Scientists?' *Isis*, 99 (2008), pp. 341–9.

Marx, K., *Capital*, 3 vols (1867–1894; New York: International Publishers, 1967).

Marx, K. and F. Engels, *The German Ideology, Part 1* (New York: International Publishers, 1970).

—, *Karl Marx and Frederick Engels: Collected Works, Volume Five* (New York: International Publishers, 1976).

Marx, L., 'Does Improved Technology Mean Progress?' in A. H. Teich (ed.), *Technology and the Future*, 9th edn (Belmont, CA: Thompson Wadsworth, 2006), pp. 3–12.

Masterman, M., 'The Nature of a Paradigm', in I. Lakatos and A. Musgrave (eds), *Criticism and the Growth of Knowledge* (Cambridge: Cambridge University Press, 1970), pp. 59–89.

Mauskopf, S. H., 'Gunpowder and the Chemical Revolution', in A. Donovan (ed.), *The Chemical Revolution: Essays in Reinterpretation, Osiris, 2nd Series, Volume 4* (Philadelphia, PA: History of Science Society, 1988), pp. 93–118.

Mayer, A.-K., 'Moralizing Science: The Uses of Science's Past in National Education in the 1920s', *British Journal for the History of Science*, 30 (1997), pp. 51–70.

—, 'Setting up a Discipline, II: British History of Science and 'the End of Ideology', 1931–1948', *Studies in History and Philosophy of Science*, 35 (2004), pp. 41–72.

—, 'When Things Don't Talk: Knowledge and Belief in the Inter-War Humanism of Charles Singer', *British Journal for the History of Science*, 38 (2005), pp. 325–47.

McCann, H. G., *Chemistry Transformed: The Paradigmatic Shift from Phlogiston to Oxygen* (Norwood, NJ: Ablex Publishing, 1978).

McEvoy, J. G., 'Joseph Priestley, Natural Philosopher: Some Comments on Professor Schofield's Views', *Ambix*, 15 (1968), pp. 115–23.

—, 'A Revolutionary Philosophy of Science: Paul K. Feyerabend and the Degeneration of Critical Rationalism into Sceptical Fallibilism', *Philosophy of Science*, 47 (1975), pp. 49–66.

—, 'Joseph Priestley: "Aerial Philosopher": Metaphysics and Methodology in Priestley's Chemical Thought, 1772–1781, Part 1', *Ambix*, 25 (1978), pp. 1–55; 'Part 2', ibid. pp. 93–116; 'Part 3', ibid., pp. 153–75; 'Part 4', *Ambix*, 26 (1979), pp. 16–38.

—, 'Enlightenment and Dissent in Science: Joseph Priestley and the Limits of Theoretical Reasoning', *Enlightenment and Dissent*, 2 (1983), pp. 47–67.

—, 'Causes and Laws, Powers and Principles: The Metaphysical Foundations of Priestley's Concept of Phlogiston', in R. G. W. Anderson and C. Lawrence (eds), *Science, Medicine, and Dissent: Joseph Priestley (1733–1804)* (London: Wellcome Trust and the Science Museum, 1987), pp. 55–71.

—, 'The Enlightenment and the Chemical Revolution', in R. S. Woolhouse (ed.), *Metaphysics and the Philosophy of Science in the Seventeenth and Eighteenth Centuries: Essays in Honor of Gerd Buchdahl* (Dordrecht: Kluwer Academic Publishers, 1988), pp. 307–25.

—, 'Continuity and Discontinuity in the Chemical Revolution', in A. Donovan (ed.), *The Chemical Revolution: Essays in Reinterpretation, Osiris, 2nd Series, Volume 4* (Philadelphia, PA: History of Science Society, 1988), pp. 195–213.

—, 'Joseph Priestley and the Chemical Revolution: A Thematic Overview', in A. T. Schwartz and J. G. McEvoy (eds), *Motion toward Perfection: The Achievement of Joseph Priestley* (Boston: Skinner House Books, 1990), pp. 129–60.

—, 'The Chemical Revolution in Context', *The Eighteenth Century: Theory and Interpretation*, 33 (1992), pp. 198–216.

—, 'Priestley Responds to Lavoisier's Nomenclature: Language, Liberty, and Chemistry in the English Enlightenment', in B. Bensaude-Vincent and F. Abbri (eds), *Lavoisier in European Context: Negotiating a New Language for Chemistry* (Canton, MA: Science History Publications, 1995), pp. 123–42.

—, 'Positivism, Whiggism and the Chemical Revolution: A Study in the Historiography of Science', *History of Science*, 35 (1997), pp. 1–33.

—, 'Postpositivist Interpretations of the Chemical Revolution', *Canadian Journal of History*, 36 (2001), pp. 453–69.

McEvoy, J. G., and J. E. McGuire, 'God and Nature: Priestley's Way of Rational Dissent', *Historical Studies in the Physical Sciences*, 6 (1975), pp. 325–404.

McKie, D., *Antoine Lavoisier: The Father of Modern Chemistry* (Philadelphia, PA: J.B. Lippincott, 1935).

Meinel, C., 'Demarcation Debates: Lavoisier in Germany', in M. Beretta (ed.), *Lavoisier in Perspective* (Munich: Deutsches Museum, 2005), pp. 153–66.

Melhado, E., 'Chemistry, Physics, and the Chemical Revolution', *Isis*, 76 (1985), pp. 195–211.

—, 'Toward an Understanding of the Chemical Revolution', *Knowledge and Society: Studies in the Sociology of Culture Past and Present*, 8 (1989), pp. 123–37.

—, 'Metzger, Kuhn, and Eighteenth-Century Disciplinary History', in G. Freudenthal (ed.), *Etudes Sur/Studies on Hélène Metzger* (Leiden: E.J. Brill, 1990), pp. 111–35.

—, 'On the Historiography of Science: A Reply to Perrin', *Isis*, 81 (1990).

Miller, D. P., *Discovering Water: James Watt, Henry Cavendish, and the Nineteenth-Century "Water Controversy"* (Aldershot and Burlington, VT: Ashgate, 2004).

Montrose, L. A., 'Professing the Renaissance: The Poetics and Politics of Culture', in H. Veeser (ed.), *The New Historicism* (New York: Routledge, 1989), pp. 15–36.

Morris, R. J., 'Lavoisier and the Caloric Theory', *British Journal for the History of Science*, 6 (1972), pp. 1–38.

Motzkin, G., 'The Catholic Response to Secularization and the Rise of the History of Science as a Discipline', *Science in Context*, 3 (1989), pp. 203–22.

Muir, M., *Heroes of Science: Chemists* (New York: E. & J.B. Young, 1883).

Mulkay, M., *Science and the Sociology of Knowledge* (London: Allen and Unwin, 1979).

Multhauf, R. P., 'On the Use of the Balance in Chemistry', *Proceedings of the American Philosophical Society*, 106 (1962), pp. 210–8.

Musgrave, A., 'Why did Oxygen Supplant Phlogiston? Research Programmes in the Chemical Revolution', in C. Howson (ed.), *Method and Appraisal in the Physical Sciences: The Critical Background to Modern Science, 1800–1905* (Cambridge: Cambridge University Press, 1976), pp. 181–209.

Nickles, T., 'Remarks on the Uses of History as Evidence', *Synthese*, 69 (1986), pp. 253–66.

—, 'Reconstructing Science: Dscovery and Experiment ', in D. Batens and J. P. V. Bendegem (eds), *Theory and Experiment: Recent Insights and Perspectives on Their Relation* (Dordrecht: D. Reidel, 1988), pp. 33–53.

Nordmann, A., 'The Passion for Truth: Lavoisier's and Litchenberg's Enlightenments', in M. Beretta (ed.), *Lavoisier in Perspective* (Munich: Deutsches Museum, 2005), pp. 109–28.

Oldroyd, D., 'An Examination of G. E. Stahl's *Philosophical Problems of Universal Chemistry*', *Ambix*, 20 (1973), pp. 36–52.

—, 'Mineralogy and the Chemical Revolution', *Centaurus*, 21 (1975), pp. 54–71.

—, 'The Doctrine of Property-Conferring Principles in Chemistry: Origins and Antecedents', *Organon*, 12–13 (1976/7), pp. 441–62.

—, 'Grid/Group Analysis for Historians of Science', *History of Science*, 24 (1986), pp. 145–71.

Ophir, A., and S. Shapin, 'The Place of Knowledge. A Methodological Survey', *Science in Context*, 4 (1991), pp. 3–21.

Parascandola, J., and A. J. Ihde, 'History of the Pneumatic Trough', *Isis*, 55 (1969), pp. 351–61.

Partington, J. R., *A Short History of Chemistry* (London: Macmillan, 1937).

Partington, J. R., and D. McKie, 'Historical Studies on the Phlogiston Theory – IV. Last Phases of the Theory', *Annals of Science*, 4 (1939), pp. 113–49.

Perkins, J., 'Creating Chemistry in Provincial France before the Revolution: The Examples of Nancy and Metz. Part I. Nancy', *Ambix*, 50 (2003), pp. 145–81.

—, 'Creating Chemistry in Provincial France before the Revolution: The Examples of Nancy and Metz. Part II. Metz', *Ambix*, 51 (2004), pp. 43–76.

Perrin, C. E., 'Early Opposition to the Phlogiston Theory: Two Anonymous Attacks', *British Journal for the History of Science*, 5 (1970), pp. 128–44.

—, 'Lavoisier's Table of the Elements: A Reappraisal', *Ambix*, 20 (1973), pp. 95–105.

—, 'A Reluctant Catalyst: Joseph Black and the Edinburgh Reception of Lavoisier's Chemistry', *Ambix*, 29 (1982), pp. 141–76.

—, 'Revolution or Reform: The Chemical Revolution and Eighteenth-Century Concepts of Scientific Change', *History of Science*, 25 (1987), pp. 395–423.

—, 'Research Traditions, Lavoisier, and the Chemical Revolution', in A. Donovan (ed.), *The Chemical Revolution: Essays in Reinterpretation, Osiris, 2nd Series, Volume, 4* (Philadelphia, PA: History of Science Society, 1988), pp. 53–81.

—, 'The Chemical Revolution: Shifts in Guiding Assumptions', in A. Donovan, L. Laudan and R. Laudan (eds), *Scrutinizing Science: Empirical Studies of Scientific Change* (Dordrecht: Kluwer Academic Publishers, 1988), pp. 105–24.

—, 'Document, Text, and Myth: Lavoisier's Crucial Year Revisited', *British Journal for the History of Science*, 22 (1989), pp. 3–25.

—, 'Chemistry as Peer of Physics: A Response to Donovan and Melhado on Lavoisier', *Isis*, 81 (1990), pp. 259–70.

Pickering, A., *Constructing Quarks: A Sociological History of Particle Physics* (Edinburgh: Edinburgh University Press, 1984).

—, 'Knowledge, Practice, and Mere Construction', *Social Studies of Science*, 20 (1990), pp. 682–729.

—, 'From Science as Knowledge to Science as Practice', in A. Pickering (ed.), *Science as Practice and Culture* (Chicago, IL: University of Chicago Press, 1992), pp. 1–26.

—, *The Mangle of Practice: Time, Agency, and Science* (Chicago, IL: University of Chicago Press, 1995).

Pickering, M., *Auguste Comte: An Intellectual Biography* (Cambridge: Cambridge University Press, 1983).

Pippin, R. B., *Modernism as a Philosophical Problem: On the Disatisfactions of European High Culture* (Oxford: Blackwell, 1991).

Polanyi, M., *Personal Knowledge: Towards a Post-Critical Philosophy* (Chicago, IL: University of Chicago Press, 1958).

Popper, K. R., *The Poverty of Historicism* (Boston: Beacon Press, 1957).

—, *The Logic of Scientific Discovery*, 3rd edn (London: Hutchinson, 1968).

—, *Conjectures and Refutations: The Growth of Scientific Knowledge* (New York: Harper Torchbooks, 1968).

—, *Objective Knowledge: An Evolutionary Approach* (Oxford: Oxford University Press, 1972).

Porter, R., 'Science, Provincial Culture, and Public Opinion in Enlightenment England', *British Journal for Eighteenth-Century Studies*, 3 (1980), pp. 20–46.

—, 'The Enlightenment in England', in R. Porter and M. Teich (eds), *The Enlightenment in National Context* (Cambridge: Cambridge Univesity Press, 1981), pp. 1–19.

Porter, T. M., 'The Promotion of Mining and the Advancement of Science: The Chemical Revolution of Mineralogy', *Annals of Science*, 38 (1981), pp. 543–70.

Priestley, J., 'Experiments and Observations Relating to Phlogiston and the Seeming Conversion of Water into Air', *Philosophical Transactions*, 73 (1783), pp. 398–14.

—, 'Experiments and Observations Relating to the Principle of Acidity, the Composition of Water, and Phlogiston', *Philosophical Transactions*, 78 (1788), pp. 147–57.

—, *Experiments and Observations on Different Kinds of Air, and Other Branches of Natural Philosophy Concerned with the Subject, in Three Volumes, Being the Former Six Volumes Abridged and Methodized, with Many Additions* (Birmingham: J. Johnson, 1790).

—, 'Further Experiments Relating to the Decomposition of Dephlogisticated Air and Inflammable Air', *Philosophical Transactions*, 81 (1791), pp. 213–22.

—, *Heads of Lectures on a Course of Experimental Philosophy, Particularly Including Chemistry, Delivered at the New College in Hackney* (London: J. Johnson, 1794).

—, *Experiments and Observations Relating to the Analysis of Atmospherical Air ... To Which Are Added Considerations on the Doctrine of Phlogiston and the Decomposition of Water* (London: J. Johnson, 1796).

—, *Considerations on the Doctrine of Phlogiston and the Decomposition of Water, Part Ii* (Philadelphia, PA: T. Dobson, 1797).

—, 'Experiements on the Production of Air by the Freezing of Water', *New York Medical Repository*, 4 (1801), pp. 17–21.

—, *The Doctrine of Phlogiston Established and that of the Decomposition of Water Refuted* (Philadelphia: P. Byrne, 1803).

Prinz, J. P., 'Lavoisier's Experimental Method of His Research on Human Respiration', in M. Beretta (ed.), *Lavoisier in Perspective* (Munich: Deutsches Museum, 2005), pp. 43–51.

Quine, W. V., 'Two Dogmas of Empiricism', in M. Curd and J. A. Cover (eds), *Philosophy of Science: The Central Issues* (1953; New York: W.W. Norton, 1998), pp. 280–301.

Richardson, A., 'Scientific Philosophy as a Topic for History of Science', *Isis*, 99 (2008), pp. 88–96.

Roberts, L., 'A Word and the World: The Significance of Naming the Calorimeter', *Isis*, 82 (1991), pp. 199–222.

—, 'Setting the Table: The Disciplinary Development of Eighteenth-Century Chemistry as Read through the Changing Structure of Its Tables', in P. Dear (ed.), *The Literary Structure of Scientific Argument: Historical Studies* (Philadelphia, PA: University of Pennsylvania Press, 1991), pp. 99–132.

—, 'Understanding Science: Beyond the Antics of Ontics (Symposium on the Possibilities of a Postmodern Philosophy of Science)', *Social Epistemology*, 5 (1991), pp. 247–55.

—, 'Condillac, Lavoisier, and the Instrumentalization of Science', *Eighteenth Century: Theory and Interpretation*, 33 (1992), pp. 252–71.

—, 'Filling the Space of Possibilities: Eighteenth-Century Chemistry's Transition from Art to Science', *Science in Context*, 6 (1993), pp. 511–53.

—, 'The Death of the Sensuous Chemist: The "New" Chemistry and the Transformation of Sensuous Technology', *Studies in History and Philosophy of Science*, 26 (1995), pp. 503–29.

—, 'Going Dutch: Situating Science in the New Enlightenment', in W. Clarke, J. Golinski and S. Schaffer (eds), *The Sciences in Enlightened Europe* (Chicago, IL: University of Chicago Press, 1999), pp. 350–88.

Rodwell, G. F., 'On the Theory of Phlogiston', *Philosophical Magazine*, 35 (1868), pp. 1–32.

Rorty, R., *Philosophy and the Mirror of Nature* (Princeton, NJ: Princeton University Press, 1979).

—, 'Habermas and Lyotard on Postmodernity', in R. J. Bernstein (ed.), *Habermas and Modernity* (Cambridge, MA: M.I.T. Press, 1985), pp. 161–75.

—, 'Solidarity or Objectivity', in L. Cahoone (ed.), *From Modernism to Postmodernism: An Anthology* (1985; Oxford: Blackwell Publishers, 2003), pp. 447–56.

Rosemann, P. W., 'Heidegger's Transcendental History', *Journal of the History of Philosophy*, 40 (2002), pp. 501–23.

Rosenau, P. M., *Postmodernism and the Social Sciences: Insights, Inroads, and Intrusions* (Princeton, NJ: Princeton University Press, 1992).

Rouse, J., *Knowledge and Power: Towards a Political Philosophy of Science* (Ithaca, NY: Cornell University Press, 1987).

—, 'Philosophy of Science and the Persistent Narratives of Modernity', *Studies in History and Philosophy of Science*, 22 (1991), pp. 141–62.

—, 'Foucault and the Natural Sciences', in J. Caputo and M. Yount (eds), *Foucault and the Critique of Institutions* (University Park, PA: Pennsylvania University Press, 1993), pp. 137–62.

—, 'Power/Knowledge', in G. Gutting (ed.), *The Cambridge Companion to Foucault* (Cambridge: Cambridge University Press, 1994), pp. 92–114.

Sarton, G., 'The History of Science', *Monist*, 31 (1916), pp. 321–65.

—, *Introduction to the History of Science*, 3 vols (Baltimore, MD: Williams and Wilkins, 1927–48).

Schaffer, S., 'Natural Philosophy', in G. S. Rousseau and R. Porter (eds), *The Ferment of Knowledge: Studies in the Historiography of Eighteenth-Century Science* (Cambridge: Cambridge University Press, 1980), pp. 55–91.

—, 'Natural Philosophy and Public Spectacle in the Eighteenth Century', *History of Science*, 21 (1983), pp. 1–43.

—, 'Making Certain (Essay Review of J. Shapiro, *Probability and Certainty in Seventeenth-Century England*)', *Social Studies of Science*, 14 (1984), pp. 137–52.

—, 'Priestley Questions: An Historiographic Survey', *History of Science*, 22 (1984), pp. 151–83.

—, 'Scientific Discoveries and the End of Natural Philosophy', *Social Studies of Science*, 16 (1986), pp. 387–420.

—, 'Priestley and the Politics of Spirit', in R. G. W. Anderson and C. Lawrence (eds), *Science, Medicine, and Dissent: Joseph Priestley (1733–1804)* (London: Wellcome Trust and the Science Museum, 1987), pp. 39–53.

—, 'Measuring Virtue: Eudiometry, Enlightenment, and Pneumatic Medicine', in A. Cunningham and R. French (eds), *The Medical Enlightenment of the Eighteenth Century* (Cambridge: Cambridge University Press, 1990), pp. 281–318.

—, 'Machine Philosophy: Demonstration Devices in Georgian Mechanics', in A. van Helden and T. L. Hankins (eds), *Instruments (Osiris, 2nd Series, Volume 9)* (Philadelphia, PA: History of Science Society, 1994), pp. 157–82.

—, 'Enlightened Automota', in W. Clark, J. Golinski and S. Schaffer (eds), *The Sciences in Enlightened Europe* (Chicago, IL: University of Chicago Press, 1999), pp. 126–65.

Schlick, M., 'The Foundations of Knowledge', in A. J. Ayer (ed.), *Logical Positivism* (1934; New York: Free Press, 1959), pp. 209–27.

Schofield, R. E., 'The Scientific Background of Joseph Priestley', *Annals of Science*, 13 (1957), pp. 148–63.

—, *A Scientific Autobiography of Joseph Priestley (1733–1804); Selected Scientific Correspondence, Edited with Commentary by Robert E. Schofield* (Cambridge, MA: MIT Press, 1966).

—, *Mechanism and Materialism: British Natural Philosophy in an Age of Reason* (Princeton, NJ: Princeton University Press, 1970).

—, 'The Counter Reformation in Eighteenth-Century Science – Last Phase', in D. H. D. Roller (ed.), *Perspectives in the History of Science and Technology* (Norman, OK: University of Oklahoma Press, 1971), pp. 39–54.

Schuster, J. A., and R. R. Yeo, 'Introduction', in J. A. Schuster and R. R. Yeo (eds), *The Politics and Rhetoric of Scientific Method: Historical Studies* (Dordrecht: Kluwer Academic Publishers, 1986), pp. ix–xxvii.

Schutz, A., 'Concept and Theory Formation in the Social Sciences', in F. Dallmayr and T. McCarthy (eds), *Understanding and Social Inquiry* (1954: Notre Dame, IN and London: University of Notre Dame Press, 1977), pp. 225–39.

Shapin, S., 'Homo Phrenologicus: Anthropological Perspectives on an Historical Problem', in B. Barnes and S. Shapin (eds), *Natural Order: Historical Studies of Scientific Culture* (Beverly Hills, CA and London: Sage Publications, 1979), pp. 41–72.

—, 'Social Uses of Science', in G. S. Rousseau and R. Porter (eds), *The Ferment of Knowledge: Studies in the Historiography of Eighteenth-Century Science* (Cambridge: Cambridge University Press, 1980a), pp. 93–139.

—, 'Of Gods and Kings: Natural Philosophy and Politics in the Leibniz-Clarke Disputes', *Isis*, 72 (1981b), pp. 187–215.

—, 'History of Science and Its Sociological Reconstructions', *History of Science*, 20 (1982), pp. 157–211.

—, 'Pump and Circumstance: Robert Boyle's Literary Technology', *Social Studies of Science*, 14 (1984), pp. 481–520.

—, 'Talking History: Reflections on Discourse Analysis', *Isis*, 75 (1984b), pp. 125–30.

—, 'Science and the Public', in R. C. Olby, G. N. Cantor, J. R. R. Christie and M. J. S. Hodge (eds), *Companion to the History of Modern Science* (London: Routledge, 1990), pp. 989–1007.

—, 'Discipline and Bounding: The History and Sociology of Science as Seen through the Externalism-Internalism Debate', *History of Science*, 30 (1992), pp. 333–69.

—, *A Social History of Truth: Civility and Science in Seventeenth-Century England* (Chicago, IL: University of Chicago Press, 1994).

—, 'Here and Everywhere: Sociolgy of Scientific Knowledge', *Annual Review of Sociology*, 21 (1995), pp. 289–321.

—, *The Scientific Revolution* (Chicago, IL: University of Chicago Press, 1996).

—, 'Placing the View from Nowhere: Historical and Sociological Problems in the Location of Science', *Transactions of the Institute of British Geographers*, 23 (1998), pp. 5–12.

Shapin, S., and S. Schaffer, *Leviathan and the Air-Pump: Hobbes, Boyle and the Experimental Life* (Princeton, NJ: Princeton University Press, 1985).

Shapin, S., and A. Thackray, 'Prosopography as a Research Tool in History of Science', *History of Science*, 12 (1974), pp. 1–28.

Siegfried, R., 'Lavoisier's View of the Gaseous State and Its Early Application to Pneumatic Chemistry', *Isis*, 63 (1972), pp. 59–78.

—, 'Lavoisier's Table of Simple Substances: Its Origin and Interpretation', *Ambix*, 29 (1982), pp. 29–48.

—, 'Lavoisier and the Phlogistic Connection', *Ambix*, 36 (1989), pp. 31–40.

Siegfried, R., and B. J. Dobbs, 'Composition: A Neglected Aspect of the Chemical Revolution', *Annals of Science*, 24 (1968), pp. 275–93.

Simon, J., 'The Chemical Revolution and Pharmacy: A Disciplinary Perspective', *Ambix*, 45 (1998), pp. 1–13.

—, 'Authority and Authorship in the Method of Chemical Nomenclature', *Ambix*, 44 (2002), pp. 2006–226.

—, 'Analysis and the Hierarchy of Nature in Eighteenth-Century Chemistry', *British Journal for the History of Science*, 35 (2002), pp. 1–16.

—, *Chemistry, Pharmacy and Revolution in France, 1777–1809* (Burlington, VT and Aldershot: Ashgate, 2005).

Starr, P., *Logics of Failed Revolts: French Theory after May '68* (Stanford, CA: Stanford University Press, 1995).

Stewart, L., 'Putting on Airs: Science, Medicine, and Polity in the Late Eighteenth Century', in T. H. Levere and G. Turner (eds), *Discussing Chemistry and Steam: The Minutes of a Coffee-House Philosophical Society, 1760–1787* (Oxford: Oxford University Press, 2002), pp. 207–55.

Stone, L., 'History and the Social Sciences in the Twentieth Century', *The Past and the Present* (1976; Boston: Routledge and Kegan Paul, 1981), pp. 3–44.

Strum, S. S., and B. Latour, 'Redefining the Social Link: From Baboons to Humans', *Information sur les Sciences Sociales*, 26 (1987), pp. 783–802.

Suppe, F., 'The Search for Philosophic Understanding of Scientific Theories', in F. Suppe (ed.), *The Structure of Scientific Theories* (Urbana, IL: University of Illinois Press, 1974), pp. 3–232.

Teich, M., 'Afterword', in R. Porter and M. Teich (eds), *The Enlightenment in National Context* (Cambridge: Cambridge University Press, 1981), pp. 215–7.

Thackray, A., *Atoms and Powers: An Essay on Newtonian Matter Theory and the Development of Chemistry* (Cambridge, MA: Harvard University Press, 1970).

—, 'Science: Has Its Present Past a Future?', in R. H. Steuwer (ed.), *Historical and Philosophical Perspectives of Science* (1970; New York: Gordon and Breach, 1989), pp. 112–33.

Thomas, B., 'The New Historicism and Other Old-Fashioned Topics', in A. H. Veeser (ed.), *The New Historicism* (New York/London: Routledge, 1989), pp. 182–203.

Thompson, E. P., 'The Poverty of Theory', *The Poverty of Theory and Other Essays* (New York: Monthly Review Press, 1978), pp. 1–210.

—, 'The Peculiarities of the English', *The Poverty of Theory and Other Essays* (1965; New York: Monthly Review Press, 1978), pp. 245–301.

Toulmin, S., 'Crucial Experiments: Priestley and Lavoisier', *Journal of the History of Ideas*, 18 (1957), pp. 205–22.

—, *Foresight and Understanding: An Inquiry into the Aims of Science* (New York: Harper and Row, 1961).

—, *The Philosophy of Science* (1953; London: Arrow Books, 1962).

—, *Human Understanding: The Collective Use and Evolution of Concepts* (Princeton, NJ: Princeton University Press, 1972).

Van Helden, A., and T. Hankins, (eds), *Instruments (Osiris, 2nd Series, Volume 9)* (Chicago, IL: University of Chicago Press, 1994).

Van Helden, A., and T. L. Hankins, 'Introduction: Instruments in the History of Science', in A. Van Helden and T. Hankins (eds), *Instruments (Osiris, 2nd Series, Volume 9)* (Chicago, IL: University of Chicago Press, 1994), pp. 1–6.

Veeser, H., (ed.) *The New Historicism* (London: Routledge, 1989).

Verbruggen, F., 'How to Explain Priestley's Defense of Phlogiston', *Janus*, 54 (1972), pp. 47–89.

Ward, S. C., *Reconfiguring Truth: Postmodernism, Science Studies, and the Search for a New Model of Knowledge* (Lanham, MD: Rowman and Littlefield, 1996).

Weeks, J., 'Foucault for Historians', *History Workshop*, 14 (1982), pp. 106–19.

Westfall, R. S., 'Marxism and the History of Science: Reflections on Ravetz's Essay', *Isis*, 72 (1981), pp. 402–5.

Westman, R. S., and D. C. Lindberg, 'Introduction', in R. Westman and D. C. Lindberg (eds), *Reappraisals of the Scientific Revolution* (Cambridge: Cambridge University Press, 1990), pp. xvii–xxvii.

White, H., 'New Historicism: A Comment', in H. Veeser (ed.), *The New Historicism* (New York: Routledge, 1989), pp. 293–302.

White, J. H., *The History of the Phlogiston Theory* (London: E. Arnold, 1932).

Whitehead, A. N., *Science and the Modern World* (New York: Macmillan, 1925).

Willey, B., *The Seventeenth-Century Background: Studies in the Thought of the Age in Relation to Poetry and Religion* (1934; Hammondsworth, Middlesex: Penguin Books, 1964).

Winch, P., *The Idea of a Social Science and Its Relation to Philosophy* (London: Routledge and Kegan Paul, 1970).

Wittgenstein, L., *Philosophical Investigations*, trans. G. E. M. Anscombe (New York: Macmillan, 1953).

Wolin, R., 'Modernism versus Postmodernism in Debates in Contemporary Culture', *Telos*, 62 (1984/5), pp. 9–29.

Woolgar, S., (ed.) *Knowledge and Refelxivity: New Frontiers in the Sociology of Knowledge* (London/Beverly Hills: Sage Publications, 1988).

Woolgar, S., and M. Ashmore, 'The Next Step: An Introduction to the Reflexive Project', in S. Woolgar (ed.), *Knowledge and Reflexivity: New Frontiers in the Sociology of Knowledge* (Beverly Hills, CA and London: Sage Publications, 1988), pp. 1–11.

Young, R., 'Post-Structuralism: An Introduction', in R. Young (ed.), *Untying the Text: A Post-Structuralist Reader* (Boston, MA: Routledge and Kegan Paul, 1981), pp. 1–28.

Young, R. M., 'Marxism and the History of Science', in R. C. Olby, G. N. Cantor, J. R. R. Christie and M. J. S. Hodge (eds), *Companion to the History of Modern Science* (London: Routledge, 1990), pp. 77–86.

INDEX